Principles of
Lightwave Communications

Principles of Lightwave Communications

Göran Einarsson
Royal Institute of Technology,
Stockholm,
Sweden

JOHN WILEY & SONS
Chichester · New York · Brisbane · Toronto · Singapore

Other Wiley Editorial Offices

John Wiley & Sons, Inc., 605 Third Avenue,
New York, NY 10158-0012, USA

Jacaranda Wiley Ltd, 33 Park Road, Milton,
Queensland 4064, Australia

John Wiley & Sons (Canada) Ltd, 22 Worcester Road,
Rexdale, Ontario M9W 1L1, Canada

John Wiley & Sons (SEA) Pte Ltd, 37 Jalan Pemimpin #05-04,
Block B, Union Industrial Building, Singapore 2057

Library of Congress Cataloging-in-Publication Data

Einarsson, G. H. (Göran H.), 1934–
    Principles of lightwave communications / Göran Einarsson.
       p. cm.
    ISBN 0 471 95297 4 : $79.65 – ISBN 0 471 95298 2 (pbk.) : $39.95
    1. Optical communications.   I. Title.
TK 5103.59.E36   1995
621.382'75—dc20   95–930
                                                           CIP

British Library Cataloguing in Publication Data

A catalogue record for this book is available from the British Library

ISBN 0 471 95297 4 (Ppc)   0 471 95298 2 (Pbk)

Typeset in 10/12pt Palatino by P&R Typesetters Ltd, Salisbury, Wilts
Printed and bound in Great Britain by Bookcraft (Bath) Ltd
This book is printed on acid-free paper responsibly manufactured from sustainable forestation,
for which at least two trees are planted for each one used for paper production.

To

Britt Elisabet

Contents

Preface

The book presents a statistical theory of optical communication along the same lines as the theory for electrical communication, which has been available for many years. A pervading characteristic is the treatment of probability statistics in the transform domain, in the form of generating functions, rather than density or distribution functions. This provides simple system characterizations and access to saddlepoint methods for accurate error probability calculations.

The results are, as far as possible, derived from first principles which makes the book useful as a text for students. But also self contained and therefore suitable as a starting point for alternative approaches. Many illustrative examples are included so that the book will work as a text for a graduate course and also as a reference for engineers in the field of optical fiber communication.

The book grew out of notes for a course in Fiber Optical Communication at the University of Lund, Sweden. Bo Bernhardsson and Richard Lundin were two students who contributed to the content. Bernhardsson is responsible for calculating the paths of light in various graded-index fibers, Figures 2.10–2.13, and Lundin for generating the characteristic diagram for cylindrical waveguides, Figure 3.8. The design procedure for optical receivers, Section 6.5, is based on work by Torbjörn Andersson.

More recently, at the Royal Institute of Technology in Stockholm, Sweden, Johan Strandberg, Magnus Sundelin, and Idelfonso Tafur Monroy have made calculations and generated many of the figures in Chapters 5–8. The frequency spectrum diagram Figure 1.5 is designed by Johan Einarsson.

I conclude with the following citation:

> *I heartily beg that what I have here done may be read with forbearance; and that my labors in a subject so difficult may be examined, not so much with the view to censure, as to remedy their defects.*

Stockholm, September 1995
Göran Einarsson

1
Introduction

1.1 A Historical Perspective

The word telecommunication comes from the Greek word 'tele' which means far away and the Latin word 'communicare', which means to share with others. The original meaning is thus communication over long distances.

Telecommunication has, ever since the beginning of history, been accomplished through messengers who travelled by foot or on horseback. The transport of written messages has the advantage that the information capacity is high but the delay which the method implies can sometimes be a disadvantage.

In situations such as war, when it is necessary for a message to reach the receiver without a time delay, optical communication in the form of beacons has been used since the beginning of mankind. The Greek tragedian Aeschylus, who lived between 525 and 456 BC, recounts in his drama *Oresteia* how the message about the fall of Troy was sent from Asia Minor to Argos (in today's Greece), a stretch of 400 km as the crow flies.

Besides the speed in transference which a chain of beacons exhibits, the method has the advantage that the message can reach several receivers simultaneously. In modern terminology this is called a multiuser system with one sender and several receivers, a 'multiuser broadcast channel'. The Swedish bishop Olaus Magnus describes such a system in his history of the Nordic people published in Rome in AD 1555, Figure 1.1.

One obvious disadvantage with signal beacons or smoke signals is that the information capacity is of necessity low. Only a limited number of predetermined messages can be transferred. This mode of communication is of an essentially binary character, in contrast with messengers who could transport long and complicated messages in the form of texts, maps, etc.

At the end of the eighteenth century, a new phase began for optical communication. The French Revolution in 1789 released much creative talent in France. During the years 1790–1794 Claude Chappe constructed an optical telegraph which was based on a chain with semaphores of the type shown in Figure 1.2.

De ignibus montanis tempore hoſtili.

C A P.   X.

Fumi.

Cuſtodia eque/
ſtrium.

Ignita pyramis

Velites eque/
ſtres.

Modus circum
ueniēdi hoſtē.

Victi.

X hac imagine duo veniunt conſyderanda : quorum al/
terum in vertice montium , ſumi ſcilicet congeſtis lignorum
ſtruibus ad claſsicos inſultus accedentium inimicorum ar/
cendos excitati : alterum verò in littoralibus anguſtiis , &
ſcopulis , ne hoſtes inſiliant , diligens equeſtrium cuſtodia
intuenda : qui montana inhabitant loca tempore hoſtili, ve/
luti celerrimi ſpeculatores , ſigna ſumo faciunt: quibus viſis,
alii reliquos montes incolentes , itidem alongè poſitis igni/
ta pyramide demonſtrant , vt quilibet armatus iuxta nume/
rum Principis , & patriæ lege decretum , ex campeſtribus locis pro littorum cuſto/
dia ſine mora deſcendat . Inter quos velites equeſtres celerrimi adſunt , vt hoſtibus
portuum , vel riparum acceſſum præcluſuri , ſagittariis plebeiæ multitudinis locum
ſtatuant , vbi commodius excipiant , ac conficiant hoſtem,omnino aggredi conten/
dentem : vt ſcilicet in vallibus , aut cauernis expectent , vel ad iniqua & ignota ho/
ſtibus loca , quaſi fugitiui declinent , ne forte eos taliter inſequentes educere valeant
in agmina robuſtiora : quę vrgente neceſsitate , vſφ ad infinitam multitudinem au/
geri ſolent . Nec deſunt exploratores quoquo verſum emiſsi , qui renuntiant , qua
ex parte adhuc immineant hoſtes , vt his celerius, & copioſius occurren/
tes, vel conſilio , vel virtute , vel inſidiis , vel neceſsitate, vel
deſperatione, vel loci ſecuritate nutantibus,nedum
victoriam adimant,ſed et victos , vt im/
perata perficiant, militari le/
ge conſtringant .

Figure 1.1   'On fires on the mountains in times of war'. From O. Magnus: *Historia de gentibus septentrionalibus*, AD 1555.

Forty years later the historian Thomas Carlyle wrote *The French Revolution* where he gives a lively account of Chappe's hardships and successes.

What, for example, is this that Engineer Chappe is doing, in the Park of Vincennes? In the Park of Vincennes; and onwards, they say, in the Park of Lepelletier Saint-Fargeau the assassinated Deputy; and still onwards to the Heights of Ecouen and further, he has scaffolding set up, has posts driven in;

Figure 1.2 Claude Chappe (1763–1805) and his optical telegraph. From Malmgren (1972) and Michaelis (1965). Reproduced by permission of Telia and ITU.

wooden arms with elbow-joints are jerking and fugling in the air, in the most rapid mysterious manner! Citoyens ran up, suspicious. Yes, O Citoyens, we are signalling: it is a device this, worthy of the Republic; a thing for what we will call *Far-writing* without the aid of postbags; in Greek it shall be named Telegraph. – *Télégraphe sacré !* answers Citoyenism: For writing to Traitors, to Austria? –and tears it down. Chappe had to escape, and get a new Legislative Decree. Nevertheless he has accomplished it, the indefatigable Chappe: this his *Far-writer*, with its wooden arms and elbow-joints, can intelligibly signal; and lines of them are set up, to the North Frontiers and elsewhither. On an Autumn evening of the Year Two, Far-writer having just written that Condé Town has surrendered to us, we send from the Tuileries Convention-Hall this response in the shape of Decree: 'The name of Condé is changed to *Nord-Libre*, North-Free. The Army of the North ceases not to merit well of the country.' – To the admiration of men! For lo, in some half hour, while the Convention yet debates, there arrives this new answer: 'I inform thee, *je t'annonce*, Citizen President, that the Decree of Convention, ordering change of the name Condé into *North-Free*; and the other, declaring that the Army of the North ceases not to merit well of the country; are transmitted and acknowledged by Telegraph. I have instructed my Officer at Lille to forward them to North-Free by express.
    Signed, CHAPPE.
Cited from T. Carlyle, *The French Revolution*, Vol. 2, Everyman's Library, David Campbell Publishing Ltd.

The stations in the chain were located about 10–20 km apart, a distance at which it was possible to observe the nearest station with the aid of binoculars. As stated in the above quotation, one could transfer a message between Lille and Paris, which is a distance of 230 km, in 30 minutes. This implies a signaling speed of 3–4 characters per minute.

The French optical telegraph inspired many successors in several European countries, see Holzmann and Pehrson (1994). In Sweden in 1794, Abraham Edelcrantz built an optical telegraph line between the Royal Palace in the center of Stockholm and the Summer Palace at Drottningholm. In the following years he supervised the construction of several telegraph lines in costal areas of Sweden. Edelcrantz's construction, shown in Figure 1.3, is described in his *Treatise on Telegraphs*, published in Stockholm in 1796. It consisted of a signal mast with a number of plates which could be folded or raised in order to create a pattern which represented a character in an alphabet. The position of the plates was manipulated by a keyboard which gave a transmission speed of up to 12 characters per minute.

The optical telegraph could not be used at night, and it was dependent on the weather. During the late nineteenth century it was completely replaced by the electric telegraph.

The first electric telegraph to come into practical use was constructed by the American Samuel F. B. Morse in 1838. He used the concept of electromagnetism, which was discovered by Hans Christian Ørstedt in 1820. The telephone was invented by Alexander Graham Bell in 1876, and radio communication was made possible by the contributions of James Clark Maxwell in 1873, Heinrich Hertz in 1887 and Guglielmo Marconi in 1895.

Figure 1.3   Abraham Edelcrantz (1754–1821) and his optical telegraph. From Malmgren (1972). Reproduced by permission of Telia.

A global communication network based on these inventions was quickly built up. The first successful telegraph cable between the USA and Europe opened in 1865.

Telephone signals are attenuated much more than telegraph signals when transmitted through a cable, and therefore radiotelephony was for a long time the only way to telephone between the continents. The first telephone cable at the bottom of the Atlantic between Newfoundland and Scotland was brought into use in 1956. Since 1965 intercontinental telephone traffic has been transmitted with the assistance of so-called synchronous satellites. In 1988 optical communication returned as a medium for long distance transmission. A fiber optical underwater cable was put into operation between the USA and Europe. The system, which is called TAT-8, is described in more detail in Section 1.2.

Ever since the middle of the nineteenth century, electrical signals have prevailed as the major choice for telecommunication. Despite this, there has always been interest in optical phenomena and optical signals. In 1880, Alexander Graham Bell constructed and received a patent for a 'photophone' which transmitted telephone signals which were generated by an intensity-modulated light beam. It never achieved any practical importance despite the fact that the inventor was convinced of the opposite.

During the Victorian era there was much activity and research in optics, as described by Hecht (1985). John Tyndall carried out a series of experiments before the Royal Institution of Great Britain in 1854. They are described in the notices of the meetings of the Royal Institution, Tyndall (1854). One of the things which Tyndall showed was that light sent through a jet of water stays within the beam. The light does not leak out, it is trapped within Figure 1.4. This depends, as we shall see in Chapter 2, on the physical laws for reflection of light rays. The same thing applies for glass, and total reflection constitutes the fundamental physical principle for fiber optic transmission.

One can ask why Tyndall's experiment did not lead to optical fiber communication. The reason is easily illustrated with an example.

Example 1.1   Optical system with high attenuation fiber

In order for fiber optical communication to be possible, the fibers must be made of very pure glass so that the light which arrives at the receiver is not too weak to be detected.

The purest glass that could be produced during the nineteenth century and up to 1969 had an attenuation of more than 1000 dB per kilometer.

Assume that the minimum optical signal that is needed for the receiver to detect a transmitted pulse is one photon. A photon has an energy $E = hf$ where $h = 6.625 \times 10^{-34}$ Ws^2 is Planck's constant and f is the frequency of the light. At a light wavelength $\lambda = 1$ μm

$$E_r = E_r = h\frac{c}{\lambda} = 6.626 \times 10^{-34} \times \frac{3.0 \times 10^8}{1 \times 10^{-6}} = 2.0 \times 10^{-19} \text{ Ws}$$

with c equal to the speed of light.

Figure 1.4   John Tyndall's experiment at The Royal Institution 1854.
Artist: Gunilla Johnson.

In order to accomplish this, given an attenuation of 1000 dB corresponding to
a distance of 1 km, the following initial energy is required

$$E_t = E_r 10^{1000/10} = 2.0 \times 10^{81} \text{ Ws}$$

This is an impossibly large number. The world's total yearly energy
production is of the magnitude 10^5 TWh. The magnitude of E_t corresponds
to the amount of production during 5.6×10^{60} years, that is, during a span of
time which considerably exceeds the estimate of the age of the universe. Any
optical fiber would evaporate if it were exposed to even a very small fraction
of the required energy   □

Example 1.1 shows that even if the principle for fiber optic transmission of
light was known in 1854, the method was not applicable for communication at
distances outside the laboratory.

This state of affairs, however, was to be radically altered. In the mid-
1960s K.C. Kao and G.A. Hockham (1966) published in England and A. Werts
(1966) in France journal articles in which the transmission of light through
optical fibers was analysed. It was pointed out that if it were possible to

produce glass with a sufficiently low attenuation optical fibers would be an alternative to electric wires.

Intensive research activity on optical fibers was carried out at that time at several industrial laboratories. Nippon Sheet Glass Co. succeeded in 1968 in producing glass with an attenuation of 100 dB/km. The breakthrough for the production of optical fibers was, however, when Corning Glass in USA announced in 1970 that they could produce glass with an attenuation of 20 dB/km, Kapron *et al.* (1970).

Example 1.2   Optical system with low loss fiber

How much energy is the light source in Example 1.1 required to have if a fiber with an attenuation of 20 dB/km is used?

An attenuation of 20 dB means that

$$E_t = E_r \times 10^{20/10} = 100 \times 2.0 \times 10^{-19} \text{ Ws}$$

With the same assumptions as in Example 1.1, the light source needs to transmit a pulse which contains 100 photons, which is easy to accomplish with, for example, a simple light-emitting diode.   □

The quality of optical fibers has been steadily improved. Today glass fibers with a minimum attenuation below 0.2 dB/km are available. Figure 1.9 shows the attenuation as a function of wavelength for a so-called single-mode fiber which is used in the underwater system described in Section 1.3.

The success of optical fiber communication is very much dependent on the possibility of producing glass of very high purity. Another condition is the availability of light sources and light detectors manufactured with the use of semiconductor techniques.

The first laser was demonstrated in the laboratory in 1958. Since then rapid progress has been made and in 1970, the same year in which Corning Glass had its fiber ready, a demonstration of the first semiconductor laser which did not require cooling but could be operated at room temperature was reported by Hayashi *et al.* (1970).

The area of optical fiber communication has shown a very fast technical development. Many public telecommunication administrations no longer install any new metal cables. Instead, all planning involves optical fiber cables, a development which is by no means finished. Today throughout the world, work is being done on systems with amplified light, new fiber materials and new optical semiconductor components, to name only a few developing areas.

1.2   The Electromagnetic Spectrum

Isaac Newton, in his ground breaking work on optics in 1672, considered light to be a stream of small particles travelling in straight lines. Christian Huygens

(1629–1695) asserted that this model could not explain such phenomena as interference and diffraction. Jean Fresnel proved beyond doubt in 1816 that these phenomena showed light to be a wave in motion. In 1873 James Clark Maxwell developed the mathematical theory of electromagnetic waves and claimed that light was a special case of such waves.

The particle nature of light came again into focus with the introduction of light quanta or photons in the work of Einstein in 1905, and others. To understand light, both the wave and the particle theory are needed, and we will use both in our analysis of optical fiber communication.

There is a unifying theory called quantum electrodynamics (QED). It is complex and fortunately it is not needed in this book. For the problems we study, it is adequate to use the classical electromagnetic theory for light transmission in the fiber and the particle model for the detection of light by photodiodes.

The electromagnetic spectrum and its use for telecommunication are shown in Figure 1.5. Notice that the optical spectrum, of which visible light is a small part, is located far away from the radio waves.

1.3   Long-Distance Optical Systems

Telecommunication services between continents are carried by radio (satellites) or cable systems. One application, which imposes very severe requirements on system performance, is undersea long-distance systems.

The first transatlantic telephone system TAT-1 was ready for service in 1956. It contained two separate coaxial cables, one for each direction of transmission, and was used for carrier transmission of 36 two-way voice circuits. The repeaters necessary to compensate for the cable attenuation were equipped with specially designed vacuum tubes. The repeaters were spliced into the cables at a spacing of 70 km.

During later years, six additional undersea systems with larger capacity and of more sophisticated design have been installed. The last system was completed in 1976 and has a capacity of 4200 two-way telephone channels. Its repeaters are equipped with transistors and are spaced at a distance of 9.4 km. This system is used in cable systems TAT-6 and TAT-7 installed in 1976 and 1983.

The first intercontinental optical fiber system, TAT-8 between the USA and Europe, was put into service in late 1988. The system was developed by the American company AT&T and is presented by Runge and Trischitta (1986). The cable has an underwater branching point on the European side, dividing the traffic between France and England. A similar system was installed in 1988 between the USA and Hawaii, HAW-4; and further on to Japan and Guam, TCP-3; see Amano and Iwamoto (1990).

The route of the optical cable system TAT-8 is shown in Figure 1.6. It connects the east coast of the USA with England and France, with an underwater branching point 300 km off the coast of France. The total distance spanned is 6100 km.

The construction of the cable is shown in Figure 1.7. Six optical fibers are placed around a copper-clad steel wire, called the king wire. For protection the central core is surrounded by steel wires, a copper shield and an outer layer of polyethylene. The outer diameter of the cable is 21 mm, which can be compared with the 53 mm cable in the TAT-7 system.

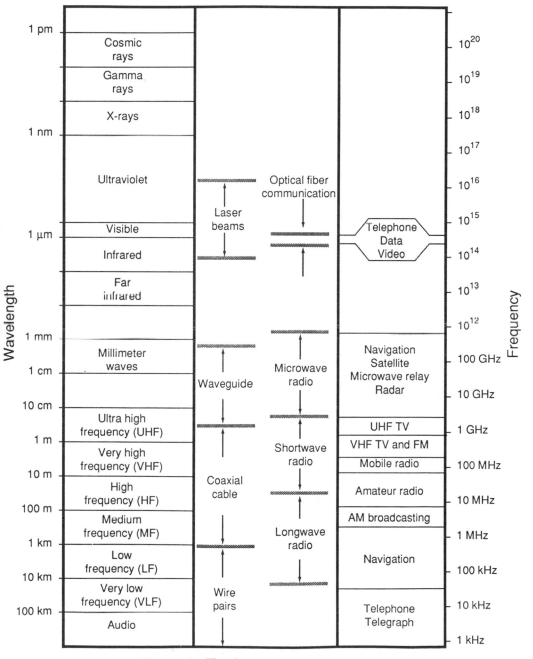

Figure 1.5   The electromagnetic spectrum.

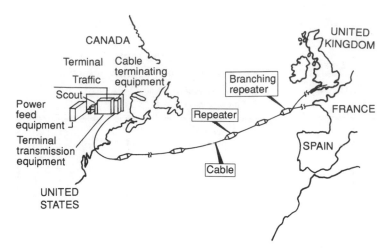

Figure 1.6 Route of the TAT-8 optical fiber undersea system. From Runge and Trischitta (1984). Reproduced by permission of IEEE.

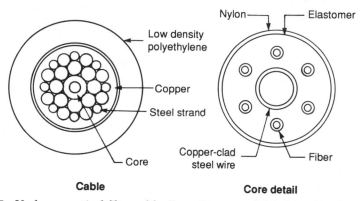

Figure 1.7 Undersea optical fiber cable. From Runge and Trischitta (1984). Reproduced by permission of IEEE.

The fibers, which support single-mode transmission, have an outer diameter of 125 μm and a core diameter of 8.3 μm. They are of a design called depressed index cladding, with an theoretical index profile shown in Figure 1.8 and attenuation according to Figure 1.9. Light with a wavelength of 1.3 μm is transmitted in the form of short pulses through the fibers.

The information transmitted is digital with a line rate of 295.6 Mbit/s which accommodates two CEPT-4 multiplex systems, each with a digital rate of 139.264 Mbit/s. A CEPT-4 multiplex can be used for simultaneous transmission of 3780 telephone channels with 64 kbit/s speech coding. Two of the three fiber pairs in the cable are used for transmission with the remaining pair as stand-by resulting in a total cable capacity of about 8000 two-way 64 kbit/s telephone channels.

The optical signals are detected and regenerated in repeaters spaced 46 km apart. A block diagram of a repeater is shown in Figure 1.10. The detector at the

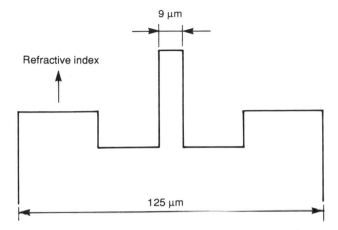

Figure 1.8 Depressed cladding single-mode fiber profile.

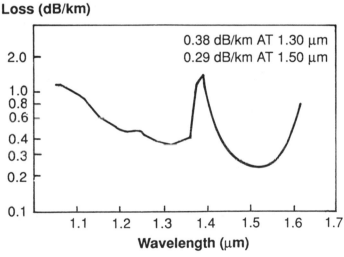

Figure 1.9 Fiber attenuation as a function of optical wavelength. From Runge and Trischitta (1983). Reproduced by permission of IEEE.

input contains an InGaAs PIN photodiode and the regenerated light pulses are produced by a buried-heterostructure laser diode.

The reliability requirements of the system are severe. No more than three repairs in the 25-year system lifespan are anticipated. The single component in the repeater with the shortest expected lifetime is the laser diode.

In contrast with coaxial cable systems, the power necessary for the repeaters cannot be supplied through the fibers. To provide power for the repeaters a direct current of 300 mA is sent through the center wire of the cable, with the aluminium shield serving as a return circuits. A supervisory system for locating faults and switching the spare lasers operates over the same circuit.

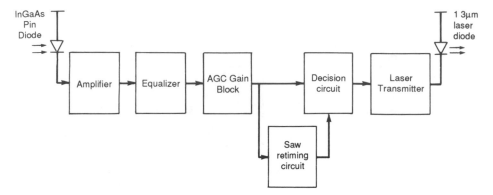

Figure 1.10 Block diagram of an optical repeater. From Runge and Trischitta (1983). Reproduced by permission of IEEE.

Today many more intercontinental optical fiber systems have been installed or are planned in various parts of the world. Figure 1.11 from Runge (1992) shows the optical systems in service or planned at that time. In the Atlantic, TAT-8 has been followed by TAT-9, -10 and -11 with the same principle of design and with improved capacity. These systems operate at a wavelength of 1.55 μm where the fiber attenuation is minimal. The transmission rate per fiber is 560 Mbit/s and the repeater distance is increased to 100 km.

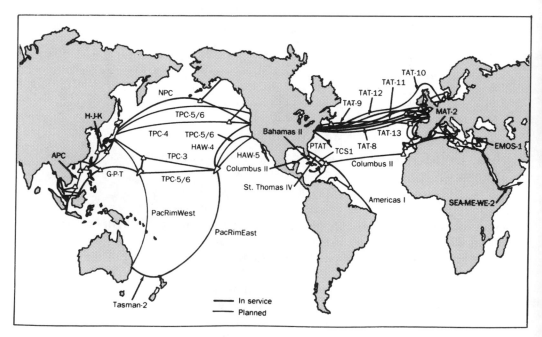

Figure 1.11 Installed and planned undersea lightwave systems (1991). From Runge (1992). Copyright 1992 AT & T. All rights reserved. Printed with permission.

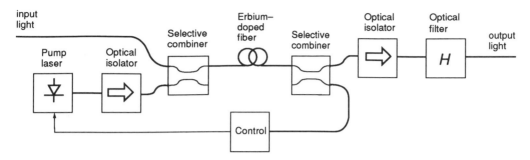

Figure 1.12   Block diagram of an Erbium Doped Fiber Amplifier (EDFA).

The next generation of undersea lightwave systems, TAT-12 and TAT-13 in the Atlantic and TPC-4 in the Pacific, will be of a new design with even greater transmission capacity. Erbium doped fiber amplifiers (EDFA) will be used instead of regenerative repeaters. The amplifying medium in an EDFA is about 100 m of silica fiber doped with erbium. Amplification of optical signals is based on stimulated photon emission which requires the erbium atoms to be excited by a pumping laser. A schematic diagram of an EDFA is shown in Figure 1.12 and the principles of its operation is presented in Chapter 7. The EDFA operates at an optical wavelength of 1.55 μm and the distance in the cable between the amplifiers is estimated at 40 km. The planned transmission rate per fiber will be 2.5 Gbit/s or 5 Gbit/s. The possibility of even higher rates by Wavelength Division Multiplexing (WDM) has been demonstrated by Taga *et al.* (1994). With two 5 Gbit/s data streams or four 2.4 Gbit/s streams transmitted simultaneously at slightly different optical wavelengths a total rate of 10 Gbit/s per fiber is created.

For the next generation of optical long-distance systems the use of soliton pulses is a possibility. Solitons are optical pulses which are stable and do not change in form during transmission, see Chapter 4, Section 4.8. The feasibility of this technique, at rates as high as 20 Gbit/s and for distances as large as 35 thousand kilometers, has been shown by Mollenauer *et al.* (1994) and by Suzuki *et al.* (1994).

2

Light Propagation in Optical Fibers

2.1 Optical Fibers

An optical fiber consists of a core surrounded by a cladding. During transmission light travels primarily in the core. The cladding, which has a refractive index smaller than that of the core, has the function of confining the light inside the fiber so that it does not leak out through the sides of the fiber. For a fiber without any coating, which is seldom the case in practice, the surrounding air acts as a cladding.

A fiber can be made of any transparent optical material such as glass or plastic. Its transmission properties depend on the optical properties of the core and cladding.

In the analysis of optical fibers, it is convenient to distinguish between two types of fibers: step-index and graded-index fibers. These are illustrated in Figure 2.1. The variation of refractive index in a graded-index fiber can have a

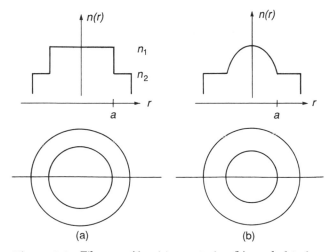

Figure 2.1 Fiber profiles (a) step-index (b) graded-index.

more complex shape than indicated in Figure 2.1. An example is the fiber in the Atlantic Undersea system presented in Section 1.3 whose index profile is shown in Figure 1.8. Note that the step-index fiber is a special case of a graded-index profile.

2.2 Step-index Fibers

In a step-index fiber the core and cladding are made of homogeneous material. The core has a constant refractive index n_1, which changes abruptly to n_2 in the cladding, as illustrated in Figure 2.1. The dimension of the core is typically 5–15 μm for so-called single-mode fibers and 50–200 μm for multimode fibers. The cladding has a thickness of 100–400 μm.

2.2.1 Light Rays in Homogeneous Media

A basic understanding and an elementary analysis of light transmission through optical fibers is obtained from the theory of light ray propagation called geometrical optics. A fundamental theorem in optics is Fermat's principle: light rays follow paths corresponding to the shortest time of propagation.

The velocity of light in a medium with refractive index n is equal to

$$v = \frac{c}{n} \tag{2.1}$$

where c is the velocity of light in a vacuum.

It is immediately clear from Fermat's principle that light rays are travelling in straight lines in a homogeneous medium. The refraction of light rays at the interface between two media with different refractive indices is also easily determined from Fermat's principle.

Example 2.1   The law of refraction

Fermat's principle can be used to determine the relation between the angles α_1 and α_2 in Figure 2.2.

The time for the light to travel from A to B (or from B to A) is

$$t = \frac{AO}{v_1} + \frac{OB}{v_2} \tag{2.2}$$

With the notations of Figure 2.2

$$t = \frac{a}{v_1 \cos \alpha_1} + \frac{b}{v_2 \cos \alpha_2} \tag{2.3}$$

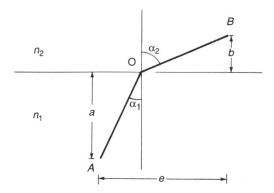

Figure 2.2   The law of refraction (Snell's law). The notation refers to Example 2.1.

and

$$a \tan \alpha_1 + b \tan \alpha_2 = e \qquad (2.4)$$

For t to be minimal, α_1 and α_2 have to satisfy the equation $dt/d\alpha_1 = 0$

$$\frac{dt}{d\alpha_1} = \frac{a \sin \alpha_1}{v_1 \cos^2 \alpha_1} + \frac{b \sin \alpha_2}{v_2 \cos^2 \alpha_2} \frac{d\alpha_2}{d\alpha_1} = 0 \qquad (2.5)$$

Derivation of (2.4) yields

$$\frac{a}{\cos^2 \alpha_1} + \frac{b}{\cos^2 \alpha_2} \frac{d\alpha_2}{d\alpha_1} = 0 \qquad (2.6)$$

Combining (2.5) and (2.6) results in

$$\frac{\sin \alpha_1}{v_1} = \frac{\sin \alpha_2}{v_2} \qquad (2.7)$$

which, with the use of (2.1), can be expressed as

$$n_1 \sin \alpha_1 = n_2 \sin \alpha_2 \qquad (2.8)$$

$\square$

The relation (2.8) is the law of refraction, also called Snell's law after its originator Willebrord Snell who discovered it experimentally in 1621. Example 2.1 shows that the law of refraction is a consequence of Fermat's principle.

Assume that in Figure 2.2 the refractive index n_1 is greater than n_2. Then for increasing values of α_1 there is a critical value $\alpha_1 = \alpha_c$ for which $\alpha_2 = \pi/2$. From (2.8)

$$n_1 \sin \alpha_c = n_2 \qquad (2.9)$$

For $\alpha_1 > \alpha_c$ the light ray is completely reflected back into the n_1-region as illustrated in Figure 2.3. Total internal reflection is the mechanism that forces the light in an optical fiber to stay confined inside the fiber.

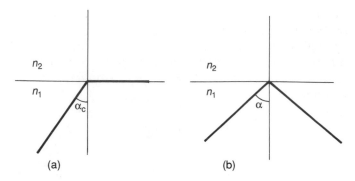

Figure 2.3 Refraction at the boundary between homogeneous media. (a) Critical angle, α_c. (b) Total reflection, $\alpha > \alpha_c$.

2.2.2   Ray Analysis of Step-Index Fibers

Figure 2.4 shows a section of a step-index fiber. The refractive index n_1 of the core is larger than n_2, the refractive index of the cladding. A light ray launched into the fiber from air $(n = 1)$ follows a path as indicated in the figure. We consider rays travelling in a plane through the axis of symmetry of the fiber, so-called meridional rays. All rays for which the angle α is greater than the critical angle α_c will be totally reflected at all points where they meet the boundary between core and cladding. Applying the law of refraction to the end face of the fiber gives

$$1 \cdot \sin \gamma = n_1 \sin \beta$$

The value of γ resulting in $\alpha = \alpha_c$ is obtained from (2.9) using the relation

$$\sin \beta = \cos \alpha = \sqrt{1 - \sin^2 \alpha}$$

The result is

$$\sin \gamma_c = n_1 \sqrt{1 - \sin^2 \alpha_c}$$

$$= n_1 \sqrt{1 - \left(\frac{n_2}{n_1}\right)^2} = \sqrt{n_1^2 - n_2^2} = \eta_A \qquad (2.10)$$

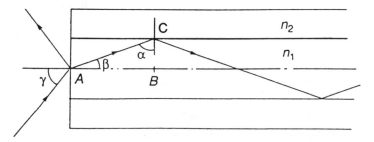

Figure 2.4   Meridional rays in a step-index fiber.

All light rays with $\gamma < \gamma_c$ satisfy the condition of total reflection inside the fiber. The quantity $\sin \gamma_c$ is called the numerical aperture η_A of the fiber. The angle γ_c is the half top angle of an acceptance cone. Rays within the cone will stay inside the fiber while rays outside the cone will be refracted into the cladding.

For a fiber to transport light, n_1 must be greater than n_2. It is convenient to introduce the quantity

$$\Delta = \frac{n_1^2 - n_2^2}{2n_1^2} \tag{2.11}$$

In practice the difference between n_1 and n_2 is often small and

$$\Delta = \frac{(n_1 - n_2)(n_1 + n_2)}{2n_1^2} \approx \frac{(n_1 - n_2)(2n_1)}{2n_1^2} = \frac{n_1 - n_2}{n_1} \tag{2.12}$$

which shows that Δ is then approximately equal to the relative index difference.

The numerical aperture (2.10) is

$$\eta_A = \sin \gamma_c = n_1 \sqrt{2\Delta} \tag{2.13}$$

Example 2.2 Numerical aperture

Silica glass has a refractive index of around 1.5. A fiber with $n_1 = 1.50$ and $n_2 = 1.48$ has

$$\Delta = \frac{n_1^2 - n_2^2}{2n_1^2} = \frac{1.50^2 - 1.48^2}{2 \cdot 1.50^2} = 0.0132$$

The relative index difference is

$$\frac{n_1 - n_2}{n_1} = \frac{1.50 - 1.48}{1.50} = 0.0133$$

The numerical aperture is, from (2.13)

$$\eta_A = n_1 \sqrt{2\Delta} = 1.50\sqrt{2 \times 0.0132} = 0.244$$

The half top angle of the acceptance cone is

$$\gamma_c = \arcsin 0.244 = 14.1°$$

The approximate value $\Delta \approx 0.0133$ gives

$$\eta_A = n_1 \sqrt{2\Delta} = 1.5\sqrt{2 \times 0.0133} = 0.245$$

□

Part of the light at the fiber core end face is reflected and lost as illustrated in Figure 2.4. The amount of loss is determined by the Fresnel reflection coefficient for normal incidence, i.e. when $\gamma = 0$ in Figure 2.4

$$\rho = \left(\frac{n_1 - n}{n_1 + n}\right)^2 \tag{2.14}$$

The quantity ρ represents the fraction of light power lost due to reflection at a boundary between two media with refractive indices n and n_1.

Example 2.3   Reflection loss

A fiber with a core having $n_1 = 1.5$ has a reflection coefficient at the boundary between glass and air $(n = 1)$

$$\rho = \left(\frac{1.5 - 1}{1.5 + 1}\right)^2 = 0.04$$

This means that 96% of the incoming light enters into the core of the fiber. The transmission loss in dB is

$$A = -10 \log(1 - \rho) = 0.18 \text{ dB}$$

which is a typical value for the loss in connectors joining two fibers.   □

2.2.3   Delay Difference and Pulse Broadening

The ray analysis shows that light can travel along different paths inside the fiber. The geometrical approach used in Section 2.2.2 indicates a continuum of possible pathways. The shortest route is a ray travelling along the axis of symmetry in the center of the core. The longest route occurs when the angle α is equal to the critical angle α_c. See Figure 2.4.

The most common way of transmitting information through optical fibers is by sending digital data using pulse modulation. Figure 2.5 shows a short light pulse transmitted over a fiber of length L. Part of the incoming light pulse travels along the shortest ray path, arriving at the receiving end at time τ_1. Other parts travel along longer paths, resulting in larger time delays. The largest delay τ_2 corresponds to the critical angle route.

It is clear from Figure 2.5 that the delay differences cause the received pulse to be broader than the transmitted pulse. This is called dispersion, and we shall show later how it impairs the information transmission capability of the fiber.

The maximal delay difference $\tau_2 - \tau_1$ can be used as a measure of the amount of dispersion.

The ratio in length of the longest and shortest routes is from Figure 2.4

$$\frac{AC}{AB} = \frac{1}{\sin \alpha_c} = \frac{n_1}{n_2}$$

The delay difference for a fiber of length L is

$$\tau_2 - \tau_1 = \frac{L}{v_1}\left(\frac{AC}{AB} - 1\right) = \frac{Ln_1}{c}\left(\frac{n_1 - n_2}{n_2}\right) \tag{2.15}$$

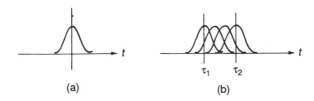

(a)   (b)

Figure 2.5 Pulse broadening (dispersion) due to delay difference of possible ray paths in the fiber. (a) input (b) output.

For small relative index differences (2.12) gives

$$\tau_2 - \tau_1 \approx \frac{Ln_1}{c}\Delta \tag{2.16}$$

Eq. (2.16) shows that a step-index fiber with small Δ shows small dispersion. This explains why optical fibers in practice are designed with Δ of 1% or less.

2.3 Graded-index Fibers

In order to treat graded-index fibers we need to study light rays in inhomogenous media.

It can be shown that Fermat's principle leads to the Euler-Lagrange equation

$$\frac{d}{ds}\left(n\frac{d\mathbf{R}}{ds}\right) = \nabla n \tag{2.17}$$

where $n = n(x, y, z)$ is equal to the media's refractive index which, in the general case, varies with position within the media. The vector $\mathbf{R}$ gives the position of a differential element ds along the light ray's path according to Figure 2.6 and ∇n is the vector $\nabla n = \left(\frac{\partial n}{\partial x}, \frac{\partial n}{\partial y}, \frac{\partial n}{\partial z}\right)$.

It is easy to convince oneself that (2.17) leads to straight line rays if n is constant.

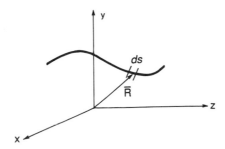

Figure 2.6 Light ray in an inhomogeneous medium.

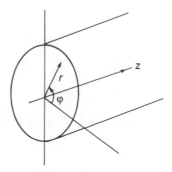

Figure 2.7 Cylindrical coordinates.

An optical fiber is conveniently analysed using cylindrical coordinates according to Figure 2.7. Using cylindrical coordinates, the Euler-Lagrange equation (2.17) becomes

$$\left.\begin{array}{c} \dfrac{d}{ds}\left(n\dfrac{dr}{ds}\right) - nr\left(\dfrac{d\varphi}{ds}\right)^2 = \dfrac{dn}{dr} \\[2mm] n\dfrac{dr\,d\varphi}{ds\,ds} + \dfrac{d}{ds}\left(nr\dfrac{d\varphi}{ds}\right) = \dfrac{dn}{d\varphi} \\[2mm] \dfrac{d}{ds}\left(n\dfrac{dz}{ds}\right) = \dfrac{dn}{dz} \end{array}\right\} \qquad (2.18)$$

We shall study a graded-index fiber with index profile $n(r)$. An example is shown in Figure 2.9 which is a so-called quadratic profile.

Because the refractive index varies only in the radial direction, we have

$$\frac{dn}{d\varphi} = 0 \quad och \quad \frac{dn}{dz} = 0$$

We restrict ourselves to studying meridional rays, that is, rays which go through the main plane through the center of the fiber. For rays in such a plane φ is constant and $d\varphi/ds = 0$.

After insertion into (2.18), two simplified equations remain.

$$\left.\begin{array}{c} \dfrac{d}{ds}\left(n\dfrac{dr}{ds}\right) = \dfrac{dn}{dr} \\[2mm] \dfrac{d}{ds}\left(n\dfrac{dz}{ds}\right) = 0 \end{array}\right\} \qquad (2.19)$$

From the second equation it follows that

$$n\frac{dz}{ds} = \text{const} = N_0 \qquad (2.20)$$

The value of the constant can be determined from an arbitrary point on the ray pathway, for example just inside the fiber's end face. The ray's position and direction at this point are easily determined from the incoming outer ray and

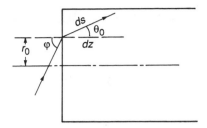

Figure 2.8   Refraction at fiber end face.

the law of refraction. From Figure 2.8 we obtain

$$\left(n\frac{dz}{ds}\right)_{r=r_0} = n(r_0)\cos[\theta(r_0)] = N_0 \tag{2.21}$$

Insertion of $ds^2 = dr^2 + dz^2$ into (2.20) gives

$$n(r)\frac{dz}{\sqrt{dz^2 + dr^2}} = N_0 \tag{2.22}$$

which can be rewritten as

$$\sqrt{n^2(r) - N_0^2}\, dz = N_0\, dr \tag{2.23}$$

Through integration we obtain

$$z - \int_{r_0}^{r} \frac{N_0}{\sqrt{n^2(r) - N_0^2}}\, dr \tag{2.24}$$

A common index profile for a graded-index fiber is

$$n^2(r) = n^2(0)[1 - 2\Delta(r/a)^2] \tag{2.25}$$

which is usually called a quadratic (also square law or parabolic) profile. See Figure 2.9.

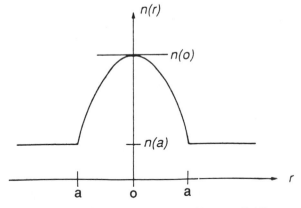

Figure 2.9   Quadratic index profile, eqn (2.25).

Insertion of (2.25) into (2.24) and using (2.21) gives, for rays in the core, i.e. for $r \le a$

$$z = \int_{r_0}^{r} \frac{\cos[\theta(r_0)]}{\left\{ [n^2(0)/n^2(r_0)][1 - 2\Delta(r/a)^2] - \cos^2[\theta(r_0)] \right\}^{1/2}} \, dr \qquad (2.26)$$

This is an integral of the type

$$\int \frac{dx}{\sqrt{\beta^2 - \alpha^2 x^2}} = \frac{1}{\alpha} \arcsin \frac{\alpha x}{\beta} \qquad (2.27)$$

Application of this to (2.26) gives z as a function of r and r_0

$$z = \frac{\cos[\theta(r_0)]}{\alpha} \arcsin \frac{\alpha r}{\beta} - \frac{\cos[\theta(r_0)]}{\alpha} \arcsin \frac{\alpha r_0}{\beta} \qquad (2.28)$$

with

$$\alpha^2 = \frac{2\Delta}{a^2} \frac{n^2(0)}{n^2(r_0)}$$

and

$$\beta^2 = \frac{n^2(0)}{n^2(r_0)} - \cos^2[\theta(r_0)]$$

Let $r_0 = 0$ which yields r as a function of z

$$r = \frac{\beta}{\alpha} \sin \frac{\alpha z}{\cos \theta_0} = \frac{a \sin \theta_0}{\sqrt{2\Delta}} \sin \left(\frac{\sqrt{2\Delta}}{\cos \theta_0} (z/a) \right) \qquad (2.29)$$

where $\theta_0 = \theta(r_0 = 0)$. This means that the rays in the fiber have a sinusoidal form with period

$$\Lambda = \frac{2\pi a \cos \theta_0}{\sqrt{2\Delta}} \qquad (2.30)$$

The period Λ depends on the ray's initial angle.

Examples of rays in a fiber with a quadratic profile are shown in Figure 2.10.

Passage time for the rays is obtained by calculating the time T for a quarter period of the ray pathway. Because

$$v(r) = \frac{ds}{dt} = \frac{c}{n(r)}$$

and

$$dt = \frac{n(r)}{c} \, ds = \frac{n(r)}{c} \sqrt{dz^2 + dr^2}$$

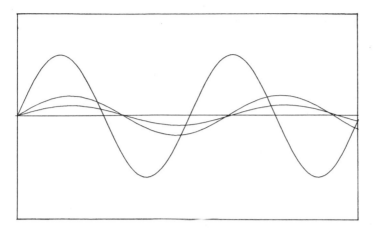

Figure 2.10   Meridional rays in a fiber with a quadratic index profile, eqn (2.25).

it follows, with the use of (2.23), that

$$T = \int dt = \frac{1}{c} \int_0^{r_1} \frac{n^2(r)}{\sqrt{n^2(r) - N_0^2}} \, dr \qquad (2.31)$$

where r_1 is the maximal value of r, i.e. the peak value of the ray pathway.
   From (2.21) it follows that r_1 is determined by

$$n(r_1) = n(0) \cos \theta_0 \qquad (2.32)$$

Insertion of the quadratic profile (2.25) into (2.31) gives

$$T = \frac{1}{c} \int_0^{r_1} \frac{n(0)[1 - \alpha^2 r^2]}{\sqrt{1 - \cos^2 \theta_0 - \alpha^2 r^2}} \, dr \qquad (2/33)$$

with $\alpha^2 = 2\Delta/a^2$.
   The integral can be solved with the help of partial integration and (2.27). The result is

$$T = \frac{n(0)}{c} \frac{1 + \cos^2 \theta_0}{2} \frac{\pi}{2\alpha} \qquad (2.34)$$

The ray's propagation velocity in the z direction is then

$$v_z = \frac{\Lambda/4}{T} = \frac{c}{n(0)} \frac{2 \cos \theta_0}{1 + \cos^2 \theta_0} \qquad (2.35)$$

It is natural to define the numerical aperture for a graded-index fiber as the largest angle of incidence for which the rays travel unbroken through the core, i.e. $r_1 \leq a$. From (2.32) we obtain the maximum value of θ_0

$$\cos \theta_{\max} = \frac{n(a)}{n(0)}$$

The numerical aperture becomes

$$\eta_A = \sin \varphi_{max} = n(0) \sin \theta_{max} = \sqrt{n^2(0) - n^2(a)} \tag{2.36}$$

Observe that the relation is valid for a graded-index fiber with an arbitrary index profile. The relation (2.36) is identical to the expression for a step-index fiber (2.10) with $n(0) = n_1$ and $n(a) = n_2$.

For the quadratic index profile (2.25) the numerical aperture is

$$\eta_A = n(0)\sqrt{2\Delta} \tag{2.37}$$

which is the same expression as (2.13).

An estimate of the dispersion of the fiber can be obtained by calculating the difference in passage time between the slowest and the fastest ray in the fiber

$$\tau_2 - \tau_1 = L\left(\frac{1}{v_{min}} - \frac{1}{v_{max}}\right) \tag{2.38}$$

where L is the length of the fiber.

The fastest ray for quadratic index profiles has $r_1 = 0$; that is, it goes straight through the center of the core

$$v_{max} = \frac{c}{n(0)}$$

The slowest ray has $r_1 = a$, and from (2.35) and (2.25) we obtain

$$\frac{v_{min}}{v_{max}} = \frac{2\cos\theta_{max}}{1 + \cos^2\theta_{max}} = \frac{\sqrt{1 - 2\Delta}}{1 - \Delta}$$

Insertion into (2.38) gives

$$\tau_2 - \tau_1 = L\,\frac{n(0)}{c}\left(\frac{1 - \Delta}{\sqrt{1 - 2\Delta}} - 1\right) \approx L\,\frac{n(0)}{c}\,\frac{\Delta^2}{2} \tag{2.39}$$

This can be compared with the corresponding value (2.16) for a step-index profile

$$\tau_2 - \tau_1 \approx L\,\frac{n_1}{c}\,\Delta \tag{2.40}$$

The passage time difference for a graded-index fiber with a quadratic index profile is thus a factor $\Delta/2$ less than for a step-index fiber.

One can ask if there exists an index profile for which all rays have the same passage time through the fiber. It can be shown, Okoshi (1982), that an index profile for which the velocity v_z is constant and independent of θ_0 for all meridional rays must have the form

$$n(r) = \frac{n(0)}{\cosh(\sqrt{2\Delta}\,r/a)} \tag{2.41}$$

The function (2.41) can also be written as

$$n(r) = n(0)\text{sech}(\sqrt{2\Delta}\ r/a)$$

which motivates the notation hyperbolic secant for this index profile.

We shall calculate the appearance of the ray pathways for this $n(r)$. Insertion into (2.24) gives, with $r_0 = 0$

$$z = \int_0^r \frac{n(0)\cos\theta_0 \cosh(\alpha r)}{\sqrt{n^2(0) - n^2(0)\cos^2\theta_0 \cosh^2(\alpha r)}}\ dr \qquad (2.42)$$

in which $\alpha = \sqrt{2\Delta}/a$.

Changing the integration variable to $t = \sinh(\alpha r)$ gives

$$z = \frac{1}{\alpha} \int_0^t \frac{\cos\theta_0}{\sqrt{1 - \cos^2\theta_0(1 + t^2)}}\ dt$$

which is an integral of the type (2.27), which gives

$$z = \frac{1}{\alpha}\arcsin(t\tan\theta_0) \qquad (2.43)$$

From (2.43) we obtain

$$\sin\alpha z = \tan\theta_0 \sinh(\alpha r)$$

that is

$$r = \text{arcsinh}\left(\frac{\sin(2\sqrt{2\Delta}/a)}{\tan\theta_0}\right) \qquad (2.44)$$

The ray pathways are, as can be seen, not sinusoidal in form but, as shown in Figure 2.11, almost sinusoidal for moderate values of Δ.

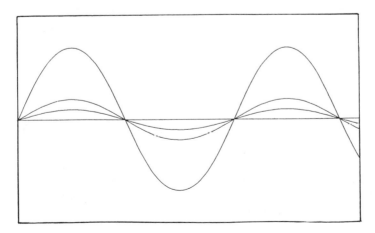

Figure 2.11 Meridional rays in a fiber with an 'optimal' index profile, eqn (2.41). All rays have same propagation time.

The period is determined by the zeros of $\sin(2\Delta z/a)$, which means that

$$\Lambda = \frac{2\pi a}{\sqrt{2\Delta}} \qquad (2.45)$$

is independent of the ray's initial angle θ_0.

Since (2.44) allows an unlimited number of rays between two intersection points on the z axis it follows from Fermat's principle that all rays must have the same propagation velocity $v_z = c/n(0)$. This can be verified by calculating T from (2.31) and showing that it becomes a constant independent of θ_0.

An alternative index profile is

$$n(r) = n(0)[1 - \Delta(r/a)^2] \qquad (2.46)$$

which gives ray pathways according to Figure 2.12.

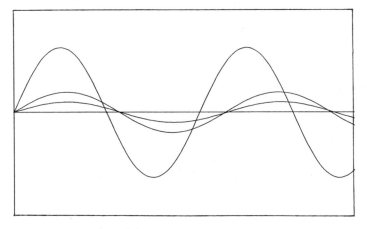

Figure 2.12　Meridional rays in a fiber with an index profile, eqn (2.46).

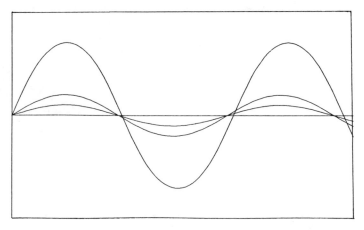

Figure 2.13　Meridional rays in a fiber with an index profile, eqn (2.47). Helical rays have same propagation time for this profile.

Another theoretically interesting case is the so called Lorentzian profile

$$n(r) = \frac{n(0)}{\left[1 + \Delta(r/a)^2\right]^{1/2}} \tag{2.47}$$

The profile (2.47) has the quality that the passage times for spiral-shaped rays are constant, Okoshi (1982) and Marcuse (1982). The ray pathways for meridional rays are shown in Figure 2.13. The fact that the rays have different periods implies that they have different passage times.

We have found that the index profile (2.41) gives a passage time difference of zero for meridional rays and therefore no dispersion for such rays. The same thing applies to the profile (2.47) regarding spiral-formed rays. Unfortunately, there is no profile where all the rays, both meridional and skewed rays, have the same passage time.

3

Optical Waveguides

3.1 Introduction

The function of an optical fiber is to transport optical signals with little loss, distortion or other impairments, i.e. to act as a reliable optical waveguide.

The fundamental property that light is confined within a fiber if the cladding has a refractive index smaller than that of the core was discussed in Chapter 2. To carry the analysis further, a refined mathematical treatment is needed, utilizing the fact that light is an electromagnetic wave which obeys Maxwell's equations. The simple analysis of Chapter 2 does not include the fundamental phenomenon that light has distinct modes of transmission.

The theory of planar waveguides is less complicated than that for optical fibers which are cylindrical structures. The wave equation for a planar structure is solvable in terms of elementary functions, and closed-form expressions for its propagation properties are available, in contrast with the cylindrical waveguide, where it is necessary to rely on numerical methods. We first present the theory of planar waveguides and determine their properties as a transmission medium. The results for the cylindrical waveguide are completely analogous and this knowledge serves as a basis for understanding the cylindrical waveguide and the optical fiber.

3.2 Wave Propagation in Planar Dielectric Waveguides

A planar optical waveguide, often called a slab waveguide, is composed of a dielectric layer surrounded by material with a lower refractive index. The central layer with refractive index n_1 is the waveguide's core. We limit the presentation to the symmetric case where both the surrounding layers have the same refractive index n_2. See Figure 3.1.

Planar optical waveguides are used, for example, in integrated optics where the dielectric layers can be made of an electro-optic material such as lithium niobate $LiNbO_3$ and the waveguide structure can be produced by vapor oxidation onto a substrate.

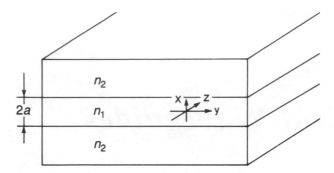

Figure 3.1 Planar optical waveguide. The core has refractive index n_1 and the surrounding material (the cladding) has refractive index n_2.

In our analysis of the transmission characteristics of planar optical waveguides, we will use the fact that light is an electromagnetic wave motion. This means that the electric and magnetic fields which describe this wave motion are determined by Maxwell's equations. For an isotropic, nonconductive medium without charges the following applies, see Cheng (1983).

$$\operatorname{curl} \mathbf{E} = -\mu \, \mu_0 \, \frac{\partial \mathbf{H}}{\partial t} \tag{3.1}$$

$$\operatorname{curl} \mathbf{H} = \varepsilon \, \varepsilon_0 \, \frac{\partial \mathbf{E}}{\partial t} \tag{3.2}$$

$$\operatorname{div} \mu \mathbf{H} = 0 \tag{3.3}$$

$$\operatorname{div} \varepsilon \mathbf{E} = 0 \tag{3.4}$$

where the vectors $\mathbf{E}$ the $\mathbf{H}$ denote the electric and magnetic field strengths, respectively.

The curl and div operators can be expressed in terms of the nabla operator

$$\nabla = \left(\frac{\partial}{\partial x}, \frac{\partial}{\partial y}, \frac{\partial}{\partial z} \right) \tag{3.5}$$

as $\operatorname{curl} \mathbf{E} = \nabla \times \mathbf{E}$ and $\operatorname{div} \mathbf{E} = \nabla \cdot \mathbf{E}$. If the curl operator is applied to (3.1) the relation

$$\nabla \times \nabla \times \mathbf{E} = \nabla (\nabla \cdot \mathbf{E}) - \nabla^2 \mathbf{E} = -\mu \mu_0 \frac{\partial (\nabla \times \mathbf{H})}{\partial t} \tag{3.6}$$

is obtained. The term $\nabla(\nabla \cdot \mathbf{E})$ contains $\nabla \cdot \mathbf{E} = \operatorname{div} \mathbf{E}$, which is equal to zero according to (3.4) because the medium is isotropic, and therefore ε is constant. Substitution of (3.2) into (3.6) gives the wave equation

$$\nabla^2 \mathbf{E} = \mu \mu_0 \varepsilon \varepsilon_0 \frac{\partial^2 \mathbf{E}}{\partial t^2} \tag{3.7}$$

where ∇^2 is the Laplace operator

$$\nabla^2 = \frac{\partial^2}{\partial x^2} + \frac{\partial^2}{\partial y^2} + \frac{\partial^2}{\partial z^2} \tag{3.8}$$

In an unlimited medium (3.7) has solutions of the type

$$\mathbf{E} = \mathbf{E}_0 \, e^{j\omega(t-z/v_f)} \tag{3.9}$$

which corresponds to a plane wave with propagation velocity v_f (phase velocity), moving in the z direction.

Substitution of (3.9) into (3.7) gives

$$\mu\mu_0\varepsilon\varepsilon_0 = 1/v_f^2 \tag{3.10}$$

In a medium with a refractive index n the velocity of light is $v_f = c/n$ and (3.10) is equivalent to

$$\mu\mu_0\varepsilon\varepsilon_0 = \left(\frac{n}{c}\right)^2 \tag{3.11}$$

Applying (3.10) to a vacuum, where the velocity of light is equal to $c = 1/\sqrt{\mu_0 \, \varepsilon_0}$, (3.11) gives the relation

$$n = \sqrt{\mu\varepsilon} \tag{3.12}$$

For nonmagnetic materials such as glass $\mu = 1$, and in connection with optical fibers, the relation is often written as

$$n = \sqrt{\varepsilon} \tag{3.13}$$

It is assumed that the waveguide has a long extension in the y and z directions and that the wave is travelling in the z direction. The electromagnetic field must therefore be constant along the y-axis and we can assume a solution of the form

$$\mathbf{E} = \mathbf{E}(x) \, e^{j(\omega t - \beta z)} \tag{3.14}$$

which is a wave in the z direction, and where the field vector $\mathbf{E}(x)$ depends only on x and not on either y or z.

Substitution into (3.7) and use of (3.11) gives

$$\frac{\partial^2}{\partial x^2} \mathbf{E}(x) + (k^2 n_i^2 - \beta^2)\mathbf{E}(x) = 0 \tag{3.15}$$

where

$$n_i, \quad i = 1, 2$$

are the refractive indices for the core and the cladding, respectively, and

$$k = \frac{\omega}{c} = \frac{2\pi}{\lambda}$$

Observe that λ in the last equation is the wavelength of light in a vacuum (air) and not the wavelength in the waveguide medium.

The quantity β is the wave's propagation parameter which determines how the light confined in the waveguide propagates. In order to determine β we must first determine the field distributions $\mathbf{E}(x)$.

Equation (3.15) is a second-order differential equation with constant coefficients and its solutions are exponential functions. For the y component of $\mathbf{E}(x)$, we assume solutions of the form

$$
\left.
\begin{aligned}
E_y(x) &= A\exp[-p(|x| - a)] \quad |x| \geq a \\
E_y(x) &= \begin{cases} B & \cos(hx) \\ B & \sin(hx) \end{cases} \quad |x| \leq a
\end{aligned}
\right\} \tag{3.16}
$$

This means that the field is concentrated in the waveguide's core and that it decreases exponentially in the areas outside the core. The light is thus contained within the core and the device acts as a waveguide.

The complete field is composed of E and H components. From (3.1) the following relations are obtained. It has been taken into consideration that $\frac{\partial}{\partial y} = 0$ and it is assumed that $\mathbf{H}$ has the same form as $\mathbf{E}$ according to (3.14).

$$
\left.
\begin{aligned}
\frac{\partial E_y}{\partial z} &= -j\omega\mu\mu_0 H_x \\
\frac{\partial E_x}{\partial z} - \frac{\partial E_z}{\partial x} &= -j\omega\mu\mu_0 H_y \\
\frac{\partial E_y}{\partial x} &= -j\omega\mu\mu_0 H_z
\end{aligned}
\right\} \tag{3.17}
$$

In the same way, the following relations are obtained from (3.2):

$$
\left.
\begin{aligned}
-\frac{\partial H_y}{\partial z} &= j\omega\varepsilon\varepsilon_0 E_x \\
\frac{\partial H_x}{\partial z} - \frac{\partial H_z}{\partial x} &= j\omega\varepsilon\varepsilon_0 E_y \\
\frac{\partial H_y}{\partial x} &= j\omega\varepsilon\varepsilon_0 E_z
\end{aligned}
\right\} \tag{3.18}
$$

These six equations can be divided into two independent groups. The first and last equations in (3.17) and the middle one in (3.18) contain only E_y together with H_x and H_z. The solutions which contain these field components are called transverse electric (TE) modes because the E field does not contain any component in the direction of propagation. The remaining equations contain only H_y, E_x and E_z, and correspond to the transverse magnetic (TM) modes for which the H field only has a transverse component.

Consider first the TE modes. For electromagnetic fields the following condition applies: the tangential components of $\mathbf{E}$ and $\mathbf{H}$ at the boundary between two regions with different material characteristics must be continuous. This gives the boundary conditions that determine which values of p and h in (3.16) are possible.

In our case E_y and H_z are tangential at the boundary surface $x = \pm a$. From

(3.16) and the last equation of (3.17) the following relations are obtained for $x = a$

$$\left.\begin{array}{l} A = B \cos(ha) \\ -pA = -Bh \sin(ha) \end{array}\right\}$$ (3.19)

and

$$\left.\begin{array}{l} A = B \sin(ha) \\ -pA = Bh \cos(ha) \end{array}\right\}$$ (3.20)

Equation (3.19) applies to the so-called even modes (cos is an even function) and (3.20) to the odd modes.

Elimination of A and B gives the following equations after multiplication with a:

$$\left.\begin{array}{ll} pa = ha \tan(ha); & \text{even modes} \\ pa = -ha \cot(ha); & \text{odd modes} \end{array}\right\}$$ (3.21)

It can easily be shown that the same equations are also obtained for the boundary surface $x = -a$.

The solutions must also satisfy the wave equation and substitution of (3.16) into (3.15) gives the relations

$$\left.\begin{array}{ll} \beta^2 = k^2 n_2^2 + p^2 & (|x| \geq a) \\ \beta^2 = k^2 n_1^2 - h^2 & (|x| \leq a) \end{array}\right\}$$ (3.22)

Combination of these equations yields the following equation for ha and pa

$$(pa)^2 + (ha)^2 = (n_1^2 - n_2^2)k^2a^2 = V^2$$ (3.23)

The quantity V is called the normalized frequency

$$V = \frac{2\pi a}{\lambda} \sqrt{n_1^2 - n_2^2}$$ (3.24)

Observe that V is inversely proportional to λ and is therefore proportional to the optical signal's frequency.

The equations (3.21) and (3.23) constitute a system of two equations in the two unknowns pa and ha, which consequently are completely determined. The equations can be solved graphically with a diagram as in Figure 3.2. The curves denoted by $m = 0, 2, 4 \ldots$ are generated by the first equation in (3.21) and those denoted by $m = 1, 3, 5 \ldots$ by the second. Equation (3.23) is a circle with radius V and the solutions are those values of pa and ha where the circle intersects the other curves. Those points on the ha axis where the curves (3.21) intersect the axis are called cutoff frequencies because they correspond to the value of V for which the corresponding mode is no longer a solution. By setting $pa = 0$ in (3.21) the cutoff frequencies are obtained:

$$V_m = m\pi/2$$ (3.25)

For $V < \pi/2$ only one intersection point corresponding to a single solution is obtained which means that only one TE mode can exist in the waveguide. This

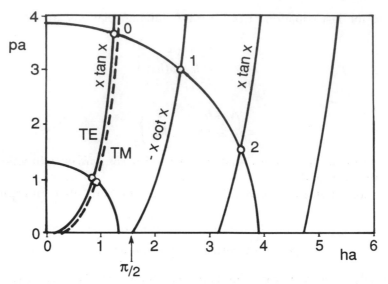

Figure 3.2   Characteristic diagram for symmetric planar waveguides.

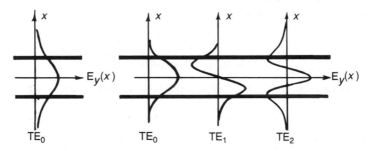

Figure 3.3   Electric field distributions for TE modes in symmetric planar waveguides.

is called single-mode or monomode. In terms of the wavelength λ the single-mode condition is

$$\lambda > 4a\sqrt{n_1^2 - n_2^2} \tag{3.26}$$

The intersection points in Figure 3.2 give the values of p and h which should be used in (3.16). Figure 3.3 shows the distributions of the electric field $E_y(x)$ for a single-mode solution (the small circle in Figure 3.2) and the three modes which are obtained for a V which corresponds to the large circle.

For TM modes the boundary condition implies that H_y and E_z must be continuous at the boundary surface $x = \pm a$. Likewise with (3.16), we assume solutions of the form

$$\left.\begin{aligned}
H_y(x) &= A\ \exp[-p(|x| - a)]; \quad |x| \geq a \\
H_y(x) &= \begin{cases} B & \cos(hx) \\ B & \sin(hx) \end{cases}; \quad |x| \leq a
\end{aligned}\right\} \tag{3.27}$$

which gives for $x = +a$ and even modes

$$A = B \cos(ha) \tag{3.28}$$

From (3.18) it follows that

$$E_z = \frac{1}{j\omega\varepsilon\varepsilon_0} \frac{\partial H_y}{\partial x} \tag{3.29}$$

which gives

$$\frac{1}{j\omega\varepsilon_2\varepsilon_0}(-p)A = -\frac{hB}{j\omega\varepsilon_1\varepsilon_0}\sin(ha) \tag{3.30}$$

Combining (3.28) and (3.30) using (3.13) gives

$$pa = \left(\frac{n_2}{n_1}\right)^2 ha \tan(ha) \tag{3.31}$$

In the same way for the odd modes

$$pa = -\left(\frac{n_2}{n_1}\right)^2 ha \cot(ha) \tag{3.32}$$

Because $n_2 < n_1$ the curves which are generated by (3.31) and (3.32) will be located to the right of the curves which are associated with the TE modes as illustrated in Figure 3.2. When the relative index difference $\Delta = (n_1 - n_2)/n_1$ is small, the intersection points for TE and TM modes will be close to each other and can be considered to have the same transmission properties.

The number of modes which can appear depends on V. Since the cutoff frequencies are given by (3.25), it follows that

$$M = 2\left(\left\lceil\frac{2V}{\pi}\right\rceil + 1\right) \tag{3.33}$$

where $\lceil\ \rceil$ denotes the integer part and the TE and TM modes are considered as separate solutions.

3.3 Propagation Constant for Planar Waveguides

The solutions obtained have the form (3.14) and constitute a wave which travels along the z-axis with phase velocity

$$v_f = \frac{\omega}{\beta} \tag{3.34}$$

The propagation constant β determines how the electromagnetic waves (the light) are transported through the waveguide. It is convenient to introduce the normalized propagation constant

$$b = \left(\frac{pa}{V}\right)^2 = 1 - \left(\frac{ha}{V}\right)^2 \tag{3.35}$$

From (3.22) and (3.23) it follows that b is equal to

$$b = \frac{\beta^2/k^2 - n_2^2}{n_1^2 - n_2^2} \qquad (3.36)$$

For even modes, substitution of the first equation in (3.21) into (3.23) yields

$$(ha)^2 \tan^2(ha) + (ha)^2 = V^2 \qquad (3.37)$$

Elimination of ha with the help of (3.35) gives

$$\tan^2(V\sqrt{1-b}) = \frac{b}{1-b} \qquad (3.38)$$

Because $\tan x$ is equal to $\tan(x + n\pi)$, (3.38) can, after taking the square root of both sides, be written as

$$\tan(V\sqrt{1-b} + \frac{m\pi}{2}) = \pm\sqrt{\frac{b}{1-b}} \qquad (3.39)$$

where $m = 0, 2, 4$, etc.

Using (3.39) and solving for V gives

$$V = \frac{1}{\sqrt{1-b}}\left[m\pi2 + \arctan\sqrt{\frac{b}{1-b}}\right] \qquad (3.40)$$

$$m = 0, 2, 4, \ldots$$

Because (3.39) is obtained by taking the square root of both sides of the equation, there is an uncertainty about the sign of the right-hand side. It can easily be shown, however, that (3.40) is the solution that satisfies the first equation in (3.21) with $pd \geq 0$ and $ha \geq 0$.

In the same way, with the help of the second equation in (3.21) and (3.23), the following is obtained for the odd modes

$$V = \frac{1}{\sqrt{1-b}}\left[\frac{(m+1)\pi}{2} - \arctan\sqrt{\frac{1-b}{b}}\right] \qquad (3.41)$$

$$m = 1, 3, 5, \ldots$$

From (3.40) and (3.41) it is easy to calculate V as a function of b. Normally the inverse function $b = f(V)$ shown in Figure 3.4 is used. In Figure 3.4 every other curve corresponds to an even (TE) mode, and the remaining alternate curves to odd modes.

Equations (3.40) and (3.41) apply to TE modes. By using (3.31) and (3.32), the function $b = f(V)$ for TM modes can be determined. The results are illustrated in Figure 3.5 for the lowest modes ($m = 0$) where $n_1 = 1.5$ and $n_2 = 1$; that is, when the waveguide consists of a sheet of glass in air. In this case, the modes show a distinct difference in propagation constants. When the relative difference between the refractive indices n_1 and n_2 is small, a few percent or less, the curves $b = f(V)$ coincide and the TE and TM modes have in practice the same value of the propagation constant b. Each curve in Figure 3.4 represents, therefore, two modes: a TE and a TM mode.

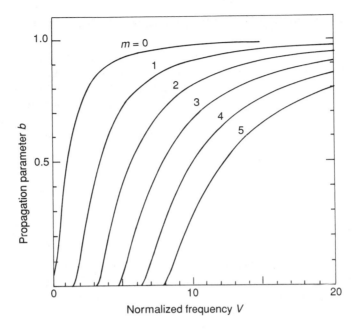

Figure 3.4   Planar waveguide. Normalized propagation parameter b as a function of the normalized optical frequency V. The curves represent TE modes.

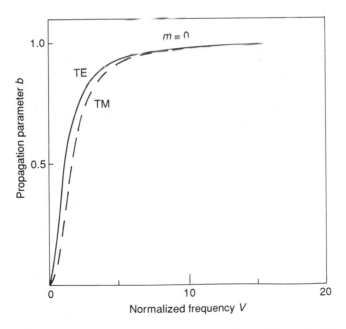

Figure 3.5   Planar waveguide with $n_1 = 1.5$ and $n_2 = 1$. Normalized propagation parameter b as a function of the normalized optical frequency V. The diagram illustrates the characteristics of TE and TM modes when the index difference is large.

3.4 Wave Propagation in Cylindrical Waveguides

The wave propagation properties of planar and cylindrical dielectric waveguides show many similarities. The cylindrical structure is more complicated in the sense that explicit solutions are not available and numerical methods are needed to obtain specific results. The basic principles of guided mode propagation are the same as for the planar waveguide, and we present the results for the cylindrical waveguide without derivations and rely on the analogy to planar waveguides. A more complete treatment can be found in Okoshi (1982) or Marcuse (1982).

A step-index fiber is a cylindrical (circular) waveguide with constant refractive indices n_1 and n_2 for the core and cladding, respectively. The mathematical analysis is most conveniently accomplished using cylindrical (polar) coordinates r, ϕ and z, and follows the same principles as for a planar waveguide. Figure 3.6 shows the expressions derived for planar dielectric waveguides, and those applying to cylindrical waveguides.

We seek solutions representing guided waves travelling in the z direction. The electromagnetic fields are assumed to be of the form

$$\mathbf{E} = \mathbf{E}(r)e^{j\nu\phi}e^{j(\omega t - \beta z)} \tag{3.42}$$

where the parameter ν is an integer called the azimuthal mode number. From Maxwell's equations in cylindrical coordinates, a wave equation analogous to (3.15) is obtained:

$$\frac{\partial^2 \mathbf{E}}{\partial r^2} + \frac{1}{r}\frac{\partial \mathbf{E}}{\partial r} + \left(k^2 n_{1,2}^2(r) - \beta^2 - \frac{\nu^2}{r^2}\right)\mathbf{E} = 0 \tag{3.43}$$

This is a Bessel differential equation and the solutions appropriate for guided modes are Bessel functions $J_\nu(x)$ and modified Bessel functions $K_\nu(x)$, see Figure 3.7. The function J_ν applies to the core and K_ν to the cladding. As seen, the same general character for the field functions is obtained as for planar waveguides with a harmonic function in the core and a monotonically decreasing function in the cladding.

The complete solution including the propagation constant β is obtained from the requirement that the tangential components of $\mathbf{E}$ and $\mathbf{H}$ must be continuous at the boundary between the core and the cladding. The principle is the same as for the planar waveguide but the algebra is considerably more tedious. The result, see e.g. Okoshi (1982), is a characteristic equation analogous to (3.21):

$$\left[\frac{J_\nu'(ua)}{ua\,J_\nu(ua)} + \frac{K_\nu'(wa)}{wa\,K_\nu(wa)}\right]\left[\frac{n_1^2}{n_2^2}\frac{J_\nu'(ua)}{ua\,J_\nu(ua)} + \frac{K_\nu'(wa)}{wa\,K_\nu(wa)}\right]$$

$$= \nu^2\left(\frac{1}{(ua)^2} + \frac{1}{(wa)^2}\right)\left(\frac{n_1^2}{n_2^2}\frac{1}{(ua)^2} + \frac{1}{(wa)^2}\right) \tag{3.44}$$

The variables ua and wa have the same interpretation as ha and pa for the planar waveguide and they are related to the normalized frequency V by an equation

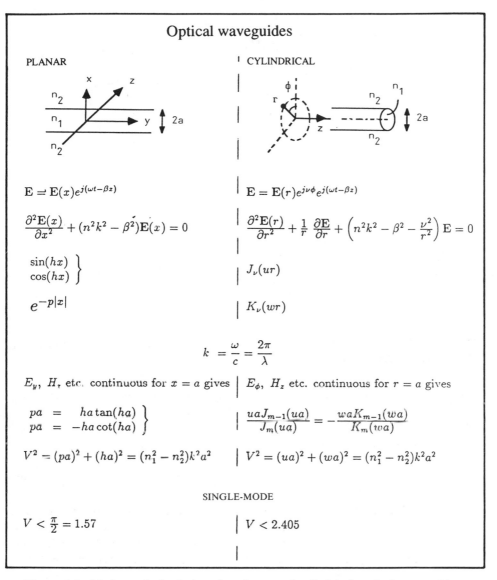

Optical waveguides

PLANAR | **CYLINDRICAL**

$E = E(x)e^{j(\omega t - \beta z)}$

$\dfrac{\partial^2 E(x)}{\partial x^2} + (n^2 k^2 - \beta^2)E(x) = 0$

$\begin{rcases} \sin(hx) \\ \cos(hx) \end{rcases}$

$e^{-p|x|}$

$E = E(r)e^{j\nu\phi}e^{j(\omega t - \beta z)}$

$\dfrac{\partial^2 E(r)}{\partial r^2} + \dfrac{1}{r}\dfrac{\partial E}{\partial r} + \left(n^2 k^2 - \beta^2 - \dfrac{\nu^2}{r^2}\right)E = 0$

$J_\nu(ur)$

$K_\nu(wr)$

$$k = \frac{\omega}{c} = \frac{2\pi}{\lambda}$$

E_y, H_z etc. continuous for $x = a$ gives | E_ϕ, H_z etc. continuous for $r = a$ gives

$\begin{rcases} pa = ha\tan(ha) \\ pa = -ha\cot(ha) \end{rcases}$

$\dfrac{ua J_{m-1}(ua)}{J_m(ua)} = -\dfrac{wa K_{m-1}(wa)}{K_m(wa)}$

$V^2 = (pa)^2 + (ha)^2 = (n_1^2 - n_2^2)k^2 a^2$ | $V^2 = (ua)^2 + (wa)^2 = (n_1^2 - n_2^2)k^2 a^2$

SINGLE-MODE

$V < \frac{\pi}{2} = 1.57$ | $V < 2.405$

Figure 3.6 Mathematical relations for planar and cylindrical optical waveguides.

equivalent to (3.23)

$$V^2 = (n_1^2 - n_2^2)k^2 a^2 = (ua)^2 + (wa)^2 \tag{3.45}$$

The solutions of the characteristic equation (3.44) in combination with (3.45) are the guided modes that can occur in the fiber. As shown in Appendix A, modes of four different types appear which are usually referred to as TE, TM, HE and EH, respectively.

We have seen that fibers for optical communication often have a small index difference between core and cladding. Such waveguides support so called

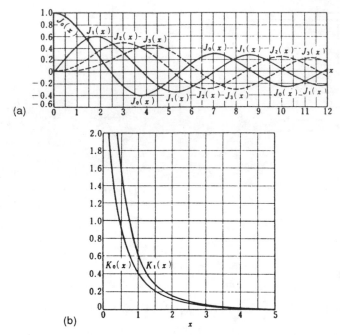

Figure 3.7 Bessel functions $J_m(x)$ and modified Bessel functions $K_m(x)$. From Okhoshi (1982). Reproduced by permission of Academic Press.

weakly guided modes with approximately the same propagation constant which can be grouped together into so-called linear polarized (LP) modes. It is shown in Appendix A that the characteristic equation (A.9) for weakly guided modes in a step-index fiber is

$$\frac{ua\, J_{m-1}(ua)}{J_m(ua)} = -\frac{wa\, K_{m-1}(wa)}{K_m(wa)} \tag{3.46}$$

The possible solutions representing LP modes are obtained by plotting the pair of values for ua and wa satisfying (3.46) in a diagram and determine the intersection of these curves with a circle of radius V generated by (3.45). See Figure 3.8. Comparison with Figure 3.2 shows that the same principle holds as for a planar waveguide.

A curve in Figure 3.8 corresponds to several degenerated modes which have the same propagation constant. With regard to propagation properties, it is therefore often unnecessary to distinguish between the degenerated modes even if they have different field patterns. A common designation LP_{mn} is used for those modes which correspond to a separate solution point in Figure 3.8.

The relations in Table 3.1 for the connections between the different modes are obtained from the derivation of (3.46).

The intensity distributions for the three lowest-order LP modes are shown in Figure 3.10 on p. 46.

The cutoff frequencies are those points on the ua-axis in Figure 3.8 where the curves start. In order to determine these, we seek the solutions to (3.46) when

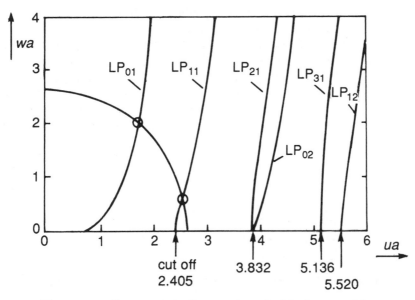

Figure 3.8 Characteristic diagram for cylindrical waveguides.

Table 3.1 Modes in cylindrical waveguides.

m	LP mode	Degenerated modes	No. of modes
$m = 0$	LP_{0n}	$2 \times HE_{1n}$	2
$m = 1$	LP_{1n}	$TE_{0n}, TM_{0n}, 2 \times HE_{2n}$	4
$m > 1$	LP_{mn}	$2 \times EH_{m-1,n}, 2 \times HE_{m+1,n}$	4

$wa = 0$. In Appendix A it is shown that the cutoff frequencies are determined by the equation (A.11)

$$J_{|m-1|}(ua) = 0 \qquad (3.47)$$

with the reservation that $ua = 0$ is a cutoff frequency only for $m = 0$, that is, for the LP_{01} mode. The roots of (3.47) can be obtained from Figure 3.7a and the corresponding cutoff frequencies are indicated in Figure 3.8. From (3.47) it follows that modes with $m = 0$ and $m = 2$ have the same cutoff frequencies.

The smallest root in Figure 3.7 which is greater than zero is given by $J_0(z) = 0$ and it is equal to $z = 2.405$. For the values of V which are smaller than 2.405 only one mode, $LP_{01} = HE_{11}$, can exist in the waveguide, and the single–mode condition for a step-index fiber is thus

$$V < 2.405 \qquad (3.48)$$

The results we have derived apply to fibers with a small index difference between the core and cladding, that is $n_1 - n_2 \ll 1$. If this condition is not fulfilled, one must proceed from the more complicated modal equation (A.1) in Appendix A. What happens is that the modes TE, TM, EH and HE, which are included in the LP modes, are separated and represented by separate curves in the characteristic diagram Figure 3.8. The result is an extended set of modes with different transmission properties. It can be shown, however, that the TE and TM modes have the same cutoff frequency, independently of whether or not the modes are weakly guided.

Because the ground mode LP_{01} consists of only one HE mode, it cannot be split up. This means that the single-mode condition (3.48) applies to optical fibers independent of the index difference $n_1 - n_2$. As we have shown in Section 3.3, the same thing does not apply to a planar optical waveguide. There the ground mode contains two components, TE and TM, which have different transmission characteristics when $n_1 - n_2$ is not small.

3.5   Propagation Constant for Cylindrical Waveguides

Our main interest is in the transmission properties and, as before, it is convenient to use a normalized propagation parameter

$$b = \left(\frac{wa}{V}\right)^2 = 1 - \left(\frac{ua}{V}\right)^2 \tag{3.49}$$

The propagation constant β is related to b by the relation (3.36).

The normalized propagation constant b is obtained, in principle, by determining the value of wa (or ua) for a particular mode at a given value of the normalized frequency V. The solutions are obtained as the intersections illustrated in Figure 3.8 and the result is the set of functions $b = f(V)$, one for each LP mode, which is shown in Figure 3.9. The diagram is from Gloge (1971) and the numbers of the curves are the LP_{mn} mode's number. With the help of Table 3.1, the LP modes can be divided up into degenerate modes, if necessary.

A comparison between Figure 3.4 and Figure 3.9 shows that the normalized propagation constant b displays a similar behavior for planar and cylindrical waveguides. At frequencies which are near the cutoff points, the modes are more poorly contained than for frequencies far from the cutoff points. This implies that a larger proportion of light leaks into the cladding. Figure 3.10 shows the distributions of the optical effect in a cross-section of the fiber for the three lowest-order LP modes for $b = 0.1$ and $b = 0.9$. Because $b = 0$ for cutoff, the light becomes less concentrated in the core for $b = 0.1$ than for $b = 0.9$.

3.6   Wave Propagation in Multimode Fibers

In Sections 3.4 and 3.5 the mode theory of step-index fibers was briefly presented. The electromagnetic field distributions turned out to be expressible in terms of Bessel functions, and the propagation constant was determined

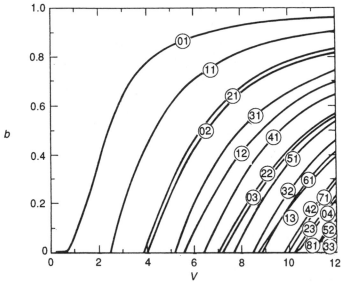

Figure 3.9 Cylindrical waveguide (step-index fiber). Normalized propagation parameter b as a function of the normalized optical frequency V for LP$_{mn}$ modes. The numbers on the curves are the indices mn of the mode. From Gloge (1971). Reproduced by permission of the Optical Society of America.

from the characteristic diagram, Figure 3.8. The results apply primarily to situations with few modes and especially to single-mode transmission. Even if a step-index fiber with many modes can be treated by solving the wave equation for each mode separately, this is cumbersome and impractical.

For a multimode fiber, where the wavelength λ is smaller than the diameter of the fiber, an approximate solution can be obtained by using the WKB method (after Wenzel, Kramers and Brillouin). The approximation can be regarded as an extension of the ray analysis of Chapter 2, where λ also was assumed to be small in comparison with the waveguide dimensions. In contrast with the simple ray theory, the WKB method includes the important notion of transmission modes. One important advantage of the WKB approximation is that it applies not only to step-index fibers but also to gradient-index fibers with more general profiles.

Our primary interest is to determine the propagation properties of the fiber and we concentrate the presentation on the propagation constant β. A simplified and somewhat heuristic derivation is given and we refer to Marcuse (1982) or Okoshi (1982) for a more stringent treatment.

The starting point is the wave equation in cylindrical coordinates (3.43). The transformation

$$r = e^x \tag{3.50}$$

reduces (3.43) to the simpler form

$$\frac{\partial^2 \mathbf{E}}{\partial x^2} + F\mathbf{E} = 0 \tag{3.51}$$

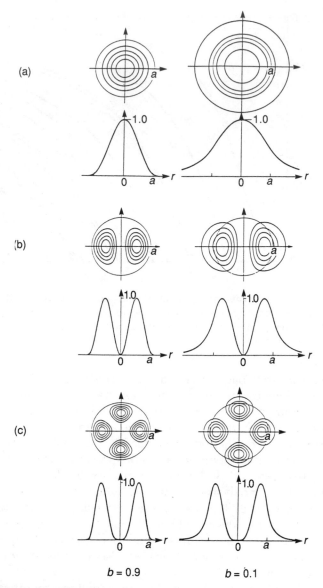

(a)

(b)

(c)

$b = 0.9$ $b = 0.1$

Figure 3.10 Normalized power density distributions of a) the LP_{01} (HE_{11}) mode b) the LP_{11} (TE_{01}, TM_{01} and HE_{21}) mode c) the LP_{21} (EH_{11} and HE_{31}) mode. From Okoshi (1982). Reproduced by permission of Academic Press.

with

$$F = [k^2 n^2(r) - \beta^2]r^2 - \nu^2 = [k^2 n^2(e^x) - \beta^2]e^{2x} - \nu^2 \tag{3.52}$$

We want to obtain approximative solutions of (3.51) representing guided modes. The study of the planar waveguide showed that a guided mode corresponds to a periodic function (3.16) in the core of the fiber. We can therefore expect the solution of (3.51) inside the core to be of the form

$$E(x) = A(x)e^{j\phi(x)} \tag{3.53}$$

where $\phi(x)$ is a real function.

Substitution of (3.53) into (3.51) gives

$$\frac{d^2A}{dx^2} + 2j\frac{dA}{dx}\frac{d\phi}{dx} + jA\frac{d^2\phi}{dx^2} - A\left(\frac{d\phi}{dx}\right)^2 + FA = 0 \tag{3.54}$$

The approximate solutions sought are obtained by neglecting the term d^2A/dx^2 which assumes that the amplitude function does not deviate too much from a linear function. This can be expected to hold for situations where λ is small compared with the fiber core radius a.

The real part of (3.54) then gives

$$\frac{d\phi}{dx} = \sqrt{F} \tag{3.55}$$

which can be integrated to give

$$\phi = \int_{x_1}^x F^{1/2}dx = \int_{r_1}^r [k^2n^2(\rho) - \beta^2 - \nu^2/\rho^2]^{1/2}d\rho \tag{3.56}$$

Since $\phi(r)$ is assumed to be real, the relation (3.56) is valid for r in the interval $r_1 \le r \le r_2$ where $F(r)$ is nonnegative. The parameters r_1 and r_2 are called the turning points of the equation. They are the roots of

$$F(r) = 0$$

and they depend on β and ν, see Figure 3.11. In optical theory the turning points have a physical interpretation as the boundaries where the optical rays change direction due to reflection.

The possible values of β for a planar waveguide were determined by a set of boundary conditions resulting in solutions illustrated in Figure 3.3. The principal feature of the field distributions is that a whole number of half-periods are contained within the core of the waveguide. We can therefore expect that the phase function $\phi(r)$ should satisfy a relation of the form $\phi(r_2) - \phi(r_1) = \mu\pi$. We refer to Marcuse (1982) or Okoshi (1982) for a more careful analysis resulting in the proper WKB equation

$$\int_{r_1}^{r_2}[k^2n^2(\rho) - \beta_{\nu\mu}^2 - \nu^2/\rho^2]^{1/2}d\rho = (\mu + 1/2)\pi \tag{3.57}$$

The propagation constant β for a mode with mode numbers ν and μ is given implicitly by (3.57). For some specific choices of index profile $n(r)$, the integral can be evaluated and an explicit expression obtained.

An important quantity that will be useful in later analysis is the number of guided modes supported by the waveguide. Let $M(\beta)$ denote the number of modes that have propagation constant above β. The analysis in Appendix A shows that each $\nu \ge 1$ generates four physical modes (HE and EH with two

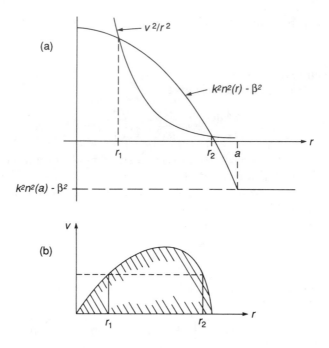

Figure 3.11   The WKB approximation. (a) Diagram showing the turning points r_1 and r_2. (b) The region of integration for the integral (3.59).

polarizations) with the exception of modes with $\nu = 0$ which are only doubly degenerate (TE and TM). $M(\beta)$ is the number of points in the ν, μ space below the dividing curve generated by $\beta = \text{const}$, see Figure 3.12. Replacing the summation by integration makes $M(\beta)$ equal to four times the area of the triangularly shaped region indicated in Figure 3.12

$$M(\beta) = 4 \int_0^{\nu(\beta)} \mu d\nu \tag{3.58}$$

Substitution of μ from (3.57) and neglecting the additive constant $1/2$ gives

$$M(\beta) = 4\pi \int_0^{\nu(\beta)} \int_{r_1}^{r_2} [k^2 n^2(r) - \beta - \nu^2/r^2]^{1/2} dr d\nu \tag{3.59}$$

The area of integration for r and ν is shown in Figure 3.11b and a change of the order of integration yields

$$M(\beta) = 4\pi \int_0^R \int_0^{\nu(r)} [k^2 n^2(r) - \beta^2 - \nu^2/r^2]^{1/2} d\nu dr \tag{3.60}$$

with $\nu(r) = r[k^2 n^2(r) - \beta^2]^{1/2}$ and R defined by

$$kn(R) = \beta \tag{3.61}$$

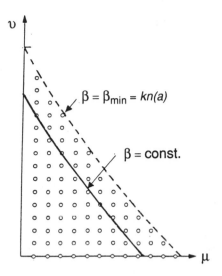

Figure 3.12 The $\mu\nu$-space for mode numbers. The modes with propagation constant above β are in the triangular region bounded by the curve $\beta = $ const.

The inner integral in (3.60) is of the type

$$\int_0^a (a^2 - x^2)^{1/2} dx = \frac{\pi}{4} a^2 \tag{3.62}$$

which when inserted into (3.60) yields

$$M(\beta) = \int_0^R r[k^2 n^2(r) - \beta^2] \, dr \tag{3.63}$$

This integral can readily be solved for the family of index profiles of the power law type

$$n(r) = n_1[1 - 2\Delta(r/a)^\alpha]^{1/2} \tag{3.64}$$

Observe that $\alpha = 2$ gives the quadratic index profile (2.25) and that $\alpha = \infty$ corresponds to a step-index fiber with

$$n_2 = n(a) = n_1[1 - 2\Delta]^{1/2} \tag{3.65}$$

This gives $\Delta = (n_1^2 - n_2^2)/2n_1^2$ which agrees with the previous definition (2.11) of Δ.

Substituting (3.64) into (3.63) and carrying out the integration yields

$$M(\beta) = k^2 n_1^2 a^2 \Delta \frac{\alpha}{\alpha + 2} \left(\frac{k^2 n_1^2 - \beta^2}{2\Delta k^2 n_1^2} \right)^{(2+\alpha)/\alpha} \tag{3.66}$$

It is convenient to express the result in terms of the normalized propagation constant b (3.39) and the normalized frequency V (3.26). The result is

$$M(b) = \frac{\alpha}{\alpha+2} \frac{V^2}{2} (1-b)^{(2+\alpha)/\alpha} \tag{3.67}$$

Because the smallest value b can have is $b = 0$, (3.67) gives the following expression for the total number of modes in a graded-index fiber with a profile according to (3.64):

$$M = M(0) = \frac{\alpha}{\alpha+2} \frac{V^2}{2} \tag{3.68}$$

A comparison with (3.33) shows that a step-index fiber ($\alpha = \infty$) with core diameter $2a$ has many more modes than a planar waveguide with a center region of thickness $2a$ at the same wavelength. This is a natural consequence of the fact that the planar waveguide constitutes a one-dimensional problem, while the fiber exhibits two-dimensional field distributions with additional degrees of freedom.

Example 3.1 Number of modes in step-index fibers

The number of modes for a planar waveguide is given exactly by (3.33). For a step-index fiber, (3.67) with $\alpha = \infty$ gives an approximate value. In Table 3.2 this is compared with the correct number from the mode analysis according to Table 3.1 in Section 3.4

Table 3.2 Number of modes in planar and cylindrical waveguides.

V	Planar waveguide M eqn (3.33)	Step-index fiber ($\alpha = \infty$) M eqn (3.68)	M exact
3	4	4.5	6
6	8	18	20
12	16	72	76

□

The example shows that the WKB analysis can gives a reasonable result even for a value of V as low as $V = 6$.

The WKB approximation was derived under the assumption that $\lambda \ll a$ and it does not apply to monomode fibers. Nevertheless, it can be used to obtain a first estimate of the cutoff frequency for monomode fibers with α-power profile.

In the derivation of relation (3.68), fourfold mode degeneracy is implied, and if we assume the formula to be valid for single modes, $M = 4$ would give the cutoff criterion

$$\frac{\alpha}{\alpha+2} \frac{V_\alpha^2}{2} = 4 \tag{3.69}$$

Table 3.3   Cutoff frequencies for a single-mode fiber with α-power law profile.

α	V_α exact	V_α eqn (3.70)	V_α eqn (3.71)
1	4.382	4.90	4.17
2	3.518	4.00	3.40
4	3.000	3.46	2.95
∞	2.405	2.83	2.40

which gives

$$V_\alpha = \sqrt{\frac{8(\alpha + 2)}{\alpha}}$$

(3.70)

The cutoff frequencies obtained from this formula for some values of α are shown in Table 3.3, together with the exact values from Oyamada and Okoshi (1980).

A somewhat better estimate is obtained if (3.70) is adjusted such that it gives correct results for a step-index fiber ($\alpha = \infty$)

$$V_\alpha = 2.405 \sqrt{\frac{\alpha + 2}{\alpha}}$$

(3.71)

Even if the estimated cutoff frequencies are not particularly accurate, the results show that V_α is decreasing with α, which means that a fiber with a graded-index profile has a larger cutoff frequency than a step index-fiber.

4

Dispersion and Attenuation

4.1 Introduction

An optical pulse transmitted through an optical fiber exhibits attenuation, delay and distortion. A constant delay of a signal due to transmission is normally not a problem, so long as the length of the delay is within reasonable limits. Through appropriate synchronization the transmitted information can be received correctly. Attenuation sets a limit on the length of the fiber and the distance between transmitter and receiver or between signal repeaters. Distortion in the form of pulse broadening can cause pulses which represent different information symbols to overlap and become inseparable at the receiver. Attenuation and distortion set the prime limits on the information transmission performance of optical fiber communication systems.

Pulse broadening, which is commonly referred to as dispersion, has several causes. It depends on the optical fiber transmission characteristics, but also on the light source.

In the following section the different mechanisms for dispersion will be described and analyzed.

4.2 System Function for Optical Waveguides

A linear system is characterized by its impulse response $h(t)$ or by its system function (frequency function) $H(\omega)$, where $\omega = 2\pi f$ is the angular frequency. See Figure 4.1. The frequency function $H(\omega)$ is the Fourier transform of $h(t)$. It is complex-valued and is usually divided up into a real-valued amplitude function $A(\omega)$ and a real-valued phase function $\phi(\omega)$

$$H(\omega) = A(\omega)e^{j\phi(\omega)} \tag{4.1}$$

If the signal $x(t) = e^{j\omega t}$ is applied to the system's input, the output signal will be $y(t) = H(\omega)e^{j\omega t}$.

The signal transmission in an optical fiber is accomplished by the

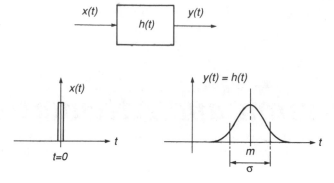

Figure 4.1   A linear system and its impulse response.

$z = O$ $z = L$

Figure 4.2   Signal transmission through an optical fiber.

transportation of optical energy by one or several modes. The analysis of optical waveguides in Section 3.4 showed that the electromagnetic field for a cylindrical structure is of the form (3.42)

$$\mathbf{E} = \mathbf{E}(r, \phi)e^{-j\beta z}e^{j\omega t} \tag{4.2}$$

where $\mathbf{E}(r, \phi)$ and β depend on which mode they represent.

Consider transmission of light by a specific mode in a fiber of length L according to Figure 4.2. Its input at $z = 0$ is

$$\mathbf{E}(z = 0) = \mathbf{E}(r, \phi)e^{j\omega t} \tag{4.3}$$

and the output at $z = L$ becomes

$$\mathbf{E}(z = L) = \mathbf{E}(r, \phi)e^{-j\beta L}e^{j\omega t} \tag{4.4}$$

The system function is the ratio between the output and the input for a $e^{j\omega t}$ signal. Dividing (4.4) by (4.3) gives the frequency function for the particular mode at fiber length L.

$$H_m(\omega) = A_m e^{-j\beta_m(\omega)L} \tag{4.5}$$

where $\beta_m(\omega)$ is the propagation constant for mode number m at the angular frequency ω. The quantity A_m is a constant which depends on how the mode is excited; that is, how large a part of the incoming optical signal is carried by that mode.

For a multimode fiber the total transmission function is obtained through summing all the contributions from the different modes.

$$H(\omega) = \sum_{m=1}^{M} A_m e^{-j\beta_m(\omega)L} \tag{4.6}$$

An optical fiber is a linear system with respect to the electromagnetic field quantities, and its frequency function is given by (4.6). In the following we assume the transmission properties of optical fibers to be modeled by (4.6). This is a consequence of the optical waveguide theory of Chapter 3 and the assumption that the waveguide is constructed from a material free from nonlinear effects.

4.3   Delay and Dispersion

Consider a linear system, i.e. an optical fiber with a narrow pulse input signal $x(t)$. The output signal $y(t)$ is then equal to the system's impulse response $h(t)$. See Figure 4.2.

If $x(t)$ is applied at $t = 0$ then the output $h(t)$ normally appears at a later time $t = \tau$, i.e. it is delayed. As a measure of the location of the delayed signal $h(t)$ we use its center of gravity or mean

$$m_t = \frac{1}{A_h} \int_{-\infty}^{\infty} t\, h(t)\, dt \tag{4.7}$$

where A_h is the area of $h(t)$

$$A_h = \int_{-\infty}^{\infty} h(t)\, dl \tag{4.8}$$

Usually the system causes a change in the shape of the input signal. We characterize the spreading in time of the output signal with σ_t, which is defined as the square root of the quadratic mean value

$$\sigma_t^2 = \frac{1}{A_h} \int_{-\infty}^{\infty} (t - m_t)^2 h(t)\, dt$$

$$= \frac{1}{A_h} \int_{-\infty}^{\infty} t^2 h(t)\, dt - m_t^2 \tag{4.9}$$

The definitions of delay (4.7) and pulse width (4.9) work well when $h(t) \geq 0$ for all the values of t. The quantities m_t and σ_t can then be interpreted as the mean value and standard deviation, respectively, for a stochastic variable with a probability density proportional to $h(t)$. For a general signal, assuming both positive and negative values, they may constitute poor estimates of signal location and broadening. The intensity of an optical signal is always nonnegative, which makes the definitions useful in the present application. They have the advantage of leading to mathematically simple relations in reference to the calculation of delay and dispersion.

In Appendix B it is shown that for a linear system with a frequency function $H(\omega) = A(\omega)e^{j\phi(\omega)}$ the following relations apply:

$$m_t = \tau(0) \tag{4.10}$$

$$\sigma_t^2 = \frac{1}{A_h}\gamma^2(0) \tag{4.11}$$

where

$$\tau(\omega) = -\frac{d\phi(\omega)}{d\omega} \tag{4.12}$$

is the system group delay and

$$\gamma^2(\omega) = -\frac{d^2 A(\omega)}{d\omega^2} \tag{4.13}$$

Notice that the area (4.8) is related to the amplitude function $A(\omega)$ by the relation $A_h = A(0)$. The quantity $\gamma^2(\omega)$ is sometimes called dispersion, but we reserve that label for σ_t.

The quantities m_t and σ_t^2 are added in the case of systems coupled in cascade, see Appendix B. This means that if the input signal to a linear system is a pulse with delay τ_1 and width σ_1 then the output signal exhibits delay τ_2 and width σ_2 determined by

$$\tau_2 = \tau_1 + \tau \tag{4.14}$$

$$\sigma_2^2 = \sigma_1^2 + \sigma^2 \tag{4.15}$$

where τ and σ are the system group delay and dispersion according to (4.10) and (4.11) respectively. This is illustrated in Figure 4.3. The additive relations (4.14) and (4.15) are valid for any input signal and for a general linear system, but observe the above mentioned limitation of σ_t as a measure of the time width for signals.

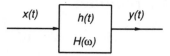

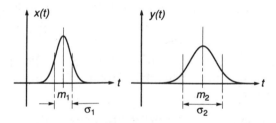

Figure 4.3   Delay and dispersion in a linear baseband system.

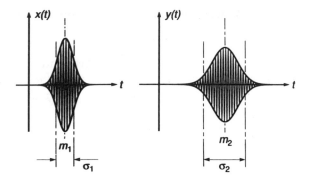

Figure 4.4   Delay and dispersion in a linear bandpass system.

For signals which are modulated on a carrier wave with frequency $f_0 = \omega_0/2\pi$, the additive relations apply approximately for the signals' envelope, as illustrated in Figure 4.4, in which

$$\tau = \tau(\omega_0) \tag{4.16}$$

$$\sigma_t^2 = \frac{1}{A_h}\, \gamma^2(\omega_0) \tag{4.17}$$

This means that the system group delay and dispersion at the carrier angular frequency ω_0 determine the delay and pulse broadening.

The relations are approximately valid if the carrier frequency f_0 is much greater than $1/\sigma_t$, i.e. if the signals are of a narrowband character. It should also be noted that they apply to the signals' mathematical envelope, which is not always the same as the outer contour of the signal.

For further details on delay and dispersion for bandpass signals see Appendix B.

4.4   Delay in Optical Waveguides

The group delay for an optical waveguide is equal to the negative derivative of the phase function according to (4.12). From (4.5) it follows that for a waveguide $\phi(\omega) = -L\beta_m(\omega)$ and

$$\tau(\omega) = L\,\frac{d\beta}{d\omega} = \frac{L}{c}\,\frac{d\beta}{dk} \tag{4.18}$$

since $k = \omega/c$.

The normalized propagation constant b is defined in (3.36). For a small index difference, i.e. $n_1 - n_2 \ll 1$ the following approximate expression is obtained by use of the fact that $n_1 k \geq \beta \geq n_2 k$.

$$b = \frac{(\beta/k - n_2)(\beta/k + n_2)}{(n_1 - n_2)(n_1 + n_2)} \approx \frac{\beta/k - n_2}{n_1 - n_2} \tag{4.19}$$

Replacing $n_1 - n_2$ by $n_1\Delta$, according to (2.12), and solving for β yields

$$\beta \approx k(n_1 b\Delta + n_2) \tag{4.20}$$

When evaluating the derivative of β it must be taken into consideration that the waveguide's material characteristics may vary with the light's wavelength, i.e. n_1 and n_2 are functions of ω or k. Derivation of (4.20) gives

$$\frac{d\beta}{dk} = n_2 + n_1\Delta\,\frac{d(kb)}{dk}$$
$$+ k\left(b\,\Delta\,\frac{dn_1}{dk} + \frac{dn_2}{dk}\right) + n_1 kb\,\frac{d\Delta}{dk} \tag{4.21}$$

The first term in (4.21) represents the delay of an ideal optical waveguide with n_1 and n_2 which are constant and independent of the parameter k. The remaining terms depend on the variation in the index of refraction as a function of the optical wavelength. The derivative of $\Delta \approx 1 - n_2/n_1$ is

$$\frac{d\Delta}{dk} = \frac{n_2\frac{dn_1}{dk} - n_1\frac{dn_2}{dk}}{n_1^2}$$

Substitution into the second part of (4.21) gives

$$\left(\frac{d\beta}{dk}\right)_{mat} = kb\left(\frac{dn_1}{dk} - \frac{dn_2}{dk}\right) + k\,\frac{dn_2}{dk}$$

If the material in the core and cladding have similar characteristics, i.e. if

$$\frac{dn_1}{dk} \approx \frac{dn_2}{dk}$$

then for $\Delta \ll 1$

$$\left(\frac{d\beta}{dk}\right)_{mat} \approx k\,\frac{dn_2}{dk} \tag{4.22}$$

The approximate expression (4.22) can be compared with what is obtained for a planar wave which propagates in an unlimited medium with refractive index n. In this case $\beta = nk$, see (3.9) - (3.14), and

$$\frac{d\beta}{dk} = n + k\,\frac{dn}{dk} \tag{4.23}$$

which agrees with (4.21) and (4.22).

By replacing n_1 with n_2 in the waveguide delay term in (4.21) and by using the approximate expression (4.22) for the material delay the final formula for the delay of a mode with index m becomes

$$\tau_m = \frac{L}{c}\,\frac{d\beta_m}{dk} \approx \frac{L}{c}\left[n_2\left(1 + \Delta\,\frac{d(Vb_m)}{dV}\right) + k\,\frac{dn_2}{dk}\right] \tag{4.24}$$

In (4.24) the derivative $d(kb)/dk$ is replaced with $d(Vb)/dV$. This is allowable because the normalized frequency V is approximately directly proportional to k see (3.23).

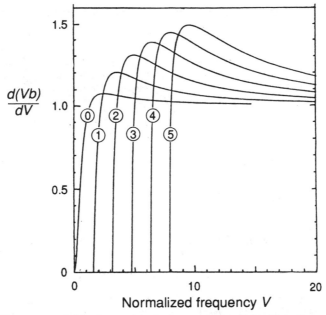

Figure 4.5 The group delay for an ideal planar waveguide. The diagram shows the quantity $d(Vb)/dV$ as a function of the normalized optical frequency V.

4.4.1 Delay in Planar Waveguides

The waveguide delay depends, according to (4.24), on the derivative $d(Vb)/dV$ of the mode in question. Derivation of (3.40) and (3.41) gives the same result

$$\frac{dV}{db} = \frac{1}{2(1-b)}\,(V + 1/\sqrt{b}) \tag{4.25}$$

from which

$$\frac{d(Vb)}{dV} = b + V\,\frac{db}{dV}$$
$$= b + 2(1-b)\,\frac{V}{V + 1/\sqrt{b}} \tag{4.26}$$

The relation (4.26) is shown in Figure 4.5 for the six lowest combined TE and TM modes.

4.4.2 Delay in Cylindrical Waveguides

For a circular waveguide the derivatives $d(Vb)/dV$ as functions of V are shown in Figure 4.6. They resemble, as can be seen, those which were calculated for planar waveguides. The mode indices in Figure 4.6 refer to LP modes.

The waveguide delay is different for different modes. It can be determined from (4.24) by reading the value of $d(Vb)/dV$ at the normalized frequency V in Figure 4.5 or in Figure 4.6 for the appropriate mode.

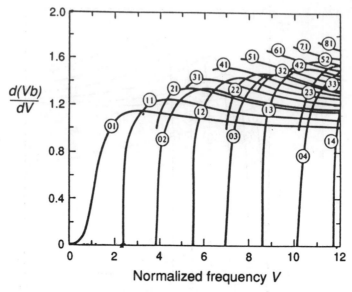

Figure 4.6 The group delay for an ideal cylindrical waveguide (step-index fiber). The diagram shows the quantity $d(Vb)/dV$ as a function of the normalized optical frequency V for LP$_{mn}$ modes. The numbers on the curves are the indices mn of the mode. From Gloge (1971). Reproduced by permission of the Optical Society of America.

4.4.3 Material Delay

In contrast to waveguide delay, material delay is equal for all modes. To calculate it the variation of refractive index with the optical frequency must be known. We illustrate this with an example

Example 4.1 Material delay

A planar waveguide has a mantle of pure (undoped) silicon glass. The core's thickness is 1.56 μm and $\Delta = 0.01$. Calculate the delay for the lowest mode at $\lambda = 1 \ \mu m$.

The refractive index for silicon glass varies with the wavelength according to Figure 4.7. For $\lambda = 1.0 \ \mu m$ the refractive index $n = 1.451$ and $dn/d\lambda$ can be approximated by 1.0, see Figure 4.7

$$\frac{dn}{d\lambda} = \frac{1.444 - 1.457}{1.5 - 0.5} = -0.013$$

The material dispersion is then, according to (4.22)

$$k\frac{dn}{dk} = k\frac{dn}{d\lambda}\frac{d\lambda}{dk} = -\lambda\frac{dn}{d\lambda}$$

The normalized frequency is

$$V = nka\sqrt{2\Delta}$$

$$= 1.451 \cdot \frac{2\pi}{10^{-6}} \cdot \left(\frac{1.56}{2}\right) \cdot 10^{-6}\sqrt{2 \cdot 0.01} = 1.0$$

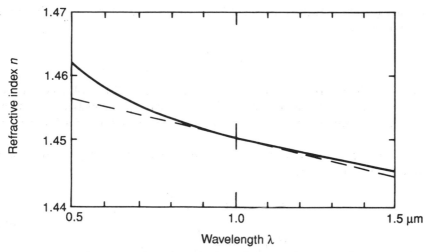

Figure 4.7   The refractive index of pure silica glass as a function of wavelength. Data from Fleming (1976). Reprinted by permission of the American Ceramic Society.

From Figure 4.5 the waveguide delay for the ground mode is obtained as

$$\frac{d(Vb)}{dV} = 0.92 \text{ for } V = 1$$

Insertion into (4.24) gives the total delay

$$\tau = \frac{L}{c}\left(n_2\left(1 + \Delta \; \frac{d(Vb)}{dV}\right) + k \; \frac{dn_2}{dk}\right)$$

$$= \frac{L}{c}\left[1.451(1 + 0.01 \cdot 0.92) + 1 \cdot 0.013\right]$$

$$= 1.477 \; \frac{L}{c} = 4.92 \; \mu s/km$$

The delay is dominated by the constant term Ln_2/c. The remaining terms are dependent on the wavelength and in this Example, they are about the same size and each is equal to about 1% of the constant delay.   □

4.4.4   Delay in Multimode Waveguides

The expression (4.24) applies to a step-index multimode waveguide by calculating the delay for each mode separately. For a fiber that supports a large number of modes this is impractical. The WKB approximation, presented in Section 3.6 offers an easier way to estimate the delays that are associated with the modes in a multimode fiber. The method also has the advantage that it is valid for step-index as well as graded-index fibers.

We start with (3.67) which expresses the number $M(b)$ of modes with normalized propagation constant exceeding b. Clearly the mode with number $m = M(b)$ will have a propagation constant close to b, and (3.67) with the use of (3.68) gives the relation

$$m = M(b) = M(1 - b)^{(2+\alpha)/\alpha} \tag{4.27}$$

By solving (4.27) for b, the normalized propagation constant for mode number m is obtained as

$$b_m = 1 - \left(\frac{m}{M}\right)^{\alpha/(\alpha+2)} \tag{4.28}$$

A step-index fiber corresponds to $\alpha = \infty$, and from (4.28) it follows that WKB analysis in this case leads to a normalized propagation constant b_m which is uniformly distributed between its smallest value $b_m = 0$ and its largest value $b_m = 1$.

The physical propagation constant β_m is obtained from (4.28) and the relation (3.36) together with (2.11):

$$\begin{aligned}
\beta_m^2 &= k^2 \left(n_1^2 - (n_1^2 - n_2^2) \left(\frac{m}{M}\right)^{\alpha/(\alpha+2)} \right) \\
&= k^2 n_1^2 \left(1 - 2\Delta \left(\frac{m}{M}\right)^{\alpha/(\alpha+2)} \right)
\end{aligned} \tag{4.29}$$

The delay for mode number m is according to (4.18)

$$\tau_m = \frac{L}{c} \frac{d\beta_m}{dk} \tag{4.30}$$

In (4.29) n_1, Δ and M depend on k. Derivation of the last expression in (4.29) gives

$$\tau_m = \frac{Lkn_1}{c\beta_m} \left[N_1 - \frac{4\Delta}{\alpha+2} \left(N_1 + \frac{n_1 k}{2\Delta} \frac{d\Delta}{dk} \right) \left(\frac{m}{M}\right)^{\alpha/(\alpha+2)} \right] \tag{4.31}$$

where

$$N_1 = \frac{d(kn_1)}{dk} = n_1 + k\frac{dn_1}{dk} \tag{4.32}$$

which is usually called the group index.

4.5   Dispersion in Multimode Waveguides

For multimode fibers a pulse spreading occurs due to the fact that the modes have different group delays.

There is, in addition, another reason for pulse broadening in fiber optical systems, namely, so-called chromatic dispersion, which will be dealt with in Section 4.6. This depends on the fact that, in practice, a light source does not contain light of only one wavelength.

4.5.1   Pulse Broadening Due to Modal Delay

Let the input to the optical waveguide be a pulse $G(t)$ which modulates the intensity of an optical signal with the frequency $f_0 = c/\lambda_0$. The intensity is

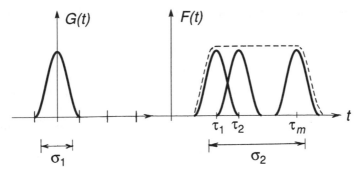

Figure 4.8 Dispersion in multi-mode fibers. The pulse broadening is caused by differences in the delays of the modes.

proportional to the square of the optical field, and the input signal to the fiber considered as a linear system is

$$x(t) = \text{const.} \sqrt{G(t)} \cdot e^{j\omega_0 t}$$

We assume that M modes can be supported in the waveguide and that these are excited uniformly by the input so that each mode contains $1/M$ of the total optical effect. The output will then consist of M pulses of the shape $G(t)$ delayed by the times τ_m, $m = 1, 2, \ldots, M$, see Figure 4.8

$$F(t) = \frac{1}{M} \sum_{m=1}^{M} G(t - \tau_m) \qquad (4.33)$$

where according to (4.18)

$$\tau_m = \tau_g(\omega_0) = L \left. \frac{d\beta}{d\omega} \right|_{\omega=\omega_0} \qquad (4.34)$$

is the group delay at the angular frequency ω_0 for mode m.

The expression (4.33) means that the intensities for the signals in the different modes are added, and not the optical field signals. In Section 4.6.2 we will study the characteristics of the light sources and show under which circumstances this applies.

Denote the area of $G(t)$ with

$$A_1 = \int G(t) \, dt \qquad (4.35)$$

Let the signal $G(t)$ be centered around $t = 0$, i.e.

$$\int t G(t) \, dt = 0 \qquad (4.36)$$

The quadratic mean width of the pulse $G(t)$ is

$$\sigma_1^2 = \frac{1}{A_1} \int t^2 G(t) \, dt \qquad (4.37)$$

For the output pulse $F(t)$ according to (4.9)

$$\sigma_2^2 = \frac{1}{A_2}\int t^2 F(t)\,dt - \left[\frac{1}{A_2}\int tF(t)\,dt\right]^2 \qquad (4.38)$$

where

$$A_2 = \int F(t)\,dt = \frac{1}{M}MA_1 = A_1$$

Substitution into (4.33) gives

$$\int tF(t)\,dt = \frac{1}{M}\sum_{m=1}^{M}\int tG(t - \tau_m)\,dt$$

$$= \frac{1}{M}\sum_{m=1}^{M}\int (x + \tau_m)G(x)\,dx = \frac{A_1}{M}\sum_{m=1}^{M}\tau_m \qquad (4.39)$$

$$\int t^2 F(t)\,dt = \frac{1}{M}\sum_{m=1}^{M}\int t^2 G(t - \tau_m)\,dt$$

$$= \frac{1}{M}\sum_{m=1}^{M}\int (x^2 + 2\tau_m x + \tau_m^2)G(x)\,dx \qquad (4.40)$$

$$= \frac{1}{M}\left[MA_1\sigma_1^2 + A_1\sum \tau_m^2\right]$$

From (4.38) follows that

$$\sigma_2^2 = \sigma_1^2 + \langle\tau_m^2\rangle - \langle\tau_m\rangle^2$$
$$= \sigma_1^2 + \sigma_M^2 \qquad (4.41)$$

where

$$\langle\tau_m^2\rangle = \frac{1}{M}\sum_{m=1}^{M}\tau_m^2 \qquad (4.42)$$

and

$$\langle\tau_m\rangle = \frac{1}{M}\sum_{m=1}^{M}\tau_m \qquad (4.43)$$

are equal to the quadratic mean value and the mean value of the modal delays τ_m, respectively.

If the input pulse is narrow, $\sigma_1 \ll \sigma_M$, the output pulse width σ_2 becomes equal to the modal dispersion σ_M, which is determined by the waveguide's delay characteristics.

4.5.2 Small Number of Modes

For an optical waveguide with a small number of modes, the modal dispersion is obtained by direct insertion into (4.41)

$$\sigma_M^2 = \langle \tau_m^2 \rangle - \langle \tau_m \rangle^2 \tag{4.44}$$

We illustrate this with an example.

Example 4.2   Planar waveguide. Small number of modes

Calculate the modal dispersion for a planar waveguide at the normalized frequency $V = 9$. The refractive index $n_1 = 1.5$ and $\Delta = 0.01$.

Solution

At $V = 9$ the waveguide supports six TE and six TM modes. From Figure 4.5 or alternately by solving (3.40), (3.41) and (4.26) numerically, the values of $d(Vb)/dV$ at $V = 9$ in Table 4.1 are obtained

Table 4.1

m	1	2	3	4	5	6
$\frac{d(Vb)}{dV}$	1.02	1.08	1.17	1.29	1.42	1.47

The group delay τ_m is, according to (4.24)

$$\tau_m = \frac{L}{c} \left[n_2 \left(1 + \Delta \, \frac{d(Vb_m)}{dV} \right) + k \, \frac{dn_2}{dk} \right] \tag{4.45}$$

This can be written as

$$\tau_m = \tau_m' + \mu \tag{4.46}$$

where

$$\tau_m' = \frac{L n_2 \Delta}{c} \frac{d(Vb_m)}{dV} \tag{4.47}$$

is the part of (4.45) which depends on m. Substitution of (4.46) into (4.44) gives

$$\sigma_M^2 = \langle \tau_m'^2 \rangle - \langle \tau_m' \rangle^2 = \left(\frac{L n_2 \Delta}{c} \right)^2 \left\{ \left\langle \left[\frac{d(Vb_m)}{dV} \right]^2 \right\rangle - \left\langle \frac{d(Vb_m)}{dV} \right\rangle^2 \right\}$$

The table above gives

$$\left\langle \left[\frac{d(Vb_m)}{dV} \right]^2 \right\rangle = \frac{1}{6} \, (1.02^2 + 1.08^2 + \cdots 1.47^2) = 1.570$$

$$\left\langle \frac{d(Vb_m)}{dV} \right\rangle = \frac{1}{6} \, (1.02 + 1.08 + \cdots 1.47) = 1.242$$

which results in

$$\sigma_M = \frac{10^3 \cdot 0.99 \cdot 1.5 \cdot 0.01}{3 \cdot 10^8} \sqrt{1.570 - 1.242^2} = 8.2 \text{ ns/km}$$

□

A cylindrical optical waveguide with a small number of modes can be treated similarly by determining $d(Vb_m)/dV$ from Figure 4.6 or calculating it from the waveguide theory.

Observe that the modal dispersion depends only on the waveguide's ideal characteristics. Variations in the refractive index do not affect σ_M because it causes an equal delay for all modes.

4.5.3   Large Number of Modes

For a multimode fiber which supports a large number of modes, it is impractical to calculate modal delay exactly for each mode. The WKB analysis in section 4.4.4 gives an approximate expression (4.31) for group delay of a step-index or graded-index fiber. Because the modal dispersion σ_M^2 contains a quadratic mean value, τ_m need not be known with great accuracy for the averaging to give the correct result if the number of modes is not too small.

We introduce a parameter

$$\varepsilon = \frac{2n_1}{N_1} \frac{k}{\Delta} \frac{\partial \Delta}{\partial k} \tag{4.48}$$

and the notations

$$\frac{m}{M} = x$$

$$\frac{\alpha}{\alpha + 2} = \gamma$$

and (4.31) can then be written as

$$\tau(x) = \frac{LN_1 k n_1}{c \beta_m} \left[1 - \frac{\Delta}{\alpha + 2} (4 + \varepsilon) x^\gamma \right] \tag{4.49}$$

Substitution of β_m from (4.29) gives

$$\tau(x) = \frac{LN_1}{c} \frac{[1 - Bx^\gamma]}{\sqrt{1 - 2\Delta x^\gamma}} \tag{4.50}$$

with

$$B = \frac{(4 + \varepsilon)\Delta}{\alpha + 2}$$

For $\Delta \ll 1$ the expression can be expanded in a Taylor series

$$\tau(x) = \frac{LN_1}{c} (1 - Bx^\gamma) \left(1 + \Delta x^\gamma + \frac{3}{2} (\Delta x^\gamma)^2 + \cdots \right)$$

If the terms up to and including Δ^2 are retained, we obtain

$$\tau(x) \simeq \frac{LN_1}{c} \left(1 + (\Delta - B)x^\gamma + \left(\frac{3}{2}\Delta^2 - \Delta B \right)x^{2\gamma} \right)$$

$$= \frac{LN_1}{c} (1 + c_1 \Delta x^\gamma + c_2 \Delta^2 x^{2\gamma})$$

(4.51)

with

$$c_1 = \frac{(\alpha - 2 - \varepsilon)}{\alpha + 2}$$

$$c_2 = \frac{3\alpha - 2(1 + \varepsilon)}{2(\alpha + 2)}$$

For a large number of modes the sums (4.42) and (4.43) can be replaced by integrals, that is

$$\frac{1}{M} \sum_{m=1}^{M} \tau_m \approx \int_0^1 \tau(x) \, dx; \quad i = 1, 2$$

(4.52)

Substitution of (4.51) into (4.52) leads to simple integrals. The result is, after substitution of $\gamma = \alpha/(\alpha + 2)$, from Olshansky and Keck (1976)

$$\sigma_M^2 = \left(\frac{LN_1 \Delta}{c} \right)^2 \frac{\alpha^2(\alpha + 2)}{4(\alpha + 1)^2(3\alpha + 2)}$$

$$\times \left[c_1^2 + \frac{4\Delta c_1 c_2(\alpha + 1)}{2\alpha + 1} + \frac{16\Delta^2 c_2^2(\alpha + 1)^2}{(5\alpha + 2)(3\alpha + 2)} \right]$$

(4.53)

A step-index profile corresponds to $\alpha = \infty$ and (4.53) gives the modal dispersion for a step-index fiber with $\Delta \ll 1$.

$$\sigma_M = \frac{LN_1 \Delta}{\sqrt{12c}}$$

(4.54)

This can be compared with the result (2.16) which was obtained by calculating the ray paths

$$\tau_2 - \tau_1 = \frac{Ln_1 \Delta}{c}$$

(4.55)

The factor $\sqrt{12}$ which distinguishes the expressions from each other is a direct consequence of the fact that σ_M represents a mean value, and $\tau_2 - \tau_1$ is the difference between two extreme delays.

For a graded-index fiber the choice of α affects the result. If $\alpha = 2 + \varepsilon$ then $c_1 = 0$ and $c_2 = 1/2$ and

$$\sigma_M^2 = \left(\frac{LN_1 \Delta}{c} \right)^2 \frac{\Delta^2 \alpha^2(\alpha + 2)}{(3\alpha + 2)^2(5\alpha + 2)}$$

(4.56)

If $\Delta \ll 1$ then normally also $\varepsilon \ll 1$ and $\alpha \simeq 2$. Insertion of $\alpha = 2$ into (4.56) gives

$$\sigma_M = \frac{LN_1\Delta}{\sqrt{12}c}\frac{\Delta}{2} \tag{4.57}$$

Comparison with (4.54) shows that the modal dispersion for a graded-index fiber with $\alpha = 2 + \varepsilon$ is reduced by a factor $\Delta/2$, compared with a step-index fiber the same result that was obtained earlier with the help of ray analysis in (2.24) and (2.25).

Olshansky and Keck (1976) state that the value of α that minimizes (4.53) is

$$\alpha_0 = 2 + \varepsilon - \Delta\,\frac{(4 + \varepsilon)(3 + \varepsilon)}{5 + 2\varepsilon} \tag{4.58}$$

Insertion into (4.53), under the assumption that $\Delta \ll 1$ and $\varepsilon \ll 1$, gives

$$\sigma_M = \frac{LN_1\Delta}{\sqrt{12}c}\frac{\Delta}{10} \tag{4.59}$$

An index profile of type (3.64) with an optimal profile parameter close to $\alpha = 2$ and with $\Delta = 10^{-2}$, gives an improvement of the modal dispersion by a factor of 500 compared to a step-index fiber.

4.6 Chromatic Dispersion

We have until now, in our description of light as an electromagnetic wave, considered field quantities with a deterministic time variation $e^{j\omega t}$. This corresponds to an idealized situation with light of one constant wavelength $\lambda = 2\pi c/\omega$. The light from real light sources always contains a mixture of light with different wavelengths. Those sources which exhibit narrow spectral lines have the wavelengths concentrated in a smaller interval around λ than those with a broader spectrum.

The group delay $\tau(\omega)$ depends on the angular frequency ω and therefore on λ. This means that different parts of the optical signal spectrum are transported with different velocities which, as we shall show in this section, results in pulse spreading. We call this type of pulse broadening chromatic dispersion because it is caused by the fact that light does not contain only one color, but is composed of components with different wavelengths.

In order to treat chromatic dispersion we need a refined model of an optical signal.

4.6.1 A Stochastic Model of Noncoherent Light

Light from lasers and light-emitting diodes is generated by random mechanisms on the atomic level under more or less controlled conditions. It is therefore natural to represent the optical electromagnetic field as a stochastic

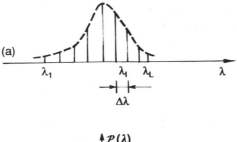

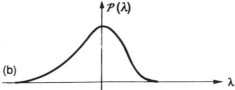

Figure 4.9 Spectral model of an optical signal: (a) discrete spectrum (b) continuous spectral density obtained by letting $\Delta\lambda \to 0$.

process. This can be done in several different ways. We have chosen a model for noncoherent optical signals which leads to simple calculations for the problems we are investigating.

We describe the electromagnetic optical field quantities E and H as stochastic processes using complex signal notation. Consider the finite sum

$$B_L(t) = \sum_{\ell=1}^{L} B_\ell \, \exp[j(\omega_\ell t + \varphi_\ell)] \tag{4.60}$$

where $\omega_\ell = 2\pi c/\lambda_\ell$. The wavelengths λ_ℓ are separated by the distance Δ_λ as is shown in Figure 4.9a. The stochastic optical signal is defined as

$$B(t) = \lim_{L \to \infty} B_L \tag{4.61}$$

The amplitudes B_ℓ and the wavelengths λ_ℓ are assumed to be deterministic, but the phase angle φ_ℓ is random with a uniform distribution $0 \le \varphi \le 2\pi$, and independent for different indices ℓ as a result of the noncoherent character of the optical signal.

The power of a sinusoidal signal with amplitude B is equal to $\mathcal{P} = B^2/2$, and the optical power (intensity) of B_L is

$$2\mathcal{P}_L = E\{|B_L(t)|^2\}$$

$$= E\left\{ \sum_\ell B_\ell e^{j(\omega_\ell t + \varphi_\ell)} \sum_n B_n e^{-j(\omega_n t + \varphi_n)} \right\} \tag{4.62}$$

$$= \sum_\ell B_\ell^2 + E\left\{ \sum_{\ell \neq n} \sum_n B_\ell B_n \, e^{j(\omega_\ell - \omega_n)t} e^{j(\varphi_\ell - \varphi_n)} \right\} = \sum_\ell B_\ell^2$$

which follows from the fact that $E\{e^{j(\varphi_\ell - \varphi_n)}\} = 0$ for $\ell \neq n$. This means power addition, i.e. that the total optical power $\mathcal{P}_L = \sum B_\ell^2/2$ is the sum of the powers of the separate frequency components.

The optical power density at $\lambda = \lambda_\ell$, measured in power per unit wavelength, is defined as

$$P(\lambda_\ell) \triangleq \frac{B_\ell^2}{2\Delta_\lambda} \tag{4.63}$$

We now let $\Delta_\lambda \to 0$ and $B_\ell^2 \to 0$ in such a way that $P(\lambda_\ell)$ approaches the optical signal spectral density $P(\lambda)$. The line spectrum in Figure 4.9a then becomes the continuous function $P(\lambda)$. See Figure 4.9b.

Insertion of (4.63) into (4.62) shows that the signal power is equal to the integral of the power density

$$P = \lim_{L \to \infty} \sum_{\ell=1}^{L} P(\lambda_\ell)\Delta_\lambda = \int_0^\infty P(\lambda)\, d\lambda \tag{4.64}$$

Note that $P(\lambda)$ is the one-sided power spectral density (intensity spectrum) of the light source.

4.6.2 Multimode Optical Signals

Modal dispersion appears because light in the optical waveguide is transmitted by modes with different group velocities. If the modes are excited by a light source according to the model (4.60), it is natural to assume that the phase angles φ will be independent between modes.

Consider an idealized multimode situation where each mode contains one single frequency ω_m.

$$B_m(t) = B_m \exp[j(\omega_m t - \varphi_m)] \tag{4.65}$$

To study a system using intensity modulation with the shape $G(t)$ consider a situation where the optical field is modulated by $\sqrt{G(t)}$. The resulting received signal is the sum over the M modes

$$B_G(t) = \sum_{m=1}^{M} \sqrt{G(t - \tau_m)} B_m(t - \tau_{fm}) \tag{4.66}$$

where τ_m is the group delay of mode m at angular frequency ω_m. The phase delay τ_{fm} for a fiber of length L_0 is related to the phase velocity (3.10)

$$\tau_{fm} = \frac{L_0}{v_f} = \frac{L_0 \beta_m}{\omega_m} \tag{4.67}$$

The time-varying optical intensity of $B_G(t)$ is

$$G_M(t) = E\{|B_G(t)|^2\}/2 = \sum_{m=1}^{M} \frac{B_m^2}{2} G(t - \tau_m) \tag{4.68}$$

Under the assumption that the modes are excited equally, each with a fraction $1/M$ of the signal power, normalization by letting $B_m^2/2 = 1/M$ yields

$$G_M(t) = \frac{1}{M} \sum_{m=1}^{M} G(t - \tau_m) \tag{4.69}$$

with $G(t)$ the optical intensity of the input signal pulse.

This is equal to the expression (4.33) used in Section 4.5.1 to calculate modal dispersion. The analysis assumes a light source composed of single frequency carriers (4.65) with random phase values with the result that the contributions from separate modes add in intensity.

A more realistic model is to let the signal in each mode be of the type (4.60)

$$B_m(t) = \sum_{\ell=1}^{L_m} B_\ell^{(m)} \exp[j(\omega_\ell^{(m)} t + \varphi_\ell^{(m)})] \tag{4.70}$$

For a signal composed of M modes

$$B(t) = \sum_{m=1}^{M} B_m(t) \tag{4.71}$$

It is easily shown with the technique used in (4.62) that the spectral density for $B(t)$ is

$$P(\lambda) = \sum_{m=1}^{M} P_m(\lambda) \tag{4.72}$$

where $P_m(\lambda)$ are the separate modes' spectral densities.

A more complete analysis using the source model (4.70) will be presented in Section 4.7.

An optical fiber with an incoherent stochastic light source is linear in intensity, i.e. the signal envelope squared adds and not the signal's amplitude. Observe that this characteristic depends on the light source's random character. The optical waveguide itself is, as we showed in Section 4.2, a linear system in the usual sense.

For single-mode transmission the idealized model (4.66) represents a coherent source with

$$B_G(t) = \sqrt{G(t - \tau)} \exp[j(\omega_0 t + \varphi)] \tag{4.73}$$

Application of (4.11), (4.13) and (4.5) gives zero dispersion

$$\sigma_t^2 = \frac{1}{A_h} \left. \frac{d^2 A(\omega)}{d\omega^2} \right|_{\omega=0} = 0 \tag{4.74}$$

This is true for a baseband system but for optical signals, which with necessity are of a bandpass character, the relation is approximate and a single-mode fiber with a coherent light source exhibits dispersion. It depends on the particular shape of the optical pulse. See Appendix B for an example of this type of pulse broadening.

The coherent dispersion in single-mode fibers is small and can often be neglected, compared with the chromatic dispersion caused by the noncoherent character of the light source.

4.6.3   *Chromatic Dispersion in Single-Mode Fibers*

Consider a single-mode fiber, where only one transmission mode is possible, in combination with a noncoherent light source. An optical pulse transmitted through the fiber is subjected to a delay τ equal to the group delay of the fiber's lowest mode. Dispersion appears because τ depends on the wavelength λ in combination with the fact that the light contains a spectrum of different wavelengths. The parts of the signal with different frequency content are transmitted with different velocities within the fiber, which results in spreading of the received pulse.

Let the input signal to the fiber be light which is intensity-modulated with a pulse $G(t)$, which means an optical field $\sqrt{G(t)}\,B(t)$ with $B(t)$ according to (4.60).

The optical field for the input signal is

$$B_G^{(1)}(t) = \sum_{\ell=1}^{L} B_\ell \sqrt{G(t)} \exp[j(\omega_\ell t + \varphi_\ell)] \tag{4.75}$$

The terms in the sum will have different delays and the output signal becomes, cf. (4.66)

$$B_G^{(2)}(t) = \sum_{\ell=1}^{L} B_\ell \sqrt{G(t - \tau_\ell)} \exp[j(\omega_\ell(t - \tau_{f\ell}) + \varphi_\ell)] \tag{4.76}$$

where $\tau_\ell = \tau(\omega_\ell)$ is the group delay and $\tau_{f\ell}$ is the phase delay at the angle frequency ω_ℓ.

The output signal's quadratic envelope is, in the same way as (4.62)

$$F(t) = E\{|\,B_G^{(2)}(t)\,|^2\} = \sum_{\ell=1}^{L} \frac{B_\ell^2}{2} G(t - \tau_\ell) \tag{4.77}$$

The expression (4.77) is similar to (4.33) and the pulse width σ_2 for $F(t)$ is analogous with (4.41)

$$\sigma_2^2 = \sigma_1^2 + \langle \tau^2 \rangle - \langle \tau \rangle^2 \tag{4.78}$$

where

$$\left. \begin{aligned} \langle \tau \rangle &= \frac{\sum B_\ell^2 \tau_\ell}{\sum B_\ell^2} \\[2mm] \langle \tau^2 \rangle &= \frac{\sum B_\ell^2 \tau_\ell^2}{\sum B_\ell^2} \end{aligned} \right\} \tag{4.79}$$

When Δ_λ approaches zero (4.79) becomes

$$\langle \tau(\lambda) \rangle = \frac{\int P(\lambda)\tau(\lambda)\, d\lambda}{\int P(\lambda)\, d\lambda} \left.\begin{matrix} \\ \\ \\ \\ \end{matrix}\right\}$$

$$\langle \tau^2(\lambda) \rangle = \frac{\int P(\lambda)\tau^2(\lambda)\, d\lambda}{\int P(\lambda)\, d\lambda} \qquad (4.80)$$

Often the light source's spectral width is small and $\tau(\lambda)$ can then be expanded in a power series around the light source's central wavelength

$$\tau(\lambda) = \tau(\lambda_0) + (\lambda - \lambda_0)\left(\frac{d\tau}{d\lambda}\right)_{\lambda=\lambda_0} \qquad (4.81)$$

with

$$\lambda_0 = \frac{\int \lambda P(\lambda)\, d\lambda}{\int P(\lambda)\, d\lambda} \qquad (4.82)$$

Insertion in (4.80) and (4.78) gives

$$\sigma_2^2 = \sigma_1^2 + \sigma_\lambda^2 \left(\frac{d\tau}{d\lambda}\right)^2 \qquad (4.83)$$

where

$$\sigma_\lambda^2 = \frac{\int (\lambda - \lambda_0)^2 P(\lambda)\, d\lambda}{\int P(\lambda)\, d\lambda} \qquad (4.84)$$

is the light source's spectral width calculated as a quadratic mean value.

4.6.4 Calculation of $d\tau/d\lambda$

The chromatic dispersion depends on the light source through the spectral width σ_λ and on the waveguide through $d\tau/d\lambda$. The group delay τ is given by (4.24) which can be written as

$$\tau = \frac{L}{c}\frac{d\beta}{dk} = \frac{Ln_2}{c}\left(1 + \Delta\frac{d(kb)}{dk}\right) + \frac{Lk}{c}\frac{dn_2}{dk} \qquad (4.85)$$

Derivation with respect to λ gives

$$\begin{aligned}
\frac{d\tau}{d\lambda} &= \frac{L}{c}\frac{dn_2}{d\lambda}\left(1 + \Delta\frac{d(kb)}{dk}\right) \\
&+ \frac{Ln_2}{c}\left(\frac{d\Delta}{d\lambda}\frac{d(kb)}{dk} + \Delta\frac{d^2(kb)}{dk^2}\frac{dk}{d\lambda}\right) \\
&+ \frac{L}{c}\frac{d}{d\lambda}\left(-\lambda\frac{dn_2}{d\lambda}\right)
\end{aligned} \qquad (4.86)$$

in which the fact $k = 2\pi/\lambda$ is used, and

$$\frac{dn}{dk} = \frac{dn}{d\lambda}\frac{d\lambda}{dk} = -\frac{\lambda^2}{2\pi}\frac{dn}{d\lambda} \tag{4.87}$$

The relations

$$\frac{dk}{d\lambda} = -\frac{2\pi}{\lambda^2} = -k/\lambda$$

and

$$\frac{d}{d\lambda}\left(-\lambda\frac{dn}{d\lambda}\right) = -\left(\frac{dn}{d\lambda} + \lambda\frac{d^2n}{d\lambda^2}\right)$$

substituted into (4.86) give

$$\frac{d\tau}{d\lambda} = -\frac{L}{c}\left[\frac{n_2\Delta}{\lambda}k\frac{d^2(kb)}{dk^2} + \lambda\frac{d^2n_2}{d\lambda^2}\right]$$

$$+\frac{L}{c}\left[\Delta\frac{dn_2}{d\lambda} + n_2\frac{d\Delta}{d\lambda}\right]\frac{d(kb)}{dk} \tag{4.88}$$

If the fiber's core and cladding are composed of glass with similar characteristics such that

$$\frac{dn_1}{d\lambda} \approx \frac{dn_2}{d\lambda}$$

then

$$\Delta\frac{dn_2}{d\lambda} + n_2\frac{d\Delta}{d\lambda} \approx 2\Delta^2\frac{dn_2}{d\lambda} \tag{4.89}$$

and the last term in (4.88) can be neglected compared with the first term for $\Delta \ll 1$. With these approximations (4.83) gives the chromatic dispersion

$$\sigma_K = \sigma_\lambda\frac{d\tau}{d\lambda} = \frac{L}{c}\sigma_\lambda\left[\frac{n_2\Delta}{\lambda}V\frac{d^2(Vb)}{dV^2} + \lambda\frac{d^2n_2}{d\lambda^2}\right] \tag{4.90}$$

in which the expression

$$k\frac{d^2(kb)}{dk^2} = V\frac{d^2(Vb)}{dV^2}$$

has been used.

The negative sign in front of the first term in (4.88) has no importance because the square of $d\tau/d\lambda$ appears in (4.83).

The first term inside the parentheses in (4.90) depends on the waveguide, and the last term on the characteristics of the material of the fiber and they represent waveguide and material dispersion, respectively. The terms may have different signs and this fact can be used in the design of the fiber. Through appropriate dimensioning and choice of material of the fiber, both terms can be made approximately the same size and of opposite signs. In this way a so-called

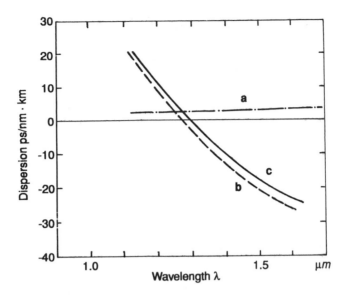

Figure 4.10 Example of dispersion-compensated fiber (a) Waveguide dispersion. (b) Material dispersion. (c) Total dispersion.

dispersion-compensated fiber with little pulse spreading can be obtained. Figure 4.10 shows an example of the waveguide dispersion and the material dispersion as functions of λ. The fiber is dispersion-compensated around $\lambda = 1.3\ \mu m$.

Example 4.3   Dispersion-compensated fiber

A step-index single-mode fiber of silica glass has a core diameter $2a = 8\ \mu m$. The refractive indices are $n_1 = 1.5$ and $n_2 = 1.495$ for core and cladding, respectively. The light source is an injection laser with spectral width $\sigma_\lambda = 1$ nm. Determine the dispersion per km fiber cable.

Solution

The relative index difference (2.12) is

$$\Delta = \frac{n_1 - n_2}{n_1} = \frac{1.5 - 1.495}{1.5} = 0.0033$$

The normalized frequency (3.24) is

$$V = \frac{2\pi n_1 a}{\lambda}\ \sqrt{2\Delta} = \frac{2\pi 1.54 \times 10^{-6}}{1.3 \times 10^{-6}}\ \sqrt{0.0067} = 2.37$$

From Figure 4.13 follows that $Vd^2(Vb)/dV^2 = 0.25$ and the material dispersion $(\lambda/c)(d^2 n/d\lambda^2) = -5$ ps/nm · km according to Figure 4.11.

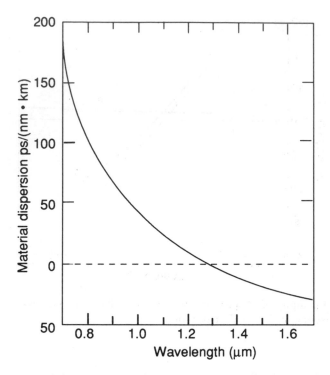

Figure 4.11 Material dispersion as a function of the optical wavelength for silica glass. Data from Fleming (1978). Reproduced by permission of McGraw Hill.

Substitution into (4.90) gives the total fiber dispersion

$$\sigma_K = \frac{L}{c} \sigma_\lambda \left[\frac{n_2 \Delta}{\lambda} V \frac{d^2(Vb)}{dV^2} + \lambda \frac{d^2 n_2}{d\lambda^2} \right]$$

$$= L\sigma_\lambda[3.15 \times 10^{-6} - 5 \times 10^{-6}] = L\sigma_\lambda[1.85 \times 10^{-6}]$$

For $\sigma_\lambda = 1$ nm and $L = 1$ km

$$\sigma_K = 10^3 \times 10^{-9} \times 1.85 \times 10^{-6} = 1.9 \text{ ps/km}$$

$\square$

Methods for overcoming the dispersion limitations are discussed in Heidemann and Veith (1993). A technique similar to dispersion compensation is dispersion shifting, used for fibers operating at $\lambda = 1.55$ μm, see papers by Reed *et al.* (1986) and Carrett *et al.* (1987). The optical wavelength $\lambda = 1.55$ μm is attractive because the fiber attenuation is minimal and optical amplifiers of the erbium-doped fiber type (EDFA) can be designed for that wavelength.

Other ways to eliminate the negative effect of dispersion is to use an optical equalizer at the receiver or solitons in the transmitter, see Section 4.8.

To upgrade existing 1.3 μm systems to 1.55 μm the addition of a special designed dispersion-compensating fiber to the installed fiber has been suggested.

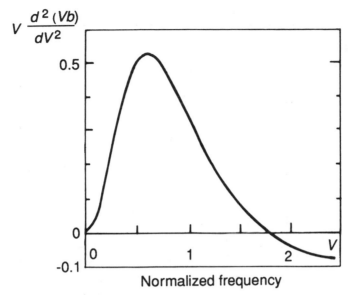

Figure 4.12   Dispersion for an ideal planar waveguide. The diagram shows the quantity $Vd^2(Vb)/dV^2$ as a function of the normalized optical frequency V.

4.6.5   Planar Waveguides

The chromatic waveguide dispersion for a planar optical waveguide depends on the second derivative of the normalized propagation constant b with respect to the normalized frequency V. Derivation of $d(Vb)/dV$ according to (4.26) gives

$$V\frac{d^2(Vb)}{dV^2} = V\frac{db}{dV}\left(1 - \frac{2V}{V + 1/\sqrt{b}} + \frac{2(1-b)(1+V/2b)}{\sqrt{b}(V+1/\sqrt{b})^2}\right) \tag{4.91}$$

The expression applies to an arbitrary mode in the waveguide. The function (4.91) is shown in Figure 4.12 for the fundamental mode of a planar waveguide.

4.6.6   Cylindrical Waveguides

For a step-index fiber no simple analytical expression for b and its derivative can be obtained. However, these quantities can, of course, be calculated numerically. Figure 4.13 shows how the second derivative of Vb varies with V. The resemblance between the curves for planar and circular waveguides is notable.

In the calculation of the chromatic dispersion, the waveguide dispersion is determined by σ_λ and $Vd^2(Vb)/dV^2$ which in turn can be read from Figure 4.13 and Figure 4.12, respectively.

In order to determine the material dispersion one must know how the refractive index of the waveguide material varies with the wavelength λ.

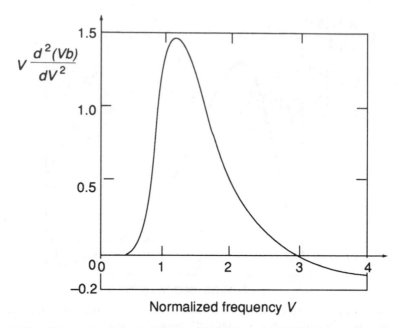

Figure 4.13 Dispersion for an ideal cylindrical waveguide (step-index fiber). The diagram shows the quantity $Vd^2(Vb)/dV^2$ as a function of the normalized optical frequency V. From Gowar (1984). Reproduced by permission of Prentice Hall International.

4.6.7   *Light Sources with Phase Noise*

The optical signal model (4.61) represents light with completely noncoherent frequency components. An alternative model, which is commonly used for light produced by a laser, is

$$B(t) = A \ \exp[j(\omega_0 t + \theta(t) + \phi)] \tag{4.92}$$

Where A is the amplitude and $\theta(t)$ is a random phase function. The phase angle ϕ is assumed to be independent of $\theta(t)$ and equally distributed in the interval $0 \le \phi \le 2\pi$.

Analysis of the laser structure supported by measurements show that the frequency noise

$$\mu(t) = \frac{1}{2\pi} \frac{d\theta(t)}{dt} \tag{4.93}$$

can be modeled as a white Gaussian stochastic process i.e. it is normally distributed and its spectral density is $R(f) = R_\mu$. For a phase process starting at $t = 0$ the phase function $\theta(t)$ is related to $\mu(t)$ by

$$\theta(t) = 2\pi \int_0^t \mu(s) \, ds \tag{4.94}$$

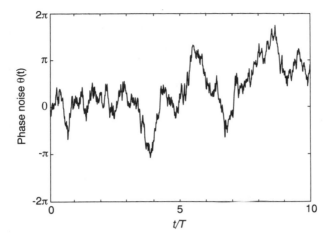

Figure 4.14 Example of a phase noise process with $B_L T = 0.1$.

This means that the phase function can be modeled as a random walk process, a so called Wiener process. How fast the phase changes in time depends on the magnitude of R_μ. Figure 4.14 shows an example of a phase noise process $\theta(t)$. The phase noise (4.94) is not a stationary stochastic process, due to the fixed starting time $t = 0$ of the random walk. However, the random phase ϕ makes $B(t)$ stationary.

The power spectral density of $\theta(t) + \phi$ is determined in (G.8) in Appendix G.

$$R(f) = \frac{A^2 R_\mu / 2}{(\pi R_\mu)^2 + (f - f_0)^2} \tag{4.95}$$

which is called a Lorentz spectrum by physicists after the Dutch scientist Hendrik Antoon Lorentz and a first order Butterworth spectrum by electrical engineers after the British engineer S. Butterworth. Its appearance is shown in Figure 4.15.

The 3-dB bandwidth B_L of the spectrum, determined by the relation $R(f_0 \pm B_L/2) = R(f_0)/2$, is

$$B_L = 2\pi R_\mu \tag{4.96}$$

In the expression (4.83) for the chromatic dispersion the light source's spectral width σ_λ is included. Insertion of (4.95) into the expression for the definition of σ_λ (4.84) with $\omega = 2\pi c/\lambda$ results in an integral which does not converge. This means that a light source with a Lorentz spectrum has a spectral width (calculated as the quadratic mean value) which is infinitely large.

This implies that pulse spreading characterized by a quadratic mean value according to (4.9) is not applicable to light sources with a spectrum of the form (4.95).

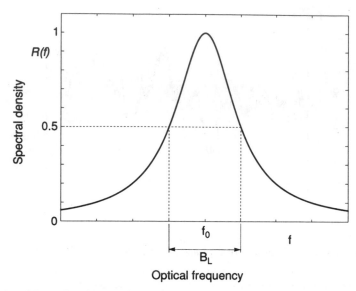

Figure 4.15 Normalized optical power spectral density (intensity spectrum) $R(f)$ of laser light with a Lorentzian spectrum.

In order to determine the pulse width for the output signal $F(t)$ from a single-mode fiber which is excited by a laser, with a Lorentz spectrum, we start with (4.77)

$$F(t) = \sum_{\ell=1}^{L} B_\ell^2 \, G(t - \tau(\lambda_\ell)) \tag{4.97}$$

If we let L approach infinity then

$$F(t) = \int_{-\infty}^{\infty} P(\lambda) \, G(t - \tau(\lambda)) \, d\lambda \tag{4.98}$$

The expansion of $\tau(\lambda)$ in a Taylor series (4.81) gives

$$F(t) = \int_{-\infty}^{\infty} P(\lambda) \, G\left(t - \lambda \frac{d\tau}{d\lambda} - \tau_0\right) d\lambda \tag{4.99}$$

with $\tau_0 = \tau(\lambda_0) - \lambda_0 d\tau/d\lambda$.

The expression (4.99) means that the output signal $F(t)$ is obtained by a convolution of $P(\lambda)$ with a function determined by the optical input signal $G(t)$.

Example 4.4   Optical system with rectangular pulses

Let $G(t)$ be a rectangular pulse with amplitude G_0 and time duration T. We assume that the light source has a Lorentz spectrum (4.95) which, transformed to the baseband, is

$$R_0(\omega) = \frac{A^2 R_\mu}{(R_\mu/2)^2 + \omega^2} \tag{4.100}$$

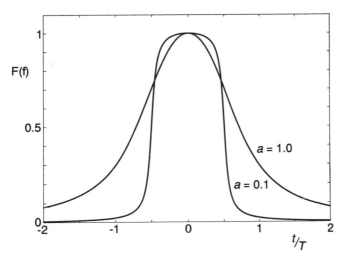

Figure 4.16 Pulse broadening of a rectangular pulse modulating a light source with Lorentzian spectrum. The diagram shows the output pulse for two values of the quantity $a = B_L\tau'/2\pi T$ where B_L is the 3 dB bandwidth of the laser and $\tau' = d\tau/d\lambda$ represents the chromatic dispersion of the fiber.

Because the spectrum (4.100) is expressed as a function of ω, it is appropriate to let τ be a function of ω and employ a Taylor expansion centered on ω_0. Analogously with (4.99)

$$F(t) = \int R_0(\omega)G\left(t - \omega\frac{d\tau}{d\omega} - c_0\right)d\omega \qquad (4.101)$$

Exchange of the integral variable $s = \omega(d\tau/d\omega) = \omega\tau'$ gives, with $x = t - c_0$,

$$F(x) = \frac{1}{\tau'}\int_{-\infty}^{\infty} R_\mu(s/\tau')G(x - s)\,ds \qquad (4.102)$$

Substitution of $G(s)$ and $R_0(s/\tau')$ gives

$$\begin{aligned}
F(x) &= \int_{x-T/2}^{x+T/2} \frac{G_0A^2 R_\mu\tau'}{(R_\mu\tau'/2)^2 + s^2}\,ds \\
&= \frac{G_0A^2}{2}\left[\arctan\left(\frac{2x+T}{R_\mu\tau'}\right) - \arctan\left(\frac{2x-T}{R_\mu\tau'}\right)\right]
\end{aligned} \qquad (4.103)$$

The light source's spectrum $R_0(\omega)$, the signal pulse $G(t)$ and the output signal $F(t)$ are shown in Figure 4.16. The mean square width of $F(t - c_0)$ is not finite, therefore we measure the pulse's dispersion as its half-value width Δ_2 defined by

$$F\left(x = \frac{\Delta_2}{2}\right) = \frac{1}{2}F(x = 0) \qquad (4.104)$$

From (4.103)

$$
\left.\begin{aligned}
F(x = 0) &= G_0 A^2 \arctan\left(\frac{T}{R_\mu \tau'}\right) \\
F\left(x = \frac{\Delta_2}{2}\right) &= \frac{G_0 A^2}{2} \arctan\left(\frac{2T/R_\mu \tau'}{1 + (\Delta_2 + T)(\Delta_2 - T)/(R_\mu \tau')^2}\right)
\end{aligned}\right\}
\tag{4.105}
$$

The expression (4.104) gives

$$
\frac{2T/R_\mu \tau'}{1 + (\Delta_2 - T)/(R_\mu \tau')^2} = \frac{T}{R_\mu \tau'}
\tag{4.106}
$$

which leads to

$$
\Delta_2 = \sqrt{T^2 + (R_\mu \tau')^2}
\tag{4.107}
$$

The input signal $G(t)$ has a half-value width

$$
\Delta_1 = T
$$

The half-value width in angular frequency for $R_0(\omega)$ is, from (4.100)

$$
\Delta_\omega = R_\mu
$$

Insertion into (4.107) gives

$$
\Delta_2^2 = \Delta_1^2 + \Delta_\omega^2 \left(\frac{d\tau}{d\omega}\right)^2
\tag{4.108}
$$

which is the same type of equation as (4.83). In contrast with (4.83), which is valid generally for arbitrary signal shapes and spectra, (4.108) assumes a definite pulse shape and a special spectral shape.   □

Dispersion analysis based on the mean square pulse width σ_t is mathematically simple. The result does not depend on the shape of the input pulse or the form of the spectral density of the light source. However, as shown by Example 4.4 there are situations where this type of analysis is not possible. The exact value of the dispersion can always be determined by calculating the form of the output pulse using (4.99) or (4.101). In the literature, pulse width and spectral width are often specified as half-value widths, which for obvious reasons always are finite quantities. If the shape is known it is easy to convert from σ_t to Δ_t.

4.7   Total Dispersion in Multimode Waveguides

4.7.1   *Small Number of Modes*

For single-mode waveguides chromatic dispersion is usually the dominating form of dispersion. With multimode transmission, both chromatic and modal dispersion must be considered.

The expression (4.77) is valid for any arbitrary mode in the waveguide. With M modes which are excited equally

$$F(t) = \frac{1}{M} \sum_{m=1}^{M} \sum_{\ell=1}^{L} B_\ell^2 G(t - \tau_{\ell m}) \tag{4.109}$$

where $\tau_{\ell m} = \tau_m(\lambda_\ell)$, is the group delay for mode m at wavelength λ_ℓ. In the same way as with the derivation of (4.78) and (4.80), the chromatic dispersion becomes

$$\sigma_k^2 = \frac{1}{M} \sum_{m=1}^{M} \sigma_m^2 \tag{4.110}$$

where

$$\sigma_m^2 = \langle \tau_m^2 \rangle - \langle \tau_m \rangle^2$$

and

$$\left.\begin{aligned} \langle \tau_m \rangle &= \frac{\int P(\lambda)\tau_m(\lambda)\, d\lambda}{\int P(\lambda)\, d\lambda} \\ \langle \tau_m^2 \rangle &= \frac{\int P(\lambda)\tau_m^2(\lambda)\, d\lambda}{\int P(\lambda)\, d\lambda} \end{aligned}\right\} \tag{4.111}$$

The dispersion for the modes is thus added quadratically and the contribution from each separate mode can be calculated separately.

If $\tau_m(\lambda)$ is replaced by the first two terms in the power series according to (4.81) then

$$\sigma_m^2 = \sigma_\lambda^2 \left(\frac{d\tau_m}{d\lambda}\right)^2 \tag{4.112}$$

where, as earlier, the derivative should be evaluated at $\lambda = \lambda_0$.

Calculation of the pulse width for $F(t)$ according to (4.109) gives the total dispersion. It is equal to the sum of the modal dispersion (4.41) and the chromatic dispersion (4.110). With the use of the power series approximation for $\tau_m(\lambda)$ the total dispersion becomes

$$\sigma_2^2 = \sigma_1^2 + E_m\{\tau_m^2(\lambda_0)\} - (E_m\{\tau_m(\lambda_0)\})^2 + \sigma_\lambda^2 E_m\{(d\tau_m/d\lambda)^2\} \tag{4.113}$$

where $E_m\{\ \}$ indicates averaging of the modes. The group delays τ_m are the values at $\lambda = \lambda_0$, the center frequency (4.82) of the light source's spectrum. The quantity σ_λ is the light source's spectral width as defined in (4.84).

4.7.2 Large Number of Modes

For a multimode waveguide with a large number of modes (4.113) is impractical to use. From the approximate expression for τ_m (4.31) which is obtained by WKB analysis, the chromatic dispersion for a multimode waveguide can be calculated.

We start from (4.51) and neglect terms of second order and higher in Δ.

$$\tau_m \simeq \frac{LN_1}{c}\left(1 + c_1\Delta\left(\frac{m}{M}\right)^\gamma\right) \tag{4.114}$$

in which $\gamma = \alpha/(\alpha+2)$ and $c_1 = (\alpha - 2 - \varepsilon)/(\alpha+2)$. The quantity N_1 is defined in (4.32) and ε in (4.48).

$$\frac{c}{L}\frac{d\tau_m}{d\lambda} = \frac{dN_1}{d\lambda}\left(1 + c_1\Delta\left(\frac{m}{M}\right)^\gamma\right) + N_1\left(\frac{d(c_1\Delta)}{d\lambda} - c_1\Delta\gamma\frac{1}{M}\frac{dM}{d\lambda}\right)\left(\frac{m}{M}\right)^\gamma$$

$$= -\lambda\frac{d^2n_1}{d\lambda^2}\left(1 + c_1\Delta\left(\frac{m}{M}\right)^\gamma\right) \tag{4.115}$$

$$+ \left[\Delta N_1\frac{2\alpha}{\alpha+2}c_1\left(\frac{1}{\lambda} - \frac{1}{n_1}\frac{dn_1}{d\lambda} - \frac{1}{N_12\alpha c_1}\frac{d\varepsilon}{d\lambda}\right) + \frac{2c_1}{\alpha+2}\frac{d\Delta}{d\lambda}\right]\left(\frac{m}{M}\right)^\gamma$$

For $\Delta \ll 1$ the second term in the first parenthesis is small and in the second parentheses $1/\lambda$ dominates the remaining terms which gives

$$\lambda\frac{d\tau_m}{d\lambda} \simeq \frac{L}{c}\left[-\lambda^2\frac{d^2n_1}{d\lambda^2} + \Delta N_1\frac{2\alpha c_1}{\alpha+2}\left(\frac{m}{M}\right)^\gamma\right] \tag{4.116}$$

Replacing the summation over m by an integral in the same way as in (4.52) and performing the integration gives the chromatic dispersion for a multimode fiber derived by Einarsson (1986).

$$\sigma_k^2 = \sigma_\lambda^2\left\langle(d\tau_md\lambda)^2\right\rangle$$

$$= \left(\frac{\sigma_\lambda}{\lambda}\right)^2\left(\frac{L}{c}\right)^2\left[\left(\lambda^2\frac{d^2n_1}{d\lambda^2}\right)^2 - 2\lambda^2\frac{d^2n_1}{d\lambda^2}\,\Delta N_1\alpha\frac{(\alpha-2-\varepsilon)}{(\alpha+2)(\alpha-1)}\right. \tag{4.117}$$

$$\left. + 4\Delta^2N_1^2\frac{2(\alpha-2-\varepsilon)^2}{(\alpha+2)^3(3\alpha+2)}\right]$$

For an index profile with $\alpha = 2 + \varepsilon$

$$\sigma_k = \frac{\sigma_\lambda}{\lambda}\frac{L}{c}\left|\lambda^2\frac{d^2n_1}{d\lambda^2}\right| \tag{4.118}$$

Comparison with (4.90) shows that this is equal to the chromatic material dispersion for a single-mode fiber.

A step-index fiber has $\alpha = \infty$, which gives

$$\sigma_k^2 = \left(\frac{\sigma_\lambda}{\lambda}\right)^2\left(\frac{L}{c}\right)^2\left[\left(\lambda^2\frac{d^2n_1}{d\lambda^2}\right)^2 - 2\lambda^2\frac{d^2n_1}{d\lambda^2}N_1\Delta + \frac{4}{3}N_1^2\Delta^2\right]$$

$$= \left(\frac{\sigma_\lambda}{\lambda}\right)^2\left(\frac{L}{c}\right)^2\left[\left(\lambda^2\frac{d^2n_1}{d\lambda^2} - N_1\Delta\right)^2 + \frac{1}{3}N_1^2\Delta^2\right] \tag{4.119}$$

The last expression shows a possibility for compensation of chromatic dispersion. If the material in the fiber is chosen such that

$$\lambda^2\frac{d^2n_1}{d\lambda^2} = N_1\Delta \tag{4.120}$$

the chromatic dispersion assumes its minimum value

$$\sigma_k = \frac{\sigma_\lambda}{\lambda} \frac{L N_1 \Delta}{c} \frac{1}{\sqrt{3}} \tag{4.121}$$

which can be compared with the modal dispersion (4.54). For the light sources, light-emitting diodes and semiconductor lasers, which are used in optical communications, $\sigma_\lambda/\lambda \ll 1$, which means that the modal dispersion for a step-index fiber normally dominates over the chromatic dispersion.

4.8   Solitons

Dispersion in a linear medium can be minimized but it cannot be eliminated. In a medium with nonlinear characteristics, it is possible under certain conditions to obtain waves which are stable in form and do not change during transmission. Such waves, called solitons, are free from dispersion and do not exhibit any pulse broadening.

For normal optical transmission the fiber can be modeled as a linear medium. At high signal amplitudes, however, nonlinear effects have to be taken into account. The refractive index of silica glass depends on the strength of the optical field, which is called the Kerr effect. For narrow single-mode pulses of sufficient amplitude the Kerr effect can compensate exactly for the chromatic dispersion (4.90). For an optical fiber without attenuation the shape $E(t, z)$ of a soliton pulse at a distance z from the transmitter is determined by the nonlinear wave equation, see e.g., Marcuse (1991).

$$\frac{\partial E}{\partial z} + \beta' \frac{\partial E}{\partial t} - \frac{j}{2} \beta'' \frac{\partial^2 E}{\partial t^2} = -jk_0 n_2 |E|^2 E \tag{4.122}$$

where β' and β'' are the first and second order derivatives of the wave propagation parameter β with respect to the angular optical frequency ω. The quantity k_0 is a constant analogous to k in the linear wave equation (3.15) and n_2 represents the nonlinear character of the refractive index.

$$n = n_0 + n_2 |E|^2 \tag{4.123}$$

To find a solution of (4.122) we try the form

$$E = Ay(x)e^{jaz} \tag{4.124}$$

where $x = (t - \beta' z)/\tau_0$.

Substitution of (4.124) into (4.122) gives the following differential equation for $y(x)$

$$y'' - \frac{2a\tau_0^2}{\beta''} y - \frac{2k_0 n_2 \tau_0^2}{\beta''} A^2 y^3 = 0 \tag{4.125}$$

A possible solution of (4.125) is

$$y(x) = \frac{1}{\cosh x} \tag{4.126}$$

Substitution of (4.126) into (4.125) and solving for the parameters a and τ_0 gives

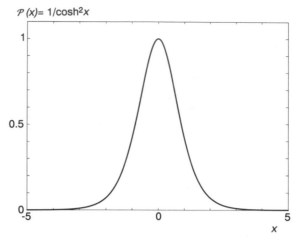

P (x)= 1/cosh²x

Figure 4.17 The shape of a soliton pulse. Its normalized optical power $P(x)$ is shown as a function of $x = (t - \beta'z)/\tau_0$.

$$a = -k_0 n_2 A^2/2 \tag{4.127}$$

and

$$\tau_0 = \sqrt{\beta''/2a} \tag{4.128}$$

The relation (4.127) shows that a is negative and from (4.128) follows that the solution (4.126) exists only when β'' is negative. This means that solitons are possible when the chromatic dispersion is negative, which for silica fiber occurs at wavelengths above 1.3 μm.

The optical power $P(t) = y^2(t)$ of the soliton pulse (4.126) has the form shown in Figure 4.17.

The area of the normalized pulse $P(x)$ is equal to unity and the mean square width (4.9) of the intensity pulse $P(t)$ is

$$\sigma_t^2 = \int_{-\infty}^{\infty} t^2 P(t)\, dt \Big/ \int_{-\infty}^{\infty} P(t)\, dt = \tau_0^2 \int_{-\infty}^{\infty} \frac{x^2}{\cosh^2 x}\, dx = \frac{\pi^2}{6} \tau_0^2 \tag{4.129}$$

A soliton is a stable solution to (4.122) which travels in the fiber without dispersion. The fiber attenuation causes the signal amplitude to decrease and the soliton condition (4.127) ceases to be satisfied. To maintain a soliton in an optical fiber it is necessary to amplify the optical signal at regular intervals to restore its amplitude. This can conveniently be done by optical amplifiers; see Chapter 7.

It has been demonstrated that soliton transmission is possible over long distances. The noise of optical amplifiers and the interaction between the soliton pulses representing adjacent digital symbols can be a problem for long distance transmission. This causes what is called the Gordon-Haus limit on the maximal transmission distance. It can be relieved by introducing optical filters or by reshaping and resynchronizing the pulses at regular intervals. With the last technique the transmission distance is practically unlimited, as shown by Nakazawa (1993) *et al.* among others.

4.9   Attenuation in Optical Fibers

The quality of an optical fiber as a transmission medium is determined by its dispersion in combination with the signal attenuation. Both these quantities increase with the fiber length, and together with the receiver sensitivity they put limits on the transmission distance possible for an optical fiber communication link.

The fiber attenuation **a** is measured in logarithmic units, usually in dB/km

$$\mathbf{a}L = 10 \log \frac{\mathcal{P}_1}{\mathcal{P}_2} \tag{4.130}$$

where $\mathcal{P}_1$ and $\mathcal{P}_2$ are the input and the output optical power, respectively, and L is the fiber length.

There are two basic mechanisms that cause attenuation of light in a glass fiber: absorption and scattering. Both are dependent on the wavelength, scattering is most pronounced at short λ and absorption dominates at long λ.

Light in an optical medium containing small particles or inhomogeneities will be subjected to Rayleigh scattering. Such a medium can be viewed as having small variations in its refractive index. If these fluctuations are on a scale much smaller than the optical wavelength, part of the light will be deflected and lost according to a theory formulated by Lord Rayleigh in 1871. This is the reason why the sky is blue. Molecules and small particles in the air scatter blue light more than red and it is this blue light, deflected from the sun, that we see when we look up into the sky.

The effect of Rayleigh scattering varies as $1/\lambda^4$. The attenuation is around 3 dB/km at $\lambda = 0.7$ μm for a silica fiber and decreases rapidly for longer wavelength. See Figure 4.18. Glass is noncrystalline or amorphous, which means that its atoms do not exhibit any long-range periodic pattern. Even if the material is extremely pure and the manufacturing process carefully controlled, solid glass will always have small fluctuations in the refractive index. Rayleigh scattering is therefore present in all glass materials and it constitutes a fundamental limit to the attenuation of silica fibers at low optical wavelengths.

At longer wavelengths, where the Rayleigh scattering can be neglected, absorption is a source of attenuation. The light causes vibrations of the crystal or molecular structures in the fiber which produces heat. This means that some of the light is transformed into thermal energy and lost. For silica glass the S_iO_2 crystal structure has a resonant wavelength, resulting in an attenuation peak at 9 μm. For lower optical wavelengths the intrinsic absorption is less severe and it can be neglected for $\lambda < 1.6$ μm.

In Figure 4.18 the broken lines show the attenuation for a silica glass fiber due to scattering and absorption respectively. The sum of these curves gives the minimum theoretical attenuation it is possible to obtain at around 1.55 μm and attenuation values as low as 0.154 dB/km at 1.55 μm are reported by Kanamori *et al.* (1986).

The discussion above assumes a perfect fiber without any defects or impurities. In the fiber production process it is difficult to eliminate water dissolved in the glass and the fiber will contain a small concentration of OH

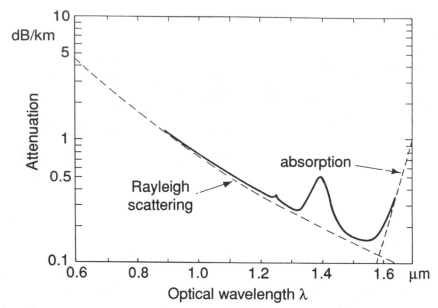

Figure 4.18   Attenuation as a function of optical wavelength for single-mode optical fibers. From Jones (1988). Reproduced by permission of Oxford University Press.

ions. This will cause attenuation due to molecular vibrations, since OH ions have a resonant wavelength of 2.73 μm and the first harmonic at $\lambda = 1.39$ μm falls in the optical transmission band. The peak in the attenuation curve around $\lambda = 1.4$ μm in Figure 4.18 is due to extrinsic absorption caused by water remnants in the fiber.

The attenuation curve of Figure 4.18 is representative for single-mode fibers. For multimode fibers the attenuation is slightly higher since they show additional attenuation due to mode conversion and other mechanisms. The attenuation of a silica glass fiber has two minima, one local minimum at $\lambda = 1.3$ μm and a global minimum at $\lambda = 1.55$ μm. The optical bands around these wavelengths are often referred to as the windows of optical fiber transmission. Operating in the band around $\lambda = 1.55$ μm gives the lowest attenuation and thereby the longest transmission span if dispersion can be neglected or controlled. The wavelength $\lambda = 1.3$ μm has the advantage that low dispersion fiber based on the principles of dispersion compensation are available at that wavelength.

The presentation above deals with the principal sources of optical fiber attenuation. In practice there are additional causes that may be needed to take into account. If the geometry of the fiber as an optical waveguide is not perfect it will cause attenuation. The cabling of the fiber may introduce additional attenuation due to microbending. A transmission line often includes splices and connectors and their attenuation must be added to the fiber attenuation.

For information on these matters and on methods of fiber manufacture we refer to Murata (1988) and also Mahlke and Gössing (1987).

5

Optical Detection Theory

5.1 Introduction

The classical problem in detection theory deals with the detection of signals in additive Gaussian noise which serves as a model for thermal and other noise sources in electronic circuitry.

In optical systems the signal exhibits random fluctuations due to the statistical nature of light as a stream of photons. This sets a lower limit to the performance of optical communication. Even in the absence of external noise it is impossible to detect an optical signal with complete certainty. In Section 5.4 we determine the quantum limit that specifies the optical power needed to guarantee a certain transmission error probability.

The optical noise caused by the random photon flow is modelled by a Poisson point process. We will start examining its effect on data transmission, and later add thermal noise which originates from the electronic circuitry in the receiver.

A complete treatment of the detection problem requires calculation of the sum of Poisson and Gaussian stochastic variables. The probability density for such a sum is easy to write down but contains an infinite summation that is difficult to handle numerically. In this chapter we show that a convenient way to deal with the situation is to utilize the characteristic functions of the densities from which bounds and accurate approximations can be derived.

In almost all textbooks on optical communication the compound density is approximated by a Gaussian probability density when detection probabilities are calculated. An alternative would be to approximate the compound density by a Poisson density. These two approaches are of about the same complexity and accuracy, as shown by Einarsson (1989). Here we use the Gaussian approximation.

The analysis presented applies to ordinary photodiodes as well as to avalanche photodiodes.

5.2   Photodetectors

In the analysis of optical waveguides in Chapters 3 and 4 we consider light to be an electromagnetic wave. A complementary description that is needed in the analysis of optical detection is the photon model. It describes light as a stream of minute particles or quanta (photons).

The theory of quantum physics states that the energy of a photon is proportional to the frequency f of the light wave:

$$E_p = hf \tag{5.1}$$

where $h = 6.6261 \times 10^{-34}$ Ws^2 is Planck's constant.

A light beam of constant optical power $\mathcal{P}$ corresponds to a stream of $\mathcal{P}/hf$ photons per second. For a modulated optical signal with power $\mathcal{P}(t)$ the instantaneous photon intensity (also called the photon flux) varies with time

$$\Gamma_p(t) = \frac{\mathcal{P}(t)}{hf} \quad \text{photons/s} \tag{5.2}$$

A conventional optical receiver contains as its first element a photodetector which converts the incoming optical signal into an electrical signal. The receiver performs signal processing operations, such as amplification and filtering, on the electrical signal in order to reach a decision about the information conveyed.

Most photodetectors are based on the principle of optical absorption. An incoming photon causes the excitation of an electric charge carrier, which under the influence of an electric field moves to the outside electric circuitry of the device. An incident optical signal generates an electric current of photoelectrons at the output of the photodetector.

An equivalent circuit diagram for an ideal semiconductor photodiode is shown in Figure 5.1. It consists of a stochastic current generator generating a signal with an intensity proportional to the power of the incident light signal.

For an ordinary photodetector (PIN diode) an optical signal with power $\mathcal{P}(t)$ generates an electron intensity

$$\Gamma_e(t) = \frac{\eta \mathcal{P}(t)}{hf} + \gamma_d \quad \text{electrons/s} \tag{5.3}$$

where η is the quantum efficiency of the device, i.e. the average number of photoelectrons produced per incoming photon. The term γ_d represents the dark current, i.e. the electrical current caused by spontaneous emission of electrons from a photodiode when no light is applied.

Figure 5.1   Equivalent circuit diagram of a photodetector.

The power $\mathcal{P}(t)$ in (5.2) is the instantaneous intensity of the optical signal. It is proportional to the square of the electromagnetic field quantities, as discussed in Section 4.6.1. The relation (5.2) provides a connection between the electromagnetic field model and the photon model of light. It constitutes what is called a semiclassical approach which proves to be sufficient for all problems of optical detection studied here. The key feature is that the time-varying intensity of the photoelectron process is proportional to the instantaneous optical signal power.

The photons and photoelectrons appear at random time instants and the stream of electrons from the photodetector is a random process. It can be shown theoretically, and it has been verified experimentally, that a valid statistical model is the Poisson process.

5.3 The Poisson Process

Consider a sequence of photoelectrons at a fixed instant of time. The time positions $\ldots t_{-1}, t_0, t_1, t_2$ of the electrons are a realization of a Poisson (point) process.

The intensity $\Gamma(t)$ of a Poisson process is a measure of the rate at which events occur. The average number of events in a time interval of length T starting at an arbitrary time t_s is

$$E\{N(T)\} = \int_{t_s}^{t_s+T} \Gamma(t)\, dt = m \tag{5.4}$$

The number $N(T)$ of events in the time interval is a random variable with a Poisson distribution

$$\Pr\{N(T) = n\} = \frac{m^n e^{-m}}{n!} \tag{5.5}$$

An example of a Poisson distribution is shown in Figure 7.3b.

The variance of $N(T)$ is

$$\mathrm{Var}\{N(T)\} = E\{(N(T) - m)^2\} = m \tag{5.6}$$

It can be shown that the time interval between two Poisson events has an exponential distribution. A property that is useful in the analysis is that the Poisson events in any fixed time interval are independent of each other if the order of their occurrences is disregarded.

5.3.1 Shot Noise

The arrivals of the photons are random events generated by a Poisson process with the intensity (5.2). There is a random relation between the absorption of photons and the emission of photoelectrons. The delay between arrival and emission times has no memory, and it can be shown that the times of occurrence of the photoelectrons can be modeled as a Poisson process with the intensity (5.3).

A stream of electrons constitutes an electric current. Each electron induces a short current pulse $g(t)$ in the electric circuit and the output electric current from the photodetector is a shot noise process

$$I(t) = \sum_{k=-\infty}^{\infty} a_k g(t - t_k) \tag{5.7}$$

where a_k is a random variable representing the multiplication mechanism of an avalanche photodetector. For an ordinary photodetector it is a constant equal to one.

The area of $g(t)$ is equal to the charge of an electron

$$\int_{-\infty}^{\infty} g(t) \, dt = q \tag{5.8}$$

The current $I(t)$ is a stochastic process generated by the Poisson events t_k. An example of a shot noise process is shown in Figure 5.2, where the pulses are triangular and $a_k = 1$.

When $\Gamma_e(t)$ is a constant, corresponding to steady incident light, $I(t)$ is a stationary stochastic process. In general, when the intensity $\Gamma_e(t)$ varies with time, it is nonstationary. The statistical properties of $I(t)$ when $\Gamma_e(t)$ is equal to a constant Γ are studied in Appendix C and the main results are summarized below.

The average of the current $I(t)$ is

$$I = E\{I(t)\} = E\{a_k\} \Gamma q \tag{5.9}$$

The power spectral density is

$$R(f) = E\{a_k^2\} \Gamma |G(f)|^2 + I^2 \delta(f) \tag{5.10}$$

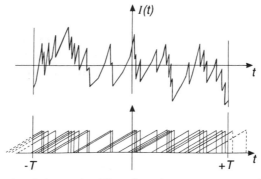

Figure 5.2　Example of shot noise. The photodetector is assumed to have a triangular impulse response.

where $G(f)$ is the Fourier transform of $g(t)$

$$G(f) = \int_{-\infty}^{\infty} g(t)e^{-j2\pi ft}\, dt \qquad (5.11)$$

The time duration of the pulses $g(t)$ is determined by the time it takes for an electron to travel through the active region of the detector under the influence of the electric field from the outside voltage applied to the device. This time is often small and $g(t)$ can then be approximated by a Dirac function. Approximation with a Dirac function $g(t) = q\delta(t)$ gives $G(f) = q$, and $R(f)$ becomes a constant spectral density, i.e. white noise

$$R(f) = \mathrm{E}\{a_k^2\}\,\Gamma q^2 + I^2\,\delta(f) \qquad (5.12)$$

If such a white shot noise process is filtered by a bandpass filter with bandwidth B, the direct current component is eliminated and the output noise signal has the variance

$$\sigma_B^2 = 2\mathrm{E}\{a_k^2\}\,\Gamma q^2 B \qquad (5.13)$$

For the special case when $a_k = 1$ corresponding to a photodiode without avalanche gain, σ_B^2 becomes

$$\sigma_B^2 = 2\Gamma q^2 B = 2IqB \qquad (5.14)$$

From (5.13) and (5.14) it follows that the photodetector shot noise process is signal-dependent. The noise variance depends on the optical signal power. Light of high intensity Γ generates a large output current I but also noise with variance proportional to I.

5.4   Ideal Receiver. The Quantum Limit

We consider binary optical transmission with on-off modulation under ideal conditions. The binary symbol 'one' is represented by an optical pulse of arbitrary photon intensity $\gamma(t)$ confined in the signaling time interval $[0, T]$ as shown in Figure 5.3. A binary 'zero' is indicated by the absence of any optical signal in the corresponding time interval. The transmitted light source is completely switched off during these intervals.

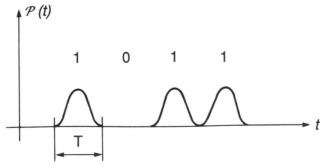

Figure 5.3   Ideal binary on–off signaling.

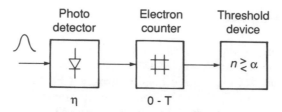

Figure 5.4   Photon-counting receiver.

The receiver is assumed to be ideal in the sense that it is capable of detecting every single photon arriving at its input. Such a receiver could in principle be realized as a photodetector, without any dark current and with quantum efficiency equal to unity, followed by a sensitive electronic counter recording the number of photoelectrons produced by the detector. See Figure 5.4.

The number of photons in the optical pulse $\gamma(t)$ is a random variable. As shown in Section 5.2, it has a Poisson distribution with a mean value obtained from (5.4).

$$m_1 = \int_0^T \gamma(t)\, dt \tag{5.15}$$

For the system to operate satisfactorily we require that the bit error probability should be at most 10^{-9}.

The receiver observes the number of photons in consecutive time intervals. For those intervals in which no photons are detected it assumes that a binary 'zero' was transmitted. The only way an error can occur is when a 'one' was transmitted but no photons are detected. With the assumption that the binary symbols are equally likely, which means that they both occur with probability $P = 1/2$, the bit error probability is, see (5.5)

$$P_e = \frac{1}{2}\, \Pr\{N(T) = 0\} = \frac{1}{2}\, e^{-m_1} \tag{5.16}$$

Letting $P_e = 10^{-9}$ gives $m_1 = 20.03$. The pulse must have an optical energy corresponding to an average of 20 photons to result in a bit error probability equal to 10^{-9}.

On average, half the signal intervals contain optical pulses, and the average number of photons per transmitted bit of information is

$$m_1/2 = 10 \text{ photons/bit} \tag{5.17}$$

This quantity of 10 photons/bit is called the quantum limit for optical detection. It is a theoretical result, assuming idealized conditions, and it represents a lower limit on the received power necessary in an direct detection optical communication system.

In the succeeding sections we will relax the idealized conditions assumed in the derivation of the quantum limit and study more realistic optical receivers.

5.5 Photon-counting Receiver

In the derivation of the quantum limit both the signaling conditions and receiver operation are idealized. In this section a photon-counting receiver is assumed but the received signal is not necessarily equal to zero for binary 'zero'.

The receiver, shown in Figure 5.4, consists of a photodetector with quantum efficiency η followed by a photoelectron counter and a threshold device.

The received optical signal is shown in Figure 5.5. The on-off pulses $\gamma(t)$ are assumed to be limited to the signaling time interval [0,T]. The received signal in Figure 5.5 has a nonzero lowest value γ_0. This corresponds, for instance, to the case when the light source is not switched off completely during the intervals representing binary 'zeroes'. An example of this is a laser operating above its threshold. See Figure 5.7 and Example 5.2.

The receiver counts the number of photoelectrons during the symbol interval $[0, T]$. If the number observed is greater than a fixed threshold it indicates that a binary 'one' was transmitted, and if it is lower or equal to the threshold a binary 'zero' is indicated.

A dark current i_d from the photodetector generates a shot noise process with intensity (electrons per second)

$$\gamma_d = \frac{i_d}{q} \tag{5.18}$$

The electron intensity depends on if a binary 'zero' or a 'one' was transmitted.

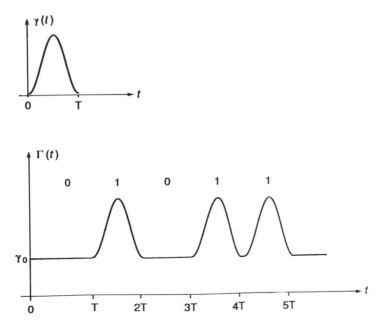

Figure 5.5 Received optical intensity for signals without intersymbol interference.

For a 'zero' the pulse $\gamma(t)$ is absent and the intensity is

$$\Gamma_0(t) = \gamma_d + \gamma_0 \tag{5.19}$$

The average number of electrons in the signaling interval $[0,T]$ is

$$m_0 = \int_0^T \Gamma_0(t)\, dt = \gamma_d T + \gamma_0 T \tag{5.20}$$

For a 'one' transmitted the electron intensity is

$$\Gamma_1(t) = \gamma_d + \gamma_0 + \gamma(t) \tag{5.21}$$

and the average number of electrons in the time interval $[0,T]$ is

$$m_1 = \int_0^T \Gamma_1(t)\, dt = m_0 + m \tag{5.22}$$

with

$$m = \int_0^T \gamma(t)\, dt \tag{5.23}$$

equal to the average number of photoelectrons in the pulse $\gamma(t)$.

The number of photoelectrons recorded by the electron counter is a random variable with a Poisson distribution (5.5). The receiver makes an erroneous decision when the observed number is greater than the threshold and a 'zero' was transmitted and when it is lower or equal than the threshold and a 'one' is transmitted. The error probability, with equal a priori probabilities for 'ones' and 'zeros', is

$$P_e = \frac{1}{2}\Pr\{N > \alpha|0\} + \frac{1}{2}\Pr\{N \le \alpha|1\}$$
$$= \frac{1}{2}\sum_{n=\alpha+1}^{\infty} \frac{m_0^n}{n!}\, e^{-m_0} + \frac{1}{2}\sum_{n=0}^{\alpha} \frac{m_1^n}{n!}\, e^{-m_1} \tag{5.24}$$

where α denotes the value of the threshold. The two terms represent the probabilities of the upper and lower tail of the relevant distribution, respectively.

The threshold α should be such that P_e is as low as possible, and it is natural to let α be the value that minimizes P_e. Consider the difference

$$\Delta P_e = P_e(\alpha) - P_e(\alpha - 1)$$
$$= \frac{1}{2}\frac{m_1^\alpha}{\alpha!}\, e^{-m_1} - \frac{1}{2}\frac{m_0^\alpha}{\alpha!}\, e^{-m_0} \tag{5.25}$$

Letting $\Delta P_e = 0$ and solving for α gives

$$\alpha = \frac{m_1 - m_0}{\ln m_1 - \ln m_0} \tag{5.26}$$

It is easy to verify that the correct integer-valued optimal threshold α is the integer part of (5.26)

$$\alpha_o = \text{int}\left(\frac{m_1 - m_0}{\ln m_1 - \ln m_0}\right) \tag{5.27}$$

Example 5.1 Quantum limit when $m_0 > 0$

The quantum limit (eq 6.33) assumes that $m_0 = 0$.

Determine the average number of photons per bit needed to achieve $P_e = 10^{-9}$ in the more realistic situation when $m_0 = 10$.

Solution

The value of m_1 needs to be determined numerically. As an example let $m_1 = 20$. The threshold (5.27) then becomes $\alpha = 14$ and evaluation of (5.24) yields $P_e = 0.94 \cdot 10^{-3}$. By successively increasing m_1 the error probability P_e decreases, and for $m_1 = 82.3$ it takes the desired value $P_e = 10^{-9}$. The corresponding threshold (5.27) is $\alpha = 34$.

The average number of photons per bit is

$$\frac{m}{2} = \frac{m_1 - m_0}{2} = \frac{82.3 - 10}{2} = 36.1$$

which is more than three times as large as the ideal quantum limit (5.17) of 10 photons per bit. □

The quantum limit defined as $(m_1 - m_0)/2$, the mean number of photons per bit in excess of the zero level m_0, required to achieve $P_e = 10^{-9}$ is shown in Figure 5.6 as a function of m_0. The uneven character of the curve is caused by the fact that the Poisson variable and the receiver threshold are integer-valued. The diagram shows that even a small amount of background noise m_0 will give a result different from the ideal limit of 10 photons per bit.

The error probability can be calculated from (5.24) using a table for the cumulative Poisson distribution, but this may be inconvenient in practice. In Appendix D upper bounds on the sums in (5.24) are studied. The following so-called improved Chernoff bounds are shown to constitute accurate approximations.

$$\Pr(N \le \alpha) = \sum_{n=0}^{\alpha} \frac{m^n}{n!} e^{-m} < \frac{1}{\sqrt{2\pi\alpha}} \frac{m}{m-\alpha} \exp[-\Theta(m,\alpha)]; \quad m > \alpha \tag{5.28}$$

$$\Pr(N \ge \alpha) = \sum_{n=\alpha}^{\infty} \frac{m^n}{n!} e^{-m} < \frac{1}{\sqrt{2\pi\alpha}} \frac{\alpha+1}{\alpha-m+1} \exp[-\Theta(m,\alpha)]; \quad m < \alpha \tag{5.29}$$

where

$$\Theta(m,\alpha) = m - \alpha[1 + \ln(m/\alpha)] \tag{5.30}$$

Note that the first term in (5.24) is $\Pr(N \ge \alpha + 1)$ and α in (5.29) should be

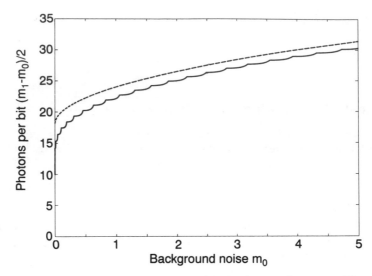

Figure 5.6 The quantum limit as a function of the background noise m_0. The solid curve is the exact solution and the dashed curve represents the Gaussian approximation.

replaced by $\alpha + 1$ in the calculation of P_e. The evaluation of error probability and the accuracy of (5.28) and (5.29) are illustrated by the following example.

Example 5.2   Optical system with on-off signaling

Consider a fiber optical system with an information rate of $B = 100$ Mb/s working at $\lambda = 0.8$ μm. The received optical signal has two intensity levels $P_0 = 10^{-9}$ W for binary 'zero' and $P_1 = 5 \cdot 10^{-9}$ W for binary 'one'. See Figure 5.7. The photodetector has a quantum efficiency $\eta = 0.9$ and produces a dark current $i_d = 1.5 \cdot 10^{-9}$ A.

Calculate the bit error probability for a photon-counting receiver.

Solution

The average number of photoelectrons occurring in the time interval $[0, T]$ at optical power P_0 is, from (5.43)

$$m_b = \frac{\eta}{hf} P_0 T = \frac{0.9 \times 0.8 \times 10^{-6}}{6.626 \times 10^{-34} \times 2.998 \times 10^8} \times 10^{-9} \cdot 10^{-8} = 36.25$$

The dark current i_d corresponds to an average number

$$m_d = \frac{i_d T}{q} = \frac{1.5 \times 10^{-9} \times 10^{-8}}{1.602 \times 10^{-19}} = 93.63$$

of electrons during the symbol interval $[0, T]$.

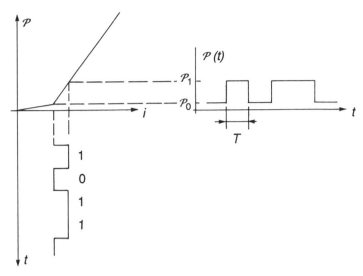

Figure 5.7 A binary optical signal produced by switching the drive current of a laser diode between two values above the lasing threshold.

The average number of electrons representing the 'zero' symbol is

$$m_0 = m_b + m_d = 129.9$$

Since $\mathcal{P}_1 = 5\mathcal{P}_0$ the average number of electrons representing the 'one' symbol is

$$m_1 = 5m_b + m_d = 274.9$$

The receiver threshold is calculated from (5.26), which gives $\alpha = 193.4$. The integer threshold for the electron counter is thus $\alpha = 193$.

The error probability calculated by the exact evaluation of the sums in formula (5.24) is

$$P_e = 1.0301 \times 10^{-7}$$

Replacing the summations in (5.24) by the upper bounds (5.28) and (5.29) results in

$$P_e < 1.054 \times 10^{-7}$$

which is a good approximation, compared to the exact value, and is easier to calculate. □

5.5.1 *Gaussian Approximation*

The basic statistical model in optical communication is the Poisson process but despite this the Gaussian density is often used to obtain an approximate expression for the error probability for optical receivers. In communication theory the Gaussian distribution is well known; it appears in many situations where additive noise is involved.

For a Gaussian-distributed decision variable a detector with threshold α has, for equal probabilities of the data symbols, the error probability

$$P_e = \frac{1}{2} \int_\alpha^\infty f_0(x)\, dx + \frac{1}{2} \int_{-\infty}^\alpha f_1(x)\, dx = \frac{1}{2}\, Q\!\left(\frac{E_1 - \alpha}{\sigma_1}\right) + \frac{1}{2}\, Q\!\left(\frac{\alpha - E_0}{\sigma_0}\right) \quad (5.31)$$

where E_1 and E_0 are the means of the decision variable σ_1 and σ_0 are the standard deviations when the symbols 'one' and 'zero', respectively, are transmitted. The function $Q(x)$ is the normalized Gaussian tail probability

$$Q(x) = \frac{1}{\sqrt{2\pi}} \int_x^\infty e^{-s^2/2} ds \quad (5.32)$$

To obtain a simple expression let α not be the optimal threshold (5.27), but be determined by making the two terms in (5.31) equal in magnitude, i.e.

$$\frac{E_1 - \alpha_g}{\sigma_1} = \frac{\alpha_g - E_0}{\sigma_0}$$

which results in

$$\alpha_g = \frac{\sigma_0 E_1 + \sigma_1 E_0}{\sigma_1 + \sigma_0} \quad (5.33)$$

The threshold (5.33) is not the best one for a Gaussian decision variable with unequal variances σ_1^2 and σ_0^2. It is introduced for computational convenience, producing a simple estimate of the error probability.

Substitution of (5.33) into (5.31) gives

$$P_e = Q(\rho) \quad (5.34)$$

where

$$\rho = \frac{E_1 - E_0}{\sigma_1 + \sigma_0} \quad (5.35)$$

is the signal-to-noise ratio (SNR) at the decision point.

A Poisson distribution with mean m has variance $\sigma^2 = m$, see (5.6). Approximating the sums in (5.24) by integrals of Gaussian density functions with the same mean and variance as the correct Poisson distributions gives

$$\rho = \frac{m_1 - m_0}{\sqrt{m_1} + \sqrt{m_0}} = \sqrt{m_1} - \sqrt{m_0} \quad (5.36)$$

The threshold (5.33) becomes

$$\alpha_g = \sqrt{m_1 m_0} \quad (5.37)$$

Substitution of (5.36) into (5.34) gives

$$P_e \approx Q(\sqrt{m_1} - \sqrt{m_0}) \quad (5.38)$$

The Q-function (5.32) can be evaluated by numerical integration or from a table over the Gaussian distribution functions. Since we are not trying to calculate the exact error probability an approximate expression for $Q(x)$ is sufficient. A simple upper bound on $Q(x)$ which is convenient to use is obtained by integrating (5.32) by parts

$$\sqrt{2\pi}Q(x) = \int_x^\infty \frac{1}{s} se^{-s^2/2}\,ds = \frac{1}{x}e^{-x^2/2} - \int_x^\infty \frac{1}{s^2}e^{-s^2/2}\,ds \qquad (5.39)$$

and

$$Q(x) \approx Q_1(x) = \frac{1}{x\sqrt{2\pi}}\exp(-x^2/2) \qquad (5.40)$$

The expression (5.40) is in fact an improved Chernoff bound for the Gaussian distribution of the same type as (5.28) and (5.29). $Q_1(x)$ is a reasonable approximation of $Q(x)$ for $x > 1$ or $Q(x) < 0.1$.

Substitution of $Q_1(\rho)$ for $Q(\rho)$ in (5.38) gives what we call the Gaussian approximation

$$P_e \approx Q_1(\rho) = Q_1(\sqrt{m_1} - \sqrt{m_0}) \qquad (5.41)$$

Application of (5.41) to Example 5.2 where $m_0 = 129.9$ and $m_1 = 274.9$ gives $\rho = \sqrt{274.9} - \sqrt{129.9} = 5.18$ and

$$P_e \approx Q_1(\rho) \approx 1.13 \times 10^{-7}$$

to be compared with the exact value $P_e = 1.03 \times 10^{-7}$.

Text books in mathematical statistics tell that the Poisson distribution asymptotically approaches the Gaussian distribution for increasing values of its mean. This is true for the overall appearance of the distribution function but in the tails the difference between the probabilities can be large. Figure 5.8 shows the cumulative probabilities for the distributions for an on-off system with $m_0 = 66$ and $m_1 = 200$. The diagram shows the upper tail $\Pr\{N > \alpha|0\}$ and the lower tail $\Pr\{N \leq \alpha|1\}$ of the true Poisson distribution together with the tails of a Gaussian distribution with the same mean and variance. The Gaussian threshold $\alpha_g = 114.9$ differs from the correct threshold $\alpha_o = 120$, but the approximate error probability 0.91×10^{-9} is close to the true value 0.76×10^{-9}.

The tail probabilities for the Poisson and the Gaussian distribution can differ with several orders of magnitude at low values. The Gaussian approximation, however, produces a reasonably accurate estimate of the error probability as illustrated in the diagram. The main reason for this is that the approximation overestimates the lower tail and underestimates the upper tail by about the same amount. The fact that the Gaussian approximation assumes a threshold (5.37) different from the optimal (5.27) helps to improve its accuracy.

Figure 5.6 shows how well the Gaussian approximation estimates the quantum limit of an on-off system. The approximation works well except for the ideal case with $m_0 = 0$.

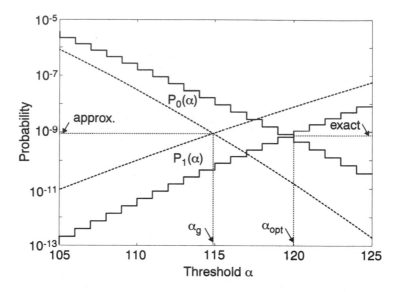

Figure 5.8   Example of detection probabilities for an optical system with $m_0 = 66$ and $m_1 = 200$. The staircase lines show the upper tail P_0 and the lower tail P_1 of the true Poisson distributions. The dashed curves are the tails of a Gaussian distributions with the same mean and variance. The optimal threshold α_{opt} and the exact error probability together with the Gaussian threshold α_g and the approximate error probability estimate are indicated in the diagram.

5.5.2   Intersymbol Interference

In this section we consider an optical system where the received pulses may overlap resulting in intersymbol interference (ISI).

The receiver is the same as before, shown in Figure 5.4. It performs symbol-by-symbol processing, which when ISI is present is suboptimal. A sequence-estimation receiver processing the entire received signal would produce a better result. In Section 5.9 below we will derive an expression for the error probability of such a receiver. Here we present a simple upper bound on the error probability in the presence of ISI.

The received optical signal is shown in Figure 5.9. The pulses are now broader than the symbol interval $[0, T]$. This is typically caused by the dispersion resulting from transmission through an optical fiber. We will find that intersymbol interference degrades receiver performance.

The received optical pulse representing binary 'one' is denoted as $\gamma(t)$. The received optical signal intensity $\Gamma(t)$ (photoelectrons per second) is

$$\Gamma(t) = \gamma_0 + \sum_{k=-\infty}^{\infty} b_k \gamma(t - kT) \tag{5.42}$$

where $b_k = 0$ or $b_k = 1$, depending on which binary symbol was transmitted. Note that (5.42) means power addition of the optical signals. Addition of two

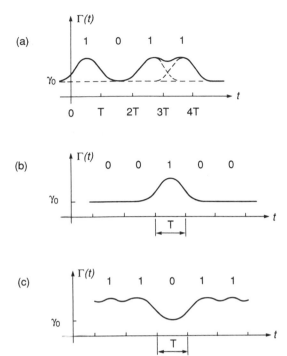

Figure 5.9 Optical signals with intersymbol interference (ISI): (a) General received signal. (b) Least favourable situation when a binary "one" is transmitted (minimal ISI). (c) Least favourable situation when a binary "zero" is transmitted (maximal ISI).

independent Poisson processes result in a Poisson process with an intensity equal to the sum of the original intensities. When optical signals are specified in terms of optical field quantities, power addition occurs for noncoherent signals. See Section 4.6.1 for an example of such signals.

In addition to (5.42) the dark current of the detector also contributes to the number of electrons produced. The number of electrons occurring in the kth symbol interval has a Poisson distribution with mean value

$$m(k) = \gamma_d T + \int_{kT}^{(k+1)T} \Gamma(t)\, dt \tag{5.43}$$

where γ_d represents the dark current electron intensity of the photodetector.

From Figure 5.9 it is evident that the part of $\Gamma(t)$ that falls into a certain symbol interval $[0, T]$ depends on the symbol in that interval, and also in general on the symbols in neighbouring intervals. This means that the detection probability will depend not only on the transmitted symbol but also on the symbols preceding and succeeding it. We deal with this situation by considering the least favourable cases for detection.

When a 'one' is transmitted in, say, time slot $k = 0$, the intersymbol interference can only increase the optical energy in the symbol interval $[0, T]$. The smallest received optical energy and the largest error probability result when the symbol $b_0 = 1$ is surrounded by 'zeros', i.e. $b_k = 0$, $k \neq 0$. See Figure 5.9b. If the receiver is designed such, that this worst case error probability is less than a specified value, it will operate at an error probability not greater than this value for all input data sequences.

It is convenient to normalize the optical pulse $\gamma(t)$. Its total optical energy expressed as the average number of photoelectrons is

$$m = \int_{-\infty}^{\infty} \gamma(t)\, dt \tag{5.44}$$

Let δ be the relative optical energy outside the signaling interval $[0, T]$

$$1 - \delta = \frac{1}{m} \int_0^T \gamma(t)\, dt \tag{5.45}$$

The average number of received photoelectrons corresponding to a transmitted 'one' during the signal interval at minimal ISI is obtained from (5.43) and (5.42) with $b_0 = 1$ and $b_k = 0$ for $k \neq 0$.

$$m_1 = \gamma_d T + \gamma_0 T + (1 - \delta)m \tag{5.46}$$

For detection of a binary 'zero' the least favourable situation is when $b_0 = 0$ and it is surrounded by $b_k = 1$, $k \neq 0$, as illustrated in Figure 5.9c which corresponds to maximal ISI.

The corresponding mean value is

$$m_0 = \gamma_d T + \gamma_0 T + \int_0^T \sum_{k \neq 0} \gamma(t - kT)\, dt \tag{5.47}$$

$$= \gamma_d T + \gamma_0 T + \delta m$$

The error probability, with equal a priori probabilities for 'ones' and 'zeros', is for all input sequences bounded by (5.24) with insertion of m_0 from (5.47) and m_1 from (5.46).

$$P_e \leq P_u = \frac{1}{2} \sum_{n=\alpha+1}^{\infty} \frac{m_0^n}{n!} e^{-m_0} + \frac{1}{2} \sum_{n=0}^{\alpha} \frac{m_1^n}{n!} e^{-m_1} \tag{5.48}$$

The tightest bound is obtained when the threshold α in (5.48) is determined from (5.27).

Since m_0 and m_1 are the least favourable values that can occur for any sequence of input symbols and since minimal and maximal ISI cannot occur simultaneously, the true error probability is certainly less than P_u when intersymbol interference is present. For a system with no ISI the error probability $P_e = P_u$. Notice that (5.48) is an upper bound on the achievable error probability in the presence of ISI with more general receiver structures as well.

Example 5.3   System with ISI

Calculate the bit error probability for a system with the same received optical power as in Example 5.2 when the received pulses have Gaussian shape

$$p(t) = A \exp(-2t^2/T^2)$$

Solution

The received signals now exhibit intersymbol interference (ISI). The received optical energy is

$$E = A \int_{-\infty}^{\infty} \exp(-2t^2/T^2)\, dt = AT\sqrt{\pi/2}$$

and the ISI parameter δ becomes (5.45)

$$\delta = 1 - \frac{1}{S} \int_{-T/2}^{T/2} p(t)\, dt$$

$$= 1 - \frac{1}{\sqrt{2\pi}} \int_{-1}^{1} \exp(-x^2/2)\, dx = 0.317$$

The same received optical symbol energy as in Example 5.2 gives

$$E = (\mathcal{P}_1 - \mathcal{P}_0)T = 4 \times 10^{-9} \times 10^{-8}\mathrm{Ws}$$

The mean value m_1 when the symbol 'one' is transmitted is from (5.46) for maximum ISI

$$m_1 = \frac{i_d T}{q} + \frac{\eta}{hf} [\mathcal{P}_0 T + (1 - \delta)E]$$

$$= 93.63 + 3.625 \times 10^{18}(10^{-9} \times 10^{-8} + 0.683 \times 4 \times 10^{-17}) = 228.9$$

The mean value m_0 when the symbol 'zero' is transmitted is from (5.47) for maximal ISI

$$m_0 = \frac{i_d T}{q} + \frac{\eta}{hf} [\mathcal{P}_0 T + \delta E]$$

$$= 93.63 + 3.625 \times 10^{18}[(10^{-9} \times 10^{-8} + 0.317 \times 4 \times 10^{-17}] = 175.8$$

The integer threshold corresponding to those values of m_1 and m_0 is, from (5.27)

$$\alpha = 201$$

which results in the bit error probability

$$P_e < P_u = 3.07 \times 10^{-2} \text{ exact}$$

$$P_e < P_u < 3.59 \times 10^{-2} \text{ approximate, using (5.28) and (5.29)}$$

The Gaussian approximation (5.41) gives

$$\rho = \sqrt{228.9} - \sqrt{175.8} = 1.87$$

and

$$P_e \approx Q_1(\rho) < 3.71 \times 10^{-2}$$

$\square$

A comparison with Example 5.2 shows a considerable degradation in performance due to the intersymbol interference.

5.6   Receiver with Rectangular Filter

The receivers studied in Sections 5.4 and 5.5 are ideal in the sense that they are assumed to be capable of detecting and counting single photons or photoelectrons. This requires a perfect detector without internal noise or any other deficiencies.

A practical receiver usually consists of a photodetector followed by a low-noise amplifier and a suitable filter before a threshold device. See Figure 5.10.

In this section we consider a receiver with a special kind of filter, the so-called integrate-and-dump filter. Such a filter consists in principle of an integrator which is reset after T seconds.

The integrate-and-dump filter has an impulse response of rectangular shape, as depicted in Figure 5.11.

$$h(t) = \begin{cases} 1; & 0 < t < T \\ 0; & t < 0,\ t > T \end{cases} \tag{5.49}$$

The integrate-and-dump filter is chosen mainly to simplify the analysis. It illustrates the properties of a realistic receiver without spending too much time on mathematical details.

We begin the analysis of the integrate-and-dump filter receiver, by assuming that the amplifier and the filter are noiseless. The photodetector sends a stream of photoelectrons to the amplifier. In the shot noise model (5.7) each electron generates a current pulse of shape $\delta(t)$. The amplified current $I(t)$ at the filter input is from (5.7) and (5.8)

$$I(t) = Aq \sum_{k=-\infty}^{\infty} \delta(t - t_k) \tag{5.50}$$

with t_k generated by a Poisson process.

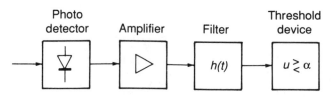

Figure 5.10   An optical receiver containing a linear filter followed by a threshold device.

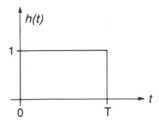

Figure 5.11   The impulse response of an integrate-and-dump (rectangular) filter.

The output of the filter at time $t = T$ is

$$u(t = T) = \int_{-\infty}^{\infty} h(T - t)I(t)\, dt = \int_{0}^{T} I(t)\, dt = NAq \qquad (5.51)$$

with $h(t)$ the current-to-voltage impulse response of the filter. The integer N is the number of photoelectrons, i.e. the number of Poisson events in the current $I(t)$ during the time interval $[0, T]$, and q is the electron charge.

An ideal integrate-and-dump filter receiver without thermal noise is thus equivalent to the photon-counting receiver analyzed in Section 5.5.

5.6.1   Thermal Noise

To study the effect of thermal noise in the receiver we introduce an additive noise component, and $I(t)$ (5.50) is now

$$I(t) = Aq \sum_{k} \delta(t - t_k) + i_n(t) \qquad (5.52)$$

The current $i_n(t)$ represents the thermal noise from the photodetector, the amplifier and the filter. It is assumed to be a zero-mean, white and Gaussian stochastic process.

Substitution of (5.52) into (5.51) gives

$$u(t = T) = \int_{0}^{T} I(t)\, dt = NAq + \xi \qquad (5.53)$$

where ξ is a zero-mean Gaussian stochastic variable. Its variance is

$$\sigma_{\xi}^2 = E\left\{ \left(\int_{0}^{T} i_n(t)\, dt \right)^2 \right\} = R_n T \qquad (5.54)$$

with R_n equal to the (two-sided) spectral density of $i_n(t)$.

The amplification factor A can be changed if the threshold is adjusted accordingly. For convenience we let $Aq = 1$ and normalize $i_n(t)$ in such a way that the decision variable (5.53) becomes

$$U = N + X \tag{5.55}$$

where N is the number of photoelectrons in the symbol interval $[0,T]$ and $X = \xi/Aq$.

The stochastic variable U is the sum of two independent variables. The component N has a Poisson distribution (5.5) with mean m, and X is Gaussian with zero mean and variance $\sigma^2 = R_n T/(Aq)^2$. Its probability density function is

$$p(x) = \frac{1}{2\pi} e^{-x^2/2\sigma^2} \tag{5.56}$$

The probability density of U is the convolution of the densities of N (5.5) and of X (5.56).

$$p(u) = \sum_{n=0}^{\infty} \frac{e^{-m}}{\sqrt{2\pi\sigma^2}} \frac{m^n}{n!} \exp[-(u-n)^2/2\sigma^2] \tag{5.57}$$

The infinite summation makes (5.57) difficult to handle analytically and numerically.

5.6.2 Saddlepoint Approximation

A more tractable way is to utilize the moment-generating function

$$\Psi(s) = E\{\exp(sU)\} = \int_{-\infty}^{\infty} p(u)e^{su} \, du \tag{5.58}$$

It can be interpreted as a double-sided Laplace transform of the density function and it contains the same information as $p(u)$.

The moment-generating function of the sum of two independent variables is the product of their moment-generating functions

$$\Psi(s) = \Psi_n(s) \cdot \Psi_x(s) \tag{5.59}$$

For the Poisson distribution,

$$\Psi_n(s) = E\{e^{sN}\} = \sum_{n=0}^{\infty} \frac{m^n e^{-m}}{n!} e^{sn}$$

$$= e^{-m} \sum_{n=0}^{\infty} \frac{(me^s)^n}{n!} = \exp[m(e^s - 1)] \tag{5.60}$$

For a Gaussian variable with zero mean and variance σ^2,

$$\Psi_x(s) = \frac{1}{\sqrt{2\pi\sigma^2}} \int_{-\infty}^{\infty} e^{sx} e^{-x^2/2\sigma^2} \, dx = \exp(s^2\sigma^2/2) \tag{5.61}$$

Substitution of (5.60) and (5.61) into (5.59) gives

$$\Psi(s) = \exp[m(e^s - 1) + s^2\sigma^2/2] \tag{5.62}$$

We will use the moment-generating function approach to derive a useful approximation of the probability that the variable U will exceed or fall below a threshold. This is what is needed to calculate the error probability of the receiver.

It is shown in Appendix D that the probability that U will exceed a threshold α,

$$q_+(\alpha) = \int_\alpha^\infty p_0(u)\, du \tag{5.63}$$

is approximately equal to

$$q_+(\alpha) \approx \frac{\exp[\psi_0(s_0)]}{\sqrt{2\pi\psi_0''(s_0)}} \tag{5.64}$$

The function $\psi_0(s)$ is related to the moment-generating function (5.62) through the expression

$$\psi_0(s) = \ln[\Psi_0(s)] - s\alpha - \ln|s| \tag{5.65}$$

where $\Psi_0(s)$ is $\Psi(s)$ for 'zero' symbol transmitted, see Appendix E.

The parameter s_0 is a positive real constant determined by the equation

$$\psi_0'(s) - 0 \tag{5.66}$$

The expression (5.64) is a saddlepoint approximation, see e.g., Carrier *et al.* (1966). The value of s determined by (eq 6.5.12) causes the integration to pass through a saddlepoint in the complex plane, resulting in a minimal value of (5.64), which results in a good approximation since q_+ is a positive quantity. See Appendix E for further details.

The saddlepoint approximation of the lower tail

$$q_-(\alpha) = \int_{-\infty}^\alpha p_1(u)\, du \tag{5.67}$$

is

$$q_-(\alpha) \approx \frac{\exp[\psi_1(s_1)]}{\sqrt{2\pi\psi_1''(s_1)}} \tag{5.68}$$

where the parameter s_1 is the negative root of

$$\psi_1'(s) = 0 \tag{5.69}$$

Substitution of (5.62) into (5.65) gives

$$\psi(s) = m(e^s - 1) + s^2\sigma^2/2 - s\alpha - \ln|s| \tag{5.70}$$

The derivative of (5.70) is

$$\psi'(s) = me^s + s\sigma^2 - \alpha - 1/s = 0 \tag{5.71}$$

The second derivative becomes

$$\psi''(s) = me^s + \sigma^2 + 1/s^2 \tag{5.72}$$

The solution of (5.71) has to be found numerically to determine the saddlepoint parameters s_0 and s_1. Since the second derivative $\psi''(s)$ is available, a suitable method is the Newton-Raphson algorithm.

We illustrate the calculation of the error probability with an example.

Example 5.4   Saddlepoint Approximation

Consider the optical system described in Example 5.2 with $m_0 = 129.9$ and $m_1 = 274.9$. Assume that Gaussian additive noise corresponding to the normalized value of $\sigma = 5$ is present in the detector.

Determine the detector threshold α and calculate the error probability.

Solution

The error probability is

$$P_e = \frac{1}{2}[\Pr\{U > \alpha|m = m_0\} + \Pr\{U < \alpha|m = m_1\}] \tag{5.73}$$

Using the saddlepoint approximations (5.64) and (5.68) for the probabilities in (5.73) gives

$$P_e \approx \frac{1}{2}[q_+(\alpha, s_0) + q_-(\alpha, s_1)] \tag{5.74}$$

where from (5.70) and (5.72)

$$q_+(\alpha, s_0) = \frac{\exp[m_0(e^{s_0} - 1) + s_0^2\sigma^2/2 - s_0\alpha - \ln|s_0|]}{\sqrt{2\pi(m_0 e^{s_0} + \sigma^2 + 1/s_0^2)}} \tag{5.75}$$

and

$$q_-(\alpha, s_1) = \frac{\exp[m_1(e^{s_1} - 1) + s_1^2\sigma^2/2 - s_1\alpha - \ln|s_1|]}{\sqrt{2\pi(m_1 e^{s_1} + \sigma^2 + 1/s_1^2)}} \tag{5.76}$$

The parameters s_0 and s_1 are determined from $\psi'(s) = 0$ with $\psi'(s)$ given by (5.71)

$$m_0 e^{s_0} + \sigma^2 s_0 - \alpha - 1/s_0 = 0 \tag{5.77}$$

and

$$m_1 e^{s_1} + \sigma^2 s_1 - \alpha - 1/s_1 = 0 \tag{5.78}$$

respectively. See (5.66) and (5.71).

So far we have considered α to be a known constant. As an example let the receiver threshold be equal to $\alpha = 200$. The solutions of (5.77) and (5.78) then result in $s_0 = 0.394$ and $s_1 = -0.298$ respectively, which should be inserted into (5.75) and (5.76) to evaluate the error probability at this particular threshold.

The best threshold is the one that minimizes P_e. In conjunction with the saddlepoint approximation it is convenient to use the threshold that minimizes the approximate expression (5.74). Setting the derivative with respect to α of (5.74) equal to zero gives

$$\frac{dP_e}{d\alpha} = -\frac{1}{2}[s_0 q_+(\alpha, s_0) + s_1 q_-(\alpha, s_1)] = 0 \qquad (5.79)$$

The three equations (5.75), (5.76) and (5.79) determine the three parameters s_0, s_1 and α. They can easily be solved numerically by the Newton-Raphson algorithm.

For $m_0 = 129.9$, $m_1 = 274.9$ and $\sigma = 5$ the best threshold turns out to be

$$\alpha = 194.06$$

resulting in an error probability (5.74) of

$$P_e \approx 5.153 \times 10^{-7}$$

An exact evaluation of (5.73) gives the same threshold and an error probability of

$$P_e = 5.156 \times 10^{-7}$$

$\Box$

The example shows that the saddlepoint approximation gives extremely accurate numerical results for an optical receiver with both Poisson and Gaussian noise.

The calculations above are simple to carry out on a digital computer but the numerical optimizations required to determine α, s_0 and s_1 may be cumbersome to perform on a calculator.

5.6.3   Gaussian Approximation

A practically useful estimate of the error probability is obtained if the distribution of the compound variable U is approximated by a Gaussian distribution. The means of U are $E_1 = m_1$ and $E_0 = m_0$, respectively. The variances are $\sigma_1^2 = m_1 + \sigma^2$ and $\sigma_0^2 = m_0 + \sigma^2$ when data symbols 'one' and 'zero', respectively, are transmitted. The signal-to-noise-ratio (5.35) is

$$\rho = \frac{m_1 - m_0}{\sqrt{m_1 + \sigma^2} + \sqrt{m_0 + \sigma^2}} = \sqrt{m_1 + \sigma^2} - \sqrt{m_0 + \sigma^2} \qquad (5.80)$$

Substitution into (5.41) gives the Gaussian approximation

$$P_e \approx Q_1\left(\sqrt{m_1 + \sigma^2} - \sqrt{m_0 + \sigma^2}\right) \qquad (5.81)$$

which constitutes a direct approximative relation between the system parameters m_0, m_1 and σ and the transmission error probability

Example 5.5   Receiver with PIN diode

Consider the optical system in Example 5.4 with $m_0 = 129.9$, $m_1 = 274.9$ and $\sigma^2 = 25$. The signal-to-noise ratio (5.80) is

$$\rho = \sqrt{274.9 + 25} - \sqrt{129.9 + 25} = 4.87$$

and the approximate error probability (5.81) is

$$P_e \approx Q_1(4.87) = 5.75 \times 10^{-7}$$

The threshold used in the approximation (5.81) is $\alpha_g = 190.5$ obtained from (5.33)

$$\alpha_g = \frac{\sigma_0 E_1 + \sigma_1 E_0}{\sigma_1 + \sigma_0} = \sqrt{(m_1 + \sigma^2)(m_0 + \sigma^2)} - \sigma^2 \tag{5.82}$$

If the the correct value $\alpha = 194.06$ obtained by the saddlepoint technique in Example 5.4 is used together with the Gaussian approximation the error probability estimate becomes

$$P_e \approx 8.24 \times 10^{-7}$$

The approximate value 5.75×10^{-7} in the Example is close to the exact value 5.16×10^{-7} calculated in Example 5.4.   □

The Gaussian approximation works well for receivers with ordinary photodetectors without avalanche gain.

Figure 5.14 illustrates that the Gaussian approximation is an acceptable estimate of the system error probability for values of $P_e < 0.1$ which clearly is sufficient in communication applications. Figure 5.14a shows a case where $\sigma^2 = 0$, i.e. with no thermal noise and Figure 5.14b illustrates a situation with both optical and thermal noise present. Notice that the Gaussian approximation in many cases is useful also for pure optical signals, despite the fact that Poisson statistics differ markedly from Gaussian.

A more detailed evaluation of the accuracy of various bounds and approximations for the Poisson distribution and the compound Poisson and Gaussian distribution is presented by Einarsson (1989).

5.7   Avalanche Photodetector

One way of improving the receiver sensitivity is to use an avalanche photodetector (APD). The primary photoelectrons produced by the detector are multiplied by the avalanche gain a_k, which means an amplification of the optical signal before the electronic signal processing in the receiver.

The shot noise current from an APD is, see (5.7) and (5.50)

$$I(t) = Aq \sum_{-\infty}^{\infty} a_k \delta(t - t_k) \tag{5.83}$$

The number of photoelectrons produced by the APD during the symbol interval [0,T] is

$$N_a = \sum_{k=1}^{N} a_k \qquad (5.84)$$

where N is the number of Poisson events during the time interval. The decision variable for a receiver with rectangular filter, in the presence of thermal noise, is analogously to (5.55)

$$U = N_a + X \qquad (5.85)$$

To calculate the error probability by the saddlepoint approximation the moment-generating function of N_a is needed. It is related to the moment-generating function of the avalanche gain $\Psi_a(s) = E\{\exp(sa_k)\}$ in the following way, cf. (5.60)

$$\Psi_{na}(s) = E\{e^{sN_a}\} = E\{\exp(s\sum_{k=1}^{N} a_k)\}$$

$$= E\left\{\prod_{k=1}^{N} \exp(sa_k)\right\} = E\{\Psi_a^N(s)\} \qquad (5.86)$$

$$= \sum_{n=0}^{\infty} \frac{m^n e^{-m}}{n!} \Psi_a^n(s) = \exp[m(\Psi_a(s) - 1)]$$

The gain factors a_k are assumed to be independent for different indices of the poisson point process.

The statistics of the avalanche gain have been studied by Personick (1971a, b). For the idealized case when the hole/electron ionization ratio $\kappa = 0$, which corresponds to an APD where only electrons cause ionizing collisions, the moment-generating function is

$$\Psi_a(s) = \frac{1}{1 - M(1 - e^{-s})} \qquad (5.87)$$

with $M = E\{a_k\}$ equal to the average avalanche gain.

For the general case with $0 < \kappa < 1$ an implicit relation is available, see Personick (1971b).

$$\boxed{s = \ln \Psi_a - \frac{1}{1 - \kappa} \ln[(1 - c)\Psi_a + c]} \qquad (5.88)$$

The parameter c in (5.88) is specified later.

Derivating both sides of (5.88) with respect to s gives

$$1 = \frac{\Psi_a'}{\Psi_a} - \frac{(1 - c)\Psi_a'}{(1 - \kappa)[(1 - c)\Psi_a + c]} \qquad (5.89)$$

Solving for Ψ_a' yields

$$\Psi_a' = \left(\frac{1}{\Psi_a} - \frac{(1-c)}{(1-\kappa)[(1-c)\Psi_a + c]} \right)^{-1} \tag{5.90}$$

From the definition of the moment generating function follows that $\Psi_a(s=0) = 1$ and that the average avalanche gain is equal to the first derivative evaluated at $s = 0$

$$M = \mathrm{E}\{a_k\} = \Psi_a'(s=0) = \frac{1-\kappa}{c-\kappa} \tag{5.91}$$

The relation (5.91) determines the constant c

$$c = \frac{1-\kappa}{M} + \kappa \tag{5.92}$$

Substituting (5.92) into (5.88) and solving for Ψ_a verifies that the moment-generating function for $\kappa = 0$ is (5.87).

The mean square value $\mathrm{E}\{a_k^2\}$ is obtained in the same way by differentiating (5.90) and letting $s = 0$

$$\mathrm{E}\{a_k^2\} = \Psi_a''(s=0) = [\kappa M + (1-\kappa)(2-M^{-1})]M^2 \tag{5.93}$$

after elimination of c by the use of (5.92).

If the avalanche gain would not be stochastic, i.e. a_k equal to a fixed value $a_k = M$ the square value (5.93) assumes its smallest value equal to M^2. The factor in front of M^2 in (5.93) is called the excess noise factor

$$F(M) = \kappa M + (1-\kappa)(2-M^{-1}) \tag{5.94}$$

It is shown in Figure 5.12 for various values of the ionization ratio κ.

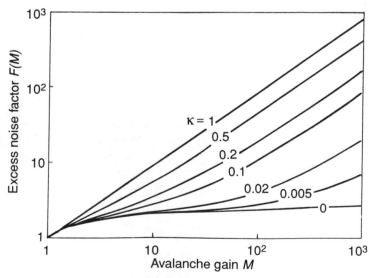

Figure 5.12 The excess noise factor for avalanche photodiodes as a function of the avalanche gain. The parameter κ is the hole/electron ionization ratio.

The probability of error can be evaluated using the saddlepoint technique described in Appendix E. The moment-generating function for the decision variable (5.85) in the presence of Gaussian noise is, analogously to (5.59)

$$\Psi_u(s) = \Psi_{na}(s) \cdot \Psi_x(s) \tag{5.95}$$

with $\Psi_x(s)$ given by (5.61).

The problem that (5.88) does not provide an explicit expression for $\Psi_a(s)$ can be handled by changing the variable in the saddlepoint approximation from s to Ψ_a as suggested by Helstrom (1984, 1988).

Substitution of (5.86) into (5.65) gives

$$\psi(s) = m[\Psi_a(s) - 1] + s^2\sigma^2/2 - s\alpha - \ln|s| \tag{5.96}$$

The saddlepoint equation $\psi'(s) = 0$ becomes

$$m\Psi_a'(s) + s\sigma^2 - \alpha - 1/s = 0 \tag{5.97}$$

Substitution of Ψ_a' from (5.90) and replacing s by (5.88) gives an equation in Ψ_a:

$$f(\Psi_a) = 0 \tag{5.98}$$

The derivative $f'(\Psi_a)$ is easily determined and (5.98) can be solved for Ψ_a numerically by e.g. the Newton-Raphson method. The saddlepoints s_0 and s_1 are obtained by insertion of the solution of (5.98) into (5.88).

Example 5.6   Receiver with APD

Consider an optical system with $m_0 = 9$, $m_1 = 418$ and large thermal noise $\sigma = 1000$. The photodetector in the receiver has an avalanche gain of $M = 110$ and an ionization ratio $\kappa = 0.04$.

Calculate the error probability using the saddlepoint approximation.

Solution

The saddlepoints are determined by (5.98) which is the following equation in the unknown $x = \Psi_a$

$$
m\left(\frac{1}{x} - \frac{(1-c)}{(1-\kappa)[(1-c)x+c]}\right)^{-1} + \left(\ln x - \frac{1}{1-\kappa}\ln[(1-c)x+c]\right)\sigma^2
$$
$$
- \left(\ln x - \frac{1}{1-\kappa}\ln[(1-c)x+c]\right)^{-1} - m - \alpha = 0
\tag{5.99}
$$

The constant c is equal to (5.92)

$$c = \frac{1-\kappa}{M} + \kappa = 0.0487$$

Solving (5.99) with $m = m_0$ and choosing the root x_0 which, when substituted into (5.88), yields a positive s, determines the saddlepoint parameter for the upper probability tail

$$s_0 = \ln x_0 - \frac{1}{1-\kappa} \ln[(1-c)x_0 + c]$$

The saddlepoint parameter for the lower tail is obtained in the same way by solving (5.99) with $m = m_1$ and using the root x_1 producing a negative s_1

$$s_1 = \ln x_1 - \frac{1}{1-\kappa} \ln[(1-c)x_1 + c]$$

The tail probabilities (5.64)

$$q_+(\alpha, s_0) \approx \frac{\exp[\psi(s_0)]}{\sqrt{2\pi\psi''(s_0)}}$$

and (5.68)

$$q_-(\alpha, s_1) \approx \frac{\exp[\psi(s_1)]}{\sqrt{2\pi\psi''(s_1)}}$$

can now be calculated. The second derivative $\psi''(s)$ is obtained by derivating (5.96) twice and substituting Ψ_a'', furnished by derivating (5.90) once more. The detection threshold is determined by simultaneously solving (5.79)

$$s_0 q_+(\alpha, s_0) + s_1 q_-(\alpha, s_1) = 0 \tag{5.100}$$

and the equations for s_0 and s_1 generated by (5.99). The resulting saddlepoint parameters are $s_0 = 9.05 \times 10^{-4}$ and $s_1 = -1.98 \times 10^{-3}$. The optimum threshold turns out to be $\alpha = 1.97 \times 10^4$. The error probability is

$$P_e \approx \frac{1}{2}[q_+(\alpha, s_0) + q_-(\alpha, s_1)] = 1.23 \times 10^{-9}$$

$\square$

The accuracy of the saddlepoint approximation for APD receivers has been investigated by Helstrom (1988). The results indicate that the relative error in the calculated error probability for a large range of values for M and κ is 15 % or better.

5.7.1   Gaussian Approximation

The Gaussian approximation is easily applied to an optical system with an avalanche photodetector receiver.

To obtain the mean and variance of the decision variable (5.85) the mean and variance of the number of secondary photoelectrons is needed. The mean of N_a (5.84) is

$$E\{N_a\} = E\{N\}E\{a_k\} = mM \tag{5.101}$$

The mean square value is

$$E\{N_a^2\} = E\left\{\sum_{j=1}^{N}\sum_{k=1}^{N}E\{a_j a_k\}\right\}$$

$$= E\left\{\frac{\sum_{k=1}^{N}E\{a_k^2\} + \sum_{j\neq k}E\{a_j\}}{E\{a_k\}}\right\} \tag{5.102}$$

$$= E\{N\}F(M)M^2 + E\{N(N-1)\}MM$$

$$= mF(M)M^2 + (m + m^2 - m)M^2$$

The variance of N_a becomes

$$\mathrm{Var}\{N_a\} = E\{N_a^2\} - (E\{N_a\})^2 = mF(M)M^2 \tag{5.103}$$

The Gaussian approximation is much simpler than the more accurate saddlepoint approximation, which is illustrated by the following example.

Example 5.7   Receiver with APD

Estimate the error probability of the APD receiver in Example 5.6 using the Gaussian approximation.

Solution

Avalanche gain $M = 110$ and $\kappa = 0.04$ give the excess noise factor from (5.94) or Figure 5.12

$$F(M) = 6.31$$

The signal-to-noise ratio is

$$\rho = \frac{m_1 M - m_0 M}{\sqrt{m_1 F(M)M^2 + \sigma^2} + \sqrt{m_0 F(M)M^2 + \sigma^2}} = 6.394$$

The error probability estimate produced by the Gaussian approximation is

$$P_e \approx Q_1(6.39) = 8.3 \times 10^{-11}$$

Comparison with the 'exact' value $P_e = 1.23 \times 10^{-9}$ from Example 5.6 shows that the Gaussian approximation is less accurate for APD receivers than for PIN diode receivers.

An ordinary photodiode would result in a detection signal-to-noise ratio (5.80)

$$\rho = \sqrt{m_1 + \sigma^2} - \sqrt{m_0 + \sigma^2} = 0.20$$

corresponding to an error probability only slightly less than 0.5, which renders the system useless for communication. □

The reason for the low accuracy of the Gaussian approximation for APD receivers is that the probability distribution of the decision variable differs

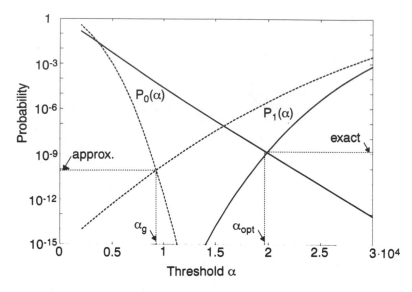

Figure 5.13 Detection probabilities for the optical system with APD receiver in Example 5.7. The solid lines show the upper and lower tails of the correct distribution calculated from the saddle point approximation. The dashed curves are the tails of a Gaussian distributions with the same mean and variance. The optimal threshold α_{opt} and the exact error probability together with the Gaussian threshold α_g and the approximate error probability estimate are shown in the diagram.

markedly from a Gaussian distribution. The deviations of the upper and lower tails show a large amount of unsymmetry and the overestimation of P_1 is not compensated for by an approximately equal underestimation of P_0. This is illustrated in Figure 5.13, which is based on the numerical data of Example 5.7.

The accuracy of the Gaussian approximation for an APD receiver has been investigated by Personick *et al.* (1977), Helstrom (1988), and also by Einarsson and Sundelin (1995).

An APD with low κ is an effective amplifier with low excess noise. At larger values of κ the decision variable contains more random noise, which helps the Gaussian approximation to function better.

5.8   Receiver with Arbitrary Filter

An optical receiver with rectangular filter is closely related to a photon counting receiver and is easy to analyze. For other types of filter the theory is more complicated. In this section we study the receiver of Figure 5.10 with an arbitrary filter.

Let $h(t)$ denote the current-to-voltage impulse response, including the gain A, of the filter in the receiver. The decision variable is analogously to (5.53)

$$U = u(t = T) = \int_0^T h(T - t)I(t) \, dt \qquad (5.104)$$

where $I(t)$ is the input current (5.52) with $A = 1$.

It is convenient to characterize the filter by the function

$$v(t) = q \, h(T - t) \qquad (5.105)$$

The part of U that comes from the optical shot noise (5.50) is then equal to, compare with (5.51)

$$Z = \int_0^T q^{-1}v(t) \sum_k q\delta(t - t_k) \, dt = \sum_{k=1}^N v(t_k) \qquad (5.106)$$

The complete decision variable is

$$U = Z + X \qquad (5.107)$$

with X a sample of filtered white zero mean Gaussian noise. The variance of X is, compare with (5.54)

$$\sigma^2 = R_n \int_0^T h^2(t) \, dt \qquad (5.108)$$

The quantity R_n is the power spectral density of the additive noise source and has physical dimension A^2/Hz. When dealing with optical detection it is convenient to represent the noise in terms of an electron intensity by introducing

$$\mathcal{N} = R_n/q^2 \qquad (5.109)$$

The noise variance (5.108) can then be expressed as

$$\sigma^2 = \mathcal{N} \int_0^T v^2(t) \, dt \qquad (5.110)$$

The components Z and X are independent random variables. The Poisson events t_k are independent and Z (5.106) is a sum of N independent variables. A complication is that N is not constant but is a random variable.

An error occurs if Z is larger than the threshold α when the 'zero' signal is transmitted or smaller than α when a 'one' is transmitted.

The exact probability density for Z is complicated. The saddlepoint approximation derived in Appendix E is therefore useful. It requires the moment-generating function of Z, which turns out to be easy to derive.

5.8.1 Moment-Generating Function

The moment-generating function (MGF) for shot noise processed by linear filter is derived in Appendix C. Substitution of (5.105) into (C.27) gives the MGF for Z

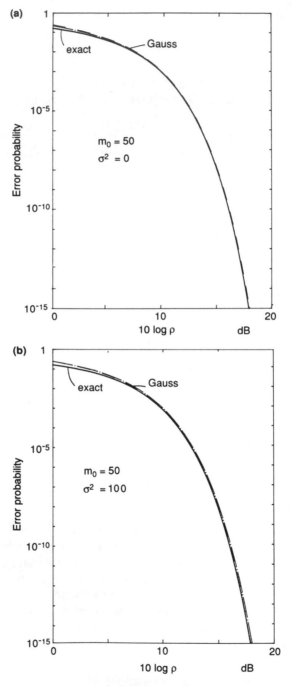

Figure 5.14   Transmission error probability for an optical system with rectangular filter receiver as a functon of the signal-to-noise ratio in dB. The exact value is shown together with the Gaussian approximation. (a) Poisson channel (no thermal noise). (b) Poisson plus Gaussian noise. From Einarsson (1989). Reproduced by permission of Springer-Verlag.

$$\Psi_z(s) = \exp\left(\int_0^T \Gamma(t) \, (\exp[sv(t)] - 1) \, dt\right) \qquad (5.111)$$

The moment-generating function (5.111) is the basis for error probability evaluation of optical receivers. The formula goes back to S. O. Rice in his famous paper from 1944 on the mathematical analysis of random noise.

When $h(t)$ is a rectangular filter the function $v(t)$ is equal to a constant, and

$$\Psi_z(s) = \exp\left(\int_0^T \Gamma(t) \, (\exp[cs] - 1) \, dt\right) \qquad (5.112)$$

$$= \exp[m(e^{cs} - 1)]$$

which for $c = 1$ is equal to the moment-generating function (5.60) for a Poisson distribution. A value $c \neq 1$ will not affect receiver performance, but the decision variable Z is Poisson for $c = 1$ only.

The moment-generating function for a detector with avalanche gain is obtained from (C.29):

$$\Psi_{za}(s) = \exp\left(\int_0^T \Gamma(t)(\Psi_a[sv(t)] - 1)dt\right) \qquad (5.113)$$

with $\Psi_a(s)$ the moment-generating function for the avalanche gain.

The complete moment-generating function, including zero mean additive Gaussian noise with variance σ^2, is obtained by multiplying (5.113) by the Gaussian moment-generating function (5.61). Using (5.110) the result is

$$\Psi_u(s) = \exp\left(\int_0^T \{\Gamma(t)(\Psi_a[sv(t)] - 1) + (\mathcal{N}/2)s^2v^2(t)\} \, dt\right) \qquad (5.114)$$

5.8.2 Saddlepoint Approximation

The moment-generating functions (5.111) and (5.114) can be used in a saddlepoint approximation in the same way as in Section 5.6. We illustrate with a simple example which does not require numerical evaluation of the integrals involved.

Example 5.8 Receiver with triangular filter

Consider a system with on-off modulation and rectangular received optical pulses, and let Γ_0 and Γ_1 denote the received optical signal levels for transmitted data symbols 'zero' and 'one', respectively.

The best receiver contains a rectangular (integrate-and-dump) filter, and the principles for calculation of the error probability of such a receiver in the presence of thermal noise are illustrated by Example 5.4.

Now let the receiver filter have a symmetric triangular filter function

$$v(t) = \begin{cases} 1 - 2|t|/T \; ; & |t| \leq T/2 \\ 0 \; ; & |t| > T/2 \end{cases} \tag{5.115}$$

Determine the optimal detector threshold and calculate the error probability.

Solution

Substitution of (5.115) into (5.111) gives the moment-generating function for the optical component of the decision variable

$$\Psi_z(s) = \exp\left\{ 2 \int_0^{T/2} \Gamma_i \left(\exp[s(1 - 2t/T)] - 1 \right) dt \right\} \tag{5.116}$$

The saddle point approximation requires the logarithm of $\Psi_z(s)$, which after changing the integration variable to $x = 1 - 2t/T$ becomes

$$\chi_i(s) = \ln[\Psi_z(s)] = \Gamma_i T \int_0^1 (e^{xs} - 1) \, dx = m_i \frac{(e^s - s - 1)}{s} \tag{5.117}$$

where m_i, $i = 0, 1$ is the average number of received photoelectrons for transmitted data symbols 'zero' and 'one', respectively. The saddlepoint function $\psi(s)$ defined by (5.65) for the complete decision variable including the Gaussian noise term (5.61) is

$$\psi_i(s) = \chi_i(s) + s^2 \sigma^2/2 - s\alpha - \ln|s| \tag{5.118}$$

The variance σ^2 is given by (5.110), and substitution of (5.115) yields

$$\sigma^2 = \mathcal{N} \int_{-T/2}^{T/2} v^2(t) \, dt = 2\mathcal{N} \int_0^{T/2} (1 - 2t/T)^2 dt = \mathcal{N}T/3 \tag{5.119}$$

To calculate the error probability the upper tail saddlepoint integral (5.63)

$$\int_\alpha^\infty p_0(u) \, du \approx q_+(\alpha, s_0) = \frac{\exp[\psi_0(s_0)]}{\sqrt{2\pi \psi_0''(s_0)}} \tag{5.120}$$

and the lower tail integral (5.67) are needed

$$\int_{-\infty}^\alpha p_1(u) \, du \approx q_-(\alpha, s_1) = \frac{\exp[\psi_1(s_1)]}{\sqrt{2\pi \psi_1''(s_1)}} \tag{5.121}$$

The quantity $\psi''(s)$ denotes the second derivative of $\psi(s)$ at the saddlepoint.

$$\psi_i''(s) = \chi_i''(s) + \sigma^2 + 1/s^2 \tag{5.122}$$

The saddlepoint parameters s_0 and s_1 are the solutions of $\psi'(s) = 0$.

$$\chi_i'(s_i) + \sigma^2 s_i - \alpha - 1/s_i^2 = 0 \; ; \quad i = 0, 1 \tag{5.123}$$

The first-order derivative of $\chi_i(s)$ is

$$\chi_i'(s) = m_i \frac{1 + (s - 1)e^s}{s^2} \tag{5.124}$$

and the second-order derivative

$$\chi_i''(s) = m_i \frac{((s-1)^2 + 1)e^s - 2}{s^3} \tag{5.125}$$

The receiver threshold α minimizing the error probability is the solution of the equation (5.79)

$$s_0 q_+(\alpha, s_0) + s_1 q_-(\alpha, s_1) = 0 \tag{5.126}$$

Numerical solution of (5.123) and (5.126) for the same system parameters $m_0 = 129.9$, $m_1 = 274.9$ and $\mathcal{N}T = 25$, as in Example 5.4, using the Newton-Raphson algorithm, gives the optimum threshold $\alpha = 97.20$ and the saddlepoint parameters $s_0 = 0.549$ and $s_1 = -0.498$. The saddlepoint estimate of the error probability is

$$P_e \approx 0.5[q_+(\alpha, s_0) + q_-(\alpha, s_1)] = 1.159 \times 10^{-5}$$

Comparison with the rectangular filter receiver in Example 5.4, which produces the error probability

$$P_e \approx 5.513 \times 10^{-7}$$

illustrates the degradation in performance for the triangular filter.   □

An analysis of optical receivers with APD detector by saddlepoint integration is presented by Helstrom (1992).

5.8.3 Channel with Intersymbol Interference

In Section 5.8 it is assumed that no intersymbol interference is present. The results apply to the detection of isolated received pulses or when the ISI is dealt with by studying worst-case situations, as in Section 5.5.2.

For a receiver with a specified filter function $v(t)$ the error probability when ISI is present can be evaluated from the moment-generating function as follows. The received photon intensity (5.42) is

$$\Gamma(t) = \gamma_0 + \sum_k b_k \gamma(t - kT) \tag{5.127}$$

Consider the case when the data symbol $b_0 = 0$ is to be detected. Substitution into (5.111) gives the moment-generating function for the decision variable (5.106)

$$\Psi_0(s) = E\left\{\exp\left(\int_0^T [\gamma_0 + \sum_{k \neq 0} b_k \gamma(t - kT)](\exp[sv(t)] - 1)\, dt\right)\right\} \tag{5.128}$$

where the expectation is to be performed over the data symbols $b_k, k \neq 0$. These are assumed to be independent variables taking the values 1 and 0 with equal probability.

Expansion of (5.128) gives

$$\Psi_0(s) = \exp\left(\int_0^T \gamma_0(\exp[sv(t)] - 1)\, dt\right)$$

$$\mathrm{E}\left\{\prod_{k \neq 0} \exp\left(b_k \int_0^T \gamma(t - kT)(\exp[sv(t)] - 1)\, dt\right)\right\}$$

$$= \exp\left(\int_0^T \gamma_0(\exp[sv(t)] - 1)\, dt\right)$$

$$\prod_{k \neq 0} \frac{1}{2}\left[1 + \exp\left(\int_0^T \gamma(t - kT)(\exp[sv(t)] - 1)\, dt\right)\right]$$

(5.129)

The moment-generating function for $b_0 = 1$ is analogously

$$\Psi_1(s) = \exp\left(\int_0^T [\gamma_0 + \gamma(t)](\exp[sv(t)] - 1)\, dt\right)$$

$$\prod_{k \neq 0} \frac{1}{2}\left[1 + \exp\left(\int_0^T \gamma(t - kT)(\exp[sv(t)] - 1)\, dt\right)\right]$$

(5.130)

The effect of additive thermal noise can be included, in the same way as in (5.114), by multiplying (5.129) and (5.130) by a Gaussian moment-generation function.

In well-designed optical systems the amount of intersymbol interference by necessity has to be small. Therefore it is often sufficient to maintain only a small number of terms in the evaluation of (5.129) and (5.130). The error probabilities when a 'zero' or a 'one' was transmitted can, for a specified receiver threshold, be calculated from the saddlepoint approximation. The optimal threshold can then be determined and the average error probability obtained.

5.8.4 Gaussian Approximation

The mean and variance of the shot noise part Z of the decision variable (5.106) are readily determined from the Campbell theorem derived in Appendix C. Substitution of (5.105) into (C.31) gives

$$\left.\begin{aligned} \mathrm{E}\{Z\} &= \int_0^T \Gamma(t)v(t)\, dt \\ \mathrm{Var}\{Z\} &= \int_0^T \Gamma(t)v^2(t)\, dt \end{aligned}\right\}$$

(5.131)

The special case of (5.131) when $\Gamma(t)$ is equal to a constant was proved by

Norman Campbell in 1909. The theorem was introduced in the engineering literature by S. O. Rice in 1944.

The Gaussian approximation is from (5.34)

$$P_e \approx Q_1(\rho) \tag{5.132}$$

with $Q_1(\ )$ defined in (5.40) and where ρ is the signal- to-noise ratio (5.35)

$$\rho = \frac{E_1 - E_0}{\sigma_1 + \sigma_0} \tag{5.133}$$

From the Campbell theorem (5.131)

$$\left.\begin{aligned} E_i &= \int_0^T \Gamma_i(t)v(t)\, dt \\ \sigma_i^2 &= \int_0^T \Gamma_i(t)v^2(t)\, dt \end{aligned}\right\} \quad i = 0,1 \tag{5.134}$$

which inserted into (5.132) and (5.133) gives a convenient, and often sufficiently accurate estimate of the error probability. The effect of additive noise is easily included by adding the variance (5.109) to the σ^2 from (5.134).

Example 5.9   Receiver with triangular filter

Consider the same system with on-off modulation and rectangular received optical pulses as in Example 5.8. Substitution of (5.87) into (5.134) gives the mean and variance of the optical component of the decision variable

$$E\{Z\} = 2\Gamma \int_0^{T/2} (1 - 2t/T)\, dt = 2\Gamma T \int_0^1 x\, dx = m/2 \tag{5.135}$$

and

$$\mathrm{Var}\{Z\} = 2\Gamma \int_0^{T/2} (1 - 2t/T)^2\, dt = 2\Gamma T \int_0^1 x^2\, dx = m/3 \tag{5.136}$$

The same result is also obtained from $\chi'(s)$ (5.124) and $\chi''(s)$ (5.125) by letting $s = 0$. After adding the thermal noise variance at the filter output $\sigma^2 = \mathcal{N}T/3$ the signal-to-noise ratio becomes

$$\rho = \frac{E_1 - E_0}{\sigma_1 + \sigma_0} = \frac{(m_1 - m_0)/2}{\sqrt{m_1/3 + \sigma^2} + \sqrt{m_0/3 + \sigma^2}} \tag{5.137}$$

The system parameters of Example 5.4 $m_0 = 129.9$, $m_1 = 274.9$ and $\mathcal{N}T = 25$ yields $\rho = 4.219$ and

$$P_e \approx Q_1(\rho) = 1.289 \times 10^{-5}$$

which is close to the more exact value

$$P_e \approx 1.159 \times 10^{-5}$$

from the saddlepoint approximation of Example 5.8.   □

5.9   Receiver for the Poisson Channel

5.9.1   Introduction

A receiver with a rectangular filter is easy to analyze and is often reasonable from a practical point of view. It is, however, not always the best choice of receiver filter. In general, the receiving filter should be designed with respect to the shape of the received optical pulses and the background noise.

An optimal receiver results in the lowest possible error probability but the problem of determining such a receiver for an optical communication system with arbitrary pulse shape and additive thermal noise is difficult. We limit our treatment to a situation with Poisson statistics which corresponds to a system without thermal noise or when the thermal noise is modeled as a shot noise process. The presentation is based on Bar-David (1969).

5.9.2   Receiver Structure

The basic assumptions are the same as in Section 5.5, but the receiver is now assumed to be general and not necessarily a photon counter.

The receiver is assumed to be capable of observing each individual photon and determining its time of occurrence. This means that the receiver extracts all information inherent in the realizations of the Poisson processes that it observes.

Let the binary symbols 'one' and 'zero' be represented by two Poisson processes with intensities $\Gamma_1(t)$ and $\Gamma_0(t)$ respectively. The receiver observes the optical signal during the time interval $[0, T]$ and decides which symbol is present. Assume that N Poisson events (photons) are observed and denote the sequence of time instances by the vector

$$\mathbf{t}_N = (t_1, t_2, \ldots, t_N)$$

A maximum a posteriori probability (MAP) receiver minimizes the error probability. It calculates the conditional probabilities $P(0|\mathbf{t}_N)$ and $P(1|\mathbf{t}_N)$ for symbols 'zero' and 'one' respectively, given the observation vector $\mathbf{t}_N$. The decision rule is to choose the symbol corresponding to the largest of these probabilities.

Let the a priori probability (at the transmitter) of the symbols 'zero' and 'one' be P_0 and P_1 respectively. The a posteriori probabilities can, using Baye's rule, be expressed as

$$P(i|\mathbf{t}_N) = \frac{P_i P_i(N) p(\mathbf{t}_N|i)}{p(\mathbf{t}_N)} \; ; \; i = 0, 1 \tag{5.138}$$

where $P_i(N)$ is the probability (5.5) that exactly N Poisson events have occurred.

The Poisson events $t_k; \; k = 1, 2, \ldots, N,$ considered individually without regard to their internal order, are statistically independent with a probability density

$$p(t_k) = \frac{\Gamma(t_k)}{m} \tag{5.139}$$

and the multidimensional probability density of the observation vector t_N is

$$p(t_N|i) = \left(\prod_{k=1}^{N} \Gamma_i(t_k) \right) / m_i^N \tag{5.140}$$

where

$$m_i = \int_0^T \Gamma_i(t) \, dt$$

From (5.138), (5.140) and (5.5) it follows that the ratio of $P(1|t_N)$ and $P(0|t_N)$ is

$$\begin{aligned} L &= \frac{P(1|t_N)}{P(0|t_N)} = \frac{P_1 P_1(N) p(t_N|1)}{P_0 P_0(N) p(t_N|0)} \\ &= \frac{P_1 e^{-m_1} \prod_{k=1}^{N} \Gamma_1(t_k)}{P_0 e^{-m_0} \prod_{k=1}^{N} \Gamma_0(t_k)} \end{aligned} \tag{5.141}$$

The decision rule is to decide the symbol 'one' when $L > 1$ and the symbol 'zero' when $L < 1$. By forming the natural logarithm of both sides of (5.141) the rule can be expressed as

$$\sum_{k=1}^{N} \ln \frac{\Gamma_1(t_k)}{\Gamma_0(t_k)} \underset{<}{\overset{>}{\underset{0}{\gtrless}}} \ln \frac{P_0}{P_1} + m_1 - m_0 \tag{5.142}$$

which we write as

$$\sum_{k=1}^{N} v^*(t_k) \underset{<}{\overset{>}{\underset{0}{\gtrless}}} \alpha^* \tag{5.143}$$

with $v^*(t) = \ln[\Gamma_1(t)/\Gamma_0(t)]$ and the threshold $\alpha^* = \ln(P_0/P_1) + m_1 - m_0$.

The decision rule (5.142) requires that at least one Poisson event is observed. The probabilities that no events are observed are

$$P_i(N = 0) = e^{-m_i} \; ; \; i = 0, 1 \tag{5.144}$$

which gives the following decision rule in the case that no Poisson events are observed:

$$\ln \frac{P_0}{P_1} + m_1 - m_0 \underset{<}{\overset{>}{\underset{1}{\gtrless}}} 0 \tag{5.145}$$

This means that the receiver, when $N = 0$, decides in favour of the symbol with the least optical power if the a priori probabilities P_0 and P_1 are not so unequal that the other alternative is more probable. The decision rule (5.145) can be included in (5.142) if the left-hand side is interpreted as equal to zero when no time events t_k are present.

A receiver operating according to the optimal decision rule (5.142) can be implemented as a linear filter followed by a threshold device. To show this let each Poisson event t_k be represented by a Dirac function at time $t = t_k$. This can be effected by a photodetector producing an output signal

$$I(t) = q \sum_{k=1}^{N} \delta(t - t_k) \tag{5.146}$$

There is no need for the pulses to be exact Dirac functions. It is sufficient that they are short compared to the functions $\Gamma_1(t)$ and $\Gamma_0(t)$.

The signal $I(t)$ is fed into a filter with impulse response

$$h(t) = q^{-1} v^*(T - t) \tag{5.147}$$

The output from the filter is

$$u(t) = \sum_{k=1}^{N} qh(t - t_k) = \sum_{k=1}^{N} v^*(T - t + t_k) \tag{5.148}$$

which at $t = T$ is equal to the sum $\sum_{k=1}^{N} v^*(t_k)$.

The receiver consists of a filter matched to the signal $v^*(t) = \ln[\Gamma_1(t)/\Gamma_0(t)]$. The filter output is sampled at the end of the signaling interval and compared with a threshold $\alpha^* = \ln(P_0/P_1) + m_1 - m_0$. The receiver structure is shown in Figure 5.10.

An important special case is when $\Gamma_0(t)$ is equal to a constant Γ_0, typically originating from background light of the light source and dark current in the photodetector. With $\Gamma_1(t) = \gamma_0 + \gamma(t)$ the function $v^*(t)$ becomes

$$v^*(t) = \ln\left(1 + \frac{\gamma(t)}{\gamma_0}\right) \tag{5.149}$$

When $\gamma(t)/\gamma_0 \ll 1$, i.e. the signal $\gamma(t)$ is weak compared to the background noise γ_0

$$v^*(t) \approx const\ \gamma(t) \tag{5.150}$$

The receiving filter is then matched to the received optical pulse in the same way as for a communication system disturbed by white additive Gaussian noise.

The shape of $v(t)$ is illustrated in Figure 5.15 when the received optical pulse $\gamma(t)$ is of triangular and of Gaussian form.

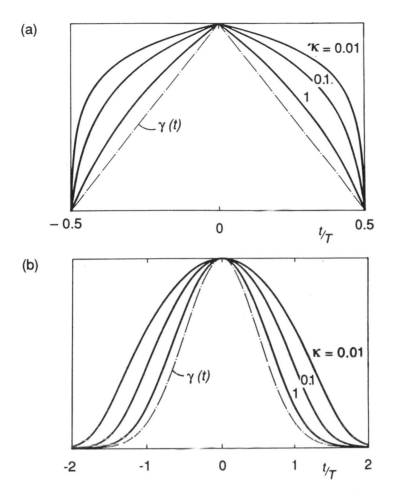

Figure 5.15   The shape $v(t)$ of the optimal (MAP) receiver filter for detection of an optical pulse $\gamma(t)$ in constant Poisson noise of intensity γ_0. The parameter $\kappa = 2\gamma_0/A$ with A equal to the peak value of $\gamma(t)$: (a) Triangular $\gamma(t)$ (b) Gaussian $\gamma(t)$ without intersymbol interference (widely separated pulses).

Example 5.10   MAP-receiver

Consider a system with on-off modulation and triangular received optical pulses

$$\gamma(t) = \begin{cases} A(1 - 2|t|/T) \; ; & |t| \leq T/2 \\ 0 \; ; & |t| > T/2 \end{cases} \tag{5.151}$$

As before let $m_0 = \gamma_0 T$ and m_1 denote the average received number of photoelectrons for the transmitted data symbols 'zero' and 'one', respectively.

Determine the optimal detector threshold and calculate the error probability.

Solution

Substitution of the MAP filter function (5.149) into (5.111) gives the logarithmic moment-generating function

$$\chi(s) = \ln[\Psi_z(s)] = \int_{-T/2}^{T/2} \Gamma(t) \left[\left(1 + \frac{\gamma(t)}{\gamma_0} \right)^s - 1 \right] dt \tag{5.152}$$

For the transmitted data symbol 'one' the received intensity $\Gamma(t) = \gamma(t) + \gamma_0$. Substitution of (5.151) and changing the integration variable to $x = 1 + \gamma(t)/\gamma_0$ yields

$$\chi_1(s) = \frac{\gamma_0^2 T}{A} \int_1^D x(x^s - 1) \, dx = \frac{m_0}{D-1} \left(\frac{D^{s+2} - 1}{s+2} - \frac{D^2 - 1}{2} \right) \tag{5.153}$$

where

$$D = 1 + A/\gamma_0 = \frac{2m_1 - m_0}{m_0}$$

In the same way, for data symbol 'zero' with $\Gamma(t) = \gamma_0$

$$\chi_0(s) = \frac{\gamma_0^2 T}{A} \int_1^D (x^s - 1) \, dx = \frac{m_0}{D-1} \left(\frac{D^{s+1} - 1}{s+1} - D + 1 \right) \tag{5.154}$$

Straightforward derivation gives the first-order derivatives

$$\chi_1'(s) = \frac{m_0}{D-1} \left(\frac{D^{s+2} \ln D}{s+2} - \frac{D^{s+2} - 1}{(s+2)^2} \right) \tag{5.155}$$

and

$$\chi_0'(s) = \frac{m_0}{D-1} \left(\frac{D^{s+1} \ln D}{s+1} - \frac{D^{s+1} - 1}{(s+1)^2} \right) \tag{5.156}$$

The second-order derivatives are

$$\chi_1''(s) = \frac{m_0}{D-1} \left(\frac{D^{s+2} (\ln D)^2}{s+2} - 2 \frac{D^{s+2} \ln D}{(s+2)^2} + 2 \frac{D^{s+2} - 1}{(s+2)^3} \right) \tag{5.157}$$

and

$$\chi_0''(s) = \frac{m_0}{D-1} \left(\frac{D^{s+1} (\ln D)^2}{s+1} - 2 \frac{D^{s+1} \ln D}{(s+1)^2} + 2 \frac{D^{s+1} - 1}{(s+1)^3} \right) \tag{5.158}$$

The saddlepoint parameter equations are

$$\chi_i'(s_i) - \alpha - 1/s_i^2 = 0 \; ; \quad i = 0, 1 \tag{5.159}$$

Numerical solution of (5.159) together with the threshold equation (5.126) for the same system parameters $m_0 = 129.9$ and $m_1 = 274.9$ as in Example 5.4, using the Newton-Raphson algorithm, results in an optimum threshold $\alpha = 145.0$ and the saddlepoint parameters $s_0 = 0.553$ and $s_1 = -0.477$. The saddlepoint estimate of the error probability is

$$P_e \approx 9.033 \times 10^{-9}$$

The Gaussian approximation requires the means (5.134)

$$\left. \begin{array}{l} E_0 = x_0'(0) = [m_0/(D-1)](D\ln D - D + 1) \\[2mm] E_1 = x_1'(0) = [m_0/4(D-1)](2D^2 \ln D - D^2 + 1) \end{array} \right\} \tag{5.160}$$

and the variances

$$\left. \begin{array}{l} \sigma_0^2 = x_0''(0) = [m_0/(D-1)][D((\ln D)^2 - 2\ln D + 2) - 2] \\[2mm] \sigma_1^2 = x_1''(0) = [m_0/4(D-1)][D^2(2(\ln D)^2 - 2\ln D + 1) - 1] \end{array} \right\} \tag{5.161}$$

The system parameters $m_0 = 129.9$ and $m_1 = 274.9$ gives $D = 3.232$ and

$$\rho = \frac{E_1 - E_0}{\sigma_1 + \sigma_0} = 5.6060$$

and

$$P_e \approx Q_1(\rho) = 1.066 \times 10^{-8}$$

For a rectangular filter receiver the shape of the optical pulses are of no consequence for system performance and such a receiver would result in the the error probability calculated in Example 5.4

$$P_e \approx 1.030 \times 10^{-7}$$

Which shows a factor of about ten in error probability between the non-optimum rectangular filter receiver and the optimal MAP filter receiver. $\square$

When $\gamma(t)$ is rectangular $v(t)$ has the same form and the receiving filter has a rectangular impulse response and can be implemented as an integrate-and-dump filter. From the derivation above it follows that the receiver in this case is equivalent to a photon- or photoelectron-counter.

To verify this equivalence let $\Gamma_1(t) = m_1/T$ and $\Gamma_0(t) = m_0/T$. The function $v^*(t) = \ln(m_1/m_0)$ for $0 \le t \le T$ and is zero otherwise, which gives

$$\sum_{k=1}^{N} v^*(t_k) = N\ln(m_1/m_0) \tag{5.162}$$

The decision rule (5.142) is for $P_0 = P_1 = 1/2$

$$N\ln(m_1/m_0) \underset{0}{\overset{1}{\gtrless}} m_1 - m_0 \tag{5.163}$$

which is equivalent to the decision rule derived in Section 5.5 for the photon-counting receiver. If the receiver filter is modified to

$$v^*(t) = \frac{\ln[1 + \gamma(t)/\gamma_0]}{\ln[1 + (m_1 - m_0)/m_0]} \qquad (5.164)$$

its output is Poisson-distributed and the threshold becomes identical to (5.26).

5.9.3   Avalanche Photodetector

In the derivation of the optimal receiver an ordinary photodetector is assumed. An avalanche photodetector (APD) produces a_k photoelectrons at time t_k and an optimal receiver observes the two vectors

$$\mathbf{t}_N = (t_1, t_2, \ldots, t_N)$$

and

$$\mathbf{a}_N = (a_1, a_2, \ldots, a_N)$$

in order to make its decision.

The compound probability density for the observation variables is

$$p(\mathbf{t}_N, \mathbf{a}_N/i) = \left(\prod_{k=1}^{N} \gamma_i(t_k) p(a_k) \right) / m_i^N \qquad (5.165)$$

under the assumption that a_k, $k = 1, 2, \ldots, N$ are independent and equally distributed with the probability density $p(a)$.

The likelihood ratio obtained from (5.165) contains the term $p(a_k)$ in both the numerator and the denominator. The terms cancel and a likelihood ratio identical to (5.141) is obtained. With the photodetector model (5.7) the avalanche gain a_k is irrelevant to an optimal receiver capable of detecting single photoelectrons at the output of the photodetector.

In a more realistic situation, however, an avalanche photodiode can improve system performance by enhancing a weak optical signal to overcome the thermal noise in the receiver.

5.10   Max SNR Receiver

Evaluation of the error probability of an optical receiver with an arbitrary filter requires fairly complicated numerical calculations, and receiver design based on the exact error probability is not practical. The Gaussian approximation is surprisingly accurate, giving estimates that are within some 10 or 20% of the exact value, and it requires much less numerical effort. It is therefore reasonable to use the error probability estimates from the Gaussian approximation as the optimization criteria. An example of this approach is the receiver design procedure presented in Chapter 6.

In direct detection PIN diode receivers thermal noise cannot be neglected, and the Poisson channel receiver (5.149) will not result in the lowest possible error probability.

In this section we investigate the shape of the receiver filter which is 'optimal' in the sense of maximizing the signal-to-noise ratio ρ in the presence of thermal noise. This means that it is the best filter when system performance is evaluated with the Gaussian approximation.

It is shown in Appendix F that with additive white noise of electron intensity $\mathcal{N}$ at the receiver input, such a filter has the form

$$v(t) = \frac{\gamma(t)}{\mathcal{N} + \gamma_0 + a\,\gamma(t)} \tag{5.166}$$

The constant a is determined by the relation

$$a = \frac{\sigma_0}{\sigma_0 + \sigma_1} \tag{5.167}$$

with σ_0 and σ_1 equal to the standard deviations of the filter output signal, i.e. the decision variable, for transmitted 'zero' and 'one', respectively.

Example 5.11   Receiver filter maximizing SNR

Determine the filter (5.166) for an optical system with received pulses of triangular shape and calculate the signal-to-noise ratio.

$$\gamma(t) = \begin{cases} A\,(1 - |2t - T|/T)\,; & 0 \le t \le T \\ 0\,; & t < 0,\ t > T \end{cases} \tag{5.168}$$

Solution

To determine the filter parameter a the variances σ_0^2 and σ_1^2 are needed. They can be determined by Campbell's Theorem (5.131) which requires the evaluation of integrals of the type

$$\int_0^T \Gamma(t)v^2(t)\,dt$$

Consider the case when $\Gamma(t) = \mathcal{N} + \gamma_0$. Substitution of (5.166) using the fact that $\gamma(t)$ is symmetrical around $t = T/2$ gives

$$\mathcal{I}_0(a) = \int_0^T bv_s^2(t)\,dt = 2b\int_0^{T/2} \left[\frac{\gamma(t)}{b + a\gamma(t)}\right]^2 dt \tag{5.169}$$

with $b = \mathcal{N} + \gamma_0$.

Insertion of (5.168) and changing the integration variable to

$$y = b + a\gamma(t) = b + 2aAt/T$$

gives

$$\mathcal{I}_0(a) = \frac{bT}{a^3 A} \int_b^{b+aA} \left(\frac{y-b}{y}\right)^2 dy$$

$$= \frac{bT}{a^3 A} \left[aA + \frac{baA}{b+aA} - 2b \ln\left(\frac{b+aA}{b}\right) \right]$$

(5.170)

In the same way

$$\mathcal{I}_1(a) = \int_0^T \gamma(t) v_s^2(t)\, dt$$

$$= \frac{T}{a^4 A} \left[\frac{(aA)^2}{2} - 2baA - \frac{b^2 aA}{b+aA} + 3b^2 \ln\left(\frac{b+aA}{b}\right) \right]$$

(5.171)

The quantities A and b are related to the system parameters by the relations

$$m_1 - m_0 = AT/2$$

and

$$\mathcal{N}T + m_0 = bT$$

The variance $\sigma_0^2 = \mathcal{I}_0(a)$ and $\sigma_1^2 = \mathcal{I}_0(a) + \mathcal{I}_1(a)$ and the filter parameter a satisfies (5.167)

$$a = \frac{\sqrt{\mathcal{I}_0(a)}}{\sqrt{\mathcal{I}_0(a)} + \sqrt{\mathcal{I}_0(a) + \mathcal{I}_1(a)}}$$

(5.172)

The solution of (5.172) is readily found by successive approximations. Start with an arbitrary value for a, i.e. $a = 0.5$, which is its maximal value. Calculate the right-hand side of (5.172) and use the result as the next value for a. The iteration process is continued until (5.172) is satisfied with sufficient accuracy. As an numerical example let $m_0 = 129.9$, $m_1 = 274.9$ and $\mathcal{N} = 0$. Then a takes the value

$$a_1 = 0.384$$

The max SNR filter (5.166) with $a = a_1$ is shown in figure 5.16. The integral

$$\mathcal{I}_2(a) = \int_0^T \gamma(t) v_s(t)\, dt = \frac{T}{a^3 A} \left[\frac{(aA)^2}{2} - baA + b^2 \ln\left(\frac{b+aA}{b}\right) \right]$$

(5.173)

is equal to $E_1 - E_0$ and the signal-to-noise ratio (5.35) of the max SNR filter becomes

$$\rho = \frac{\mathcal{I}_2(a)}{\sqrt{\mathcal{I}_0(a)} + \sqrt{\mathcal{I}_0(a) + \mathcal{I}_1(a)}}$$

(5.174)

Substitution of $a = a_1$ gives

$$\rho_{max} = 5.6061$$

A comparison with the MAP filter in Example 5.10 which resulted in $\rho = 5.6060$ shows that the filter (5.166) indeed produces a higher signal-to-

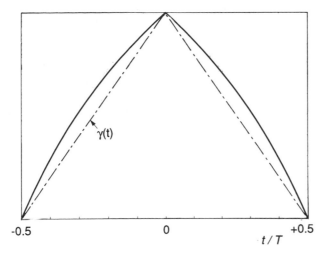

-0.5 0 $+0.5$

t / T

Figure 5.16   The shape $v(t)$ of the receiver filter maximizing the signal-to-noise ratio for an optical pulse of triangular form. The dashdot curve shows the shape of the received optical pulse.

noise ratio than the MAP filter. In this situation without thermal noise the MAP filter is of course preferable since it minimizes the true error probability. One motivation for the max SNR filter is that it specifies a receiver also in the practically important situations where thermal noise is present. Let $m_0 = 129.9$ and $m_1 = 274.9$ and $NT = 500$. Then u takes the value

$$a_2 = 0.464$$

and the signal-to-noise ratio becomes

$$\rho = 3.091$$

corresponding to a Gaussian approximation error probability estimate.

$$P_e \approx Q_1(\rho) = 1.088 \times 10^{-3}$$

A MAP filter (5.149), which is not designed for thermal noise, would result in

$$\rho = 3.083$$

and

$$P_e \approx Q_1(\rho) = 1.117 \times 10^{-3} \qquad \square$$

5.11 MSE Receiver

An alternative approach to the max SNR receiver is to base the design on a mean-square error (MSE) principle. The input data sequence $\{b_k\}$ is considered as a real-valued time discrete stochastic signal taking the values 0 and 1 with equal probability. The receiver observes the incoming signal and produces a sequence of estimates $\hat{b}_k$ such that the mean of the square difference $(b_k - \hat{b}_k)^2$ is minimized.

To obtain output data digits the real valued estimates $\hat{b}_k$ are quantized into two levels by comparing them with a threshold, a device which is sometimes called a slicer.

The MSE receiver has several interesting features. Its structure is easy to derive, and additive noise together with intersymbol interference can be incorporated in the design. However, for optical transmission it does not produce a minimum error probability receiver.

The MSE receiver for an optical channel is derived in Appendix F assuming intersymbol interference and additional (thermal) noise at the receiver.

Consider first a situation without intersymbol interference, consisting of binary transmission with nonoverlapping received optical signal intensities

$$\left.\begin{array}{ll} \Gamma_0(t) = \gamma_0 \; ; & b_k = 0 \\[2mm] \Gamma_1(t) = \gamma_0 + \gamma(t) \; ; & b_k = 1 \end{array}\right\} \tag{5.175}$$

and with additive white noise of intensity $\mathcal{N}$ (5.109).

The MSE receiver in Figure 5.22 on p. 142 consists of a linear filter $v(t) = C_0 g(t)$ sampled periodically at times $t = kT$ followed by a threshold device. The filter $g(t)$ is obtained from (F.27) and (F.17) in Appendix F

$$g(t) = \frac{\gamma(t)}{4(\mathcal{N} + \gamma_0 + \gamma(t)/2)} \tag{5.176}$$

In contrast to the MAP and the max SNR filters the amplification factor $C_0/4$ is important for the MSE filter.

Comparison with (5.149) and (5.166) reveals that the MAP, the max SNR and the MSE receivers are all different. The MSE filters for triangular and Gaussian optical pulses $\gamma(t)$ are illustrated in Figure 5.17a and b respectively. Compare with the MAP filters in Figure 5.15. The shape of the filter functions depends on the ratio $\kappa = 2\gamma_0/A$, where A is the peak value of $\gamma(t)$. The filters of Figure 5.17 assume isolated received pulses, i. e. no intersymbol interference.

The performance of the MSE receiver is illustrated by two examples representing systems with and without intersymbol interference.

Example 5.12   MSE Receiver Without ISI

Consider an optical system with triangular received pulses of time duration T:

$$\gamma(t) = A(1 - 2|t|/T) \; ; \quad |t| \leq T/2 \tag{5.177}$$

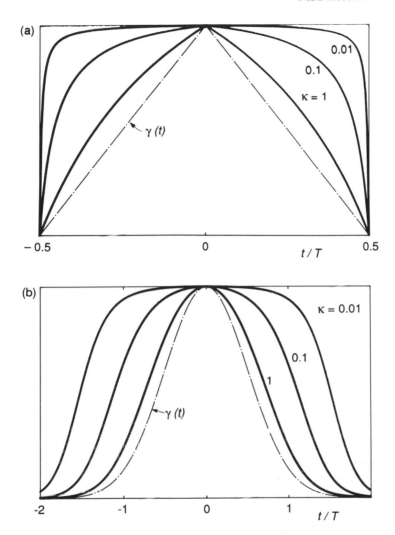

Figure 5.17 The shape $v(t)$ of the receiver filter minimizing the mean-square error (MSE) for an optical pulse $\gamma(t)$ in constant Poisson noise of intensity γ_0. The parameter $\kappa = 2\gamma_0/A$ with A equal to the peak value of $\gamma(t)$. (a) Triangular $\gamma(t)$ (b) Gaussian $\gamma(t)$ without intersymbol interference (widely separated pulses).

Without any loss of generality assume that $\mathcal{N} = 0$ (it can be included in γ_0). The optimum MSE filter obtained from (5.176) is

$$g(t) = \frac{1 - 2|t|/T}{2(2\gamma_0/A + 1 - 2|t|/T)} \tag{5.178}$$

Figure 5.17a shows $v(t) = C_0 g(t)$ for various values of $\kappa = 2\gamma_0/A$ together with the received pulse (5.177).

To determine the mean-square error at the filter output we calculate the

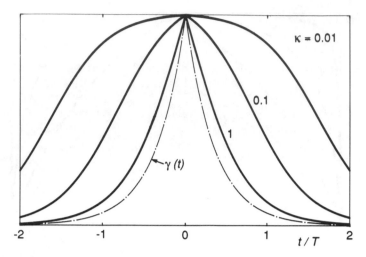

Figure 5.18 The MSE filter for an exponential received optical pulse for various values of $\kappa = 2\gamma_0/A$ without intersymbol interference (widely separated pulses).

parameter S_0 defined by (F.30) in Appendix F.

$$S_0 = \int_{-T/2}^{T/2} \gamma(t)g(t)\,dt = \frac{AT}{2}\int_0^1 \frac{(1-x)^2}{1+\kappa-x}\,dx$$
$$= \frac{AT}{2}\left[\kappa^2 \ln(\frac{1+\kappa}{\kappa}) - \kappa + 1/2\right] \qquad (5.179)$$

where $\kappa = 2\gamma_0/A$.

The mean-square error (F.34) is $\sigma_e^2 = C_0/4$ where $C_0 = 1/(1+S_0)$, see (F.32). Substitution into (F.39) gives the signal-to-noise ratio

$$\rho_{mse}^2 = S_0 \qquad (5.180)$$

which as shown in Appendix F is the average SNR over the two data symbols. Unfortunately there is no one-to-one correspondence between ρ_{mse} and the error probability. A more appropriate definition for the signal-to-noise ratio would be

$$\rho = \frac{E_1 - E_0}{\sigma_1 + \sigma_0} \qquad (5.181)$$

appearing in the Gaussian approximation (5.132). With the notations of Appendix F

$$E_1 - E_0 = r_0 = 1 - C_0$$

From (F.36) it follows that the variance σ_1^2, corresponding to the data symbol 'one', is equal to

$$\sigma_1^2 = 2\sigma_{mse}^2 - \sigma_0^2 \qquad (5.182)$$

where σ_0^2 is the variance corresponding to the data symbol 'zero' and σ_{mse} is from (F.37)

$$\sigma_{mse}^2 = \sigma_e^2 - C_0^2/4 = C_0/4 - C_0^2/4 \tag{5.183}$$

The variance σ_0^2 is produced by filtering the input signal $\Gamma_0(t) = \gamma_0$ through the MSE filter (F.26), and is equal to

$$\sigma_0^2 = \int \gamma_0 v^2(t)\, dt = C_0^2 \gamma_0 \int g^2(t)\, dt \tag{5.184}$$

The integral in (5.184) determining σ_0^2 is

$$V_0 = \gamma_0 \int_{-T/2}^{T/2} g^2(t)\, dt = \frac{AT\kappa}{8} \int_0^1 \left(\frac{1-x}{1+\kappa-x}\right)^2 dx \tag{5.185}$$

$$= \frac{AT\kappa}{8}\left[\frac{\kappa}{\kappa+1} - 2\kappa \ln(\frac{1+\kappa}{\kappa}) + 1\right] \tag{5.186}$$

The signal-to-noise ratio (5.181) is obtained from these relations as

$$\rho = \frac{1-C_0}{C_0\sqrt{V_0} + \sqrt{2C_0(1-C_0)/4 - C_0^2 V_0}}$$
$$= \frac{S_0}{\sqrt{V_0} + \sqrt{S_0/2 - V_0}} \tag{5.187}$$

The signal-to-noise ratios ρ_{mse} and ρ in dB as functions of the parameter κ are shown in Figure 5.19.   □

The signal-to-noise ratio ρ is expected to be a better measure of system performance than ρ_{mse}, since it allows the error probability to be estimated by

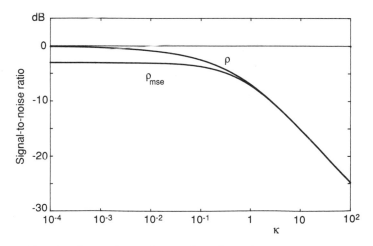

Figure 5.19   The signal-to-noise ratio ρ_{mse} based on the mean-square error and the signal-to-noise ratio ρ as functions of $\kappa = 2\gamma_0/A$. The diagram is normalized such that $\rho(0) = 0$ dB.

the Gaussian approximation, which was shown in Sections 5.5 and 5.6 to give results of sufficient accuracy. As is clear from the example and is illustrated in Figure 5.19, the maximal difference between ρ and ρ_{mse} is 3 dB, occurring when $\kappa \ll 1$, i.e. when the background noise γ_0 is low compared with the optical pulse amplitude A. For channels with a high background noise level, corresponding to $\kappa \gg 1$, the quantities are equivalent.

In contrast to the max SNR receiver, the MSE approach produces an explicit receiver configuration when intersymbol interference is present. We illustrate this with an example.

Example 5.13   MSE Receiver With ISI

Consider an optical system with data rate $B = 1/T$ using exponential pulses.

$$\gamma(t) = A \exp(-a|t|) \tag{5.188}$$

To obtain the MSE receiver filter we first determine the average received signal (F.17)

$$\psi(t) = \gamma_0 + \frac{1}{2}\sum_k \gamma(t - kT) \tag{5.189}$$

where any dark current in the photodetector is included in γ_0. In a situation with additive thermal noise γ_0 should be replaced by $\mathcal{N} + \gamma_0$. Substitution of (5.188) into (5.189) gives

$$\psi(t) = \gamma_0 + \frac{A}{2}\left(e^{-at}\sum_{k=-\infty}^{0} e^{akT} + e^{at}\sum_{k=1}^{\infty} e^{-akT}\right)$$

$$= \gamma_0 + \frac{A}{2(1-\beta)}(e^{-at} + \beta e^{at}); \quad 0 \geq t \geq T \tag{5.190}$$

where the abbreviation $\beta = e^{-aT}$ is introduced.

The function $\psi(t)$ is periodic and is depicted in Figure 5.20.

The filter function $g(t)$ obtained from (F.27) is

$$g(t) = \frac{\gamma(t)}{4\psi(t)} \tag{5.191}$$

which is shown in Figure 5.21.

Substitution of (5.188) into (5.191) gives the relation

$$g(t + mT) = \beta^m g(t); \quad 0 \leq t \leq T; \ m \geq 0 \tag{5.192}$$

The sequence S_k defined by (F.30) is divided into three parts

$$S_k = \int_{-\infty}^{-kT} \gamma(t + kT)g(t)\,dt + \int_{-kT}^{0} \cdots + \int_{0}^{\infty} \cdots \tag{5.193}$$

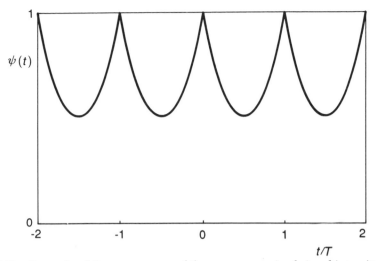

Figure 5.20 Example of the appearance of the average received signal intensity $\psi(t)$ for a system with exponential received pulses, $\kappa = 0.1$ and $B\sigma_t = 0.5$.

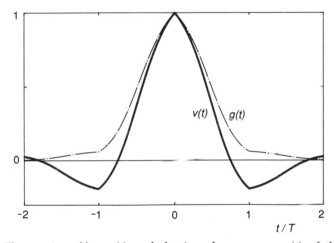

Figure 5.21 The receiver filter $g(t)$ and the impulse response $v(t)$ of the complete receiver for $\kappa = 0.1$, $B\sigma_t = 0.5$ and $\rho_{mse} = 15.6$ dB.

Assume $k \geq 0$ and consider the first integral in (5.193)

$$\int_{-\infty}^{-kT} \gamma(t + kT)g(t)\, dt = \int_0^\infty \gamma(s)g(s + kT)\, ds$$

$$= \sum_{m=0}^\infty \int_0^T \gamma(t + mT)g(t + mT + kT)\, dt$$

$$= \sum_{m=0}^\infty \int_0^T \beta^m \gamma(t)\beta^{m+k} g(t)\, dt \qquad (5.194)$$

$$= \frac{\beta^k}{1 - \beta^2} \int_0^T \gamma(t)g(t)\, dt$$

The second integral in (5.193) is in the same way

$$\int_{-kT}^{0} \gamma(t + kT)g(t)\,dt = k\beta^k \int_{0}^{T} A\exp(at)g(t)\,dt \qquad (5.195)$$

The third integral turns out to give a result identical to (5.194) and since both $\gamma(t)$ and $g(t)$ are even functions, S_k is even and equal to

$$S_k = \left(\frac{2K_1}{1 - \beta^2} + K_2|k|\right)\beta^{|k|} \qquad (5.196)$$

with

$$K_1 = \int_{0}^{T} \gamma(t)g(t)\,dt \qquad (5.197)$$

and

$$K_2 = \int_{0}^{T} A\exp(at)\gamma(t)\,dt = A^2 \int_{0}^{T} \frac{1}{4\psi(t)}\,dt \qquad (5.198)$$

The z-transform of S_k is

$$S(z) = \frac{2K_1}{(1 - \beta z)(1 - \beta/z)} + K_2\left(\frac{\beta z}{(1 - \beta z)^2} + \frac{\beta/z}{(1 - \beta/z)^2}\right) \qquad (5.199)$$

The tap gains C_k are related to $S(z)$ through (F.32)

$$C(z) = \frac{1}{1 + S(z)} \qquad (5.200)$$

The signal-to-noise ratio ρ_{mse} (F.39) is determined by the coefficient C_0 in the expansion of $C(z)$

$$\rho_{mse}^2 = \frac{1 - C_0}{C_0} \qquad (5.201)$$

The z-transform $C(z)$ is a rational function but inversion by partial fraction expansion is tedious. A particular C_k can be obtained from the general inversion formula for the z-transform

$$C_k = \frac{1}{2\pi} \int_{0}^{2\pi} C(e^{j\phi})e^{jk\phi}\,d\phi \qquad (5.202)$$

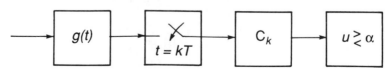

Figure 5.22   Block diagram of a MSE receiver consisting of a continous time filter $g(t)$ followed by a time discrete filter C_k (equalizer).

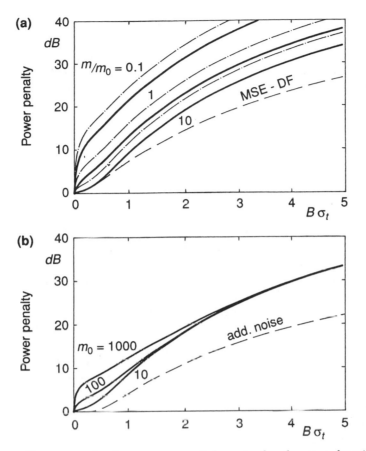

Figure 5.23 Power penalty for an exponential received pulse as a function of the dispersion parameter $B\sigma_t$. The signal-to-noise ratio $\rho_{mse} = 15.6$ dB: (a) m/m_0 constant. The dashdot curves are based on estimated values of ρ. The dashed curve shows the power penalty for a decision feedback receiver. (b) m_0 constant. The dashed curve applies to an additive noise channel.

The impulse response $v(t)$ of the complete receiver including the equalizer is (F.26)

$$v(t) = \sum C_k g(t + kT) \tag{5.203}$$

It is shown in Figure 5.21 for an output signal-to-noise ratio $\rho_{mse} = 15.6$ dB which for an additive Gaussian noise channel would correspond to an error probability $P_e = 10^{-9}$. The receiver impulse response $v(t)$ at this SNR is very close to that produced by a zero-forcing equalizer. □

The effect of intersymbol interference is often expressed as the power penalty which is the increase in the SNR at the receiver input needed to compensate for the ISI, i.e. to achieve the same SNR at the output as without ISI. Figure 5.23 shows the power penalty for exponential pulses as a function of the relative pulse width $B\sigma_t$ where $B = 1/T$ is the system data rate (or bandwidth) and σ_t^2 is the mean-square pulse width (4.9). In Figure 5.23a the ratio m/m_0 is kept fixed with m equal to the average number of received photons in the desired signal $\gamma(t)$ and m_0 is the average number of photons in the background signal, $m_0 = \gamma_0 T$. This corresponds to a situation where the background noise is generated at the transmitter and both are attenuated equally during the transmission. Figure 5.23b shows the power penalty when m_0 is kept fixed which would be the case if all the background noise is generated at the receiver. For high SNR when m/m_0 is large or m_0 is small the power penalty curves for both cases become the same.

As discussed in Example 5.12 the signal-to-noise ratio ρ_{mse} is not in any simple way related to the transmission error probability. To indicate what would result if the power penalty calculations were based on the signal-to-noise ratio ρ (5.184) the diagram Figure 5.19 has been used to calculate ρ from ρ_{mse}. The resulting power penalty curves are shown as the dashdot curves in Figure 5.23a. The MSE approach underestimates the signal-to-noise ratio and it also seems to underestimate the power penalty.

A situation with both additive, not necessarily Gaussian, and optical noise is easy to handle in the MSE model. As shown in Appendix F the parameter γ_0 should be replaced by $\gamma_0 + \mathcal{N}$ with $\mathcal{N}$ equal to the power spectral density of the noise. The MSE approach has the advantage that it is easy to apply to a situation with both optical and thermal noise present.

An optical receiver is more sensitive to intersymbol interference than a system disturbed by additive noise only. Figure 5.23b shows for comparison the power penalty of a system disturbed by additive noise at the same signal-to-noise ratio 15.6 dB. The higher sensitivity of the optical system is due to the fact that the optical noise is signal-dependent, and that overlapping signals cause an increase of the noise which is not possible to eliminate by equalization.

At high signal-to-noise ratios the detected data symbols are likely to be correct and their contribution to the intersymbol interference can be subtracted in the decision process, which is called decision feedback. It has be shown by Messerschmitt (1978) that C_0 for a decision feedback equalizer is

$$C_0 = 4\exp\left\{ -\frac{1}{2\pi} \int_0^{2\pi} \ln[4\{S(e^{j\phi}) + 1\}]\, d\phi \right\} \tag{5.204}$$

and that the signal-to-noise ratio ρ_{mse} is given by (5.201).

The power penalty for an optical decision feedback receiver at $m/m_0 = 10$ is illustrated by the dashed curve in Figure 5.23a.

5.12   Alternative Modulation Methods

On-off modulation (OOK) is used almost universally for optical systems with direct detection. Other modulation methods are also possible and some examples are presented here. They have received increasing attention together with optical preamplification.

The analysis presented is brief. The effect of thermal noise is not dealt with; it can easily be included in the same way as for OOK in Section 5.6.

5.12.1   Multilevel ASK

By letting the optical intensity of the transmitted signal attain more than two intensity levels, more than one bit of information per signaling interval can be transferred. Figure 5.24 shows amplitude shift keying (ASK) with $K = 4$ intensity levels. The information rate is two bits per transmission.

A receiver of the photoelectron counting type bases its decision on the number N of observed photoelectrons during the signaling interval $[0,T]$. The optimal decision rule, minimizing the error probability, is the maximum a posteriori probability (MAP) concept introduced in Section 5.9. The decision variable N has a Poisson distribution and the probability that intensity level k was transmitted, conditionally on $N = n$ photoelectrons being observed at the receiver can, using Baye's rule, be expressed as

$$P(k|n) = \frac{P_k P(n|k)}{P(n)} = \frac{P_k}{P(n)} \frac{m_k^n e^{-m_k}}{n!} \tag{5.205}$$

where m_k is the average number of photoelectrons corresponding to level k.

With the assumption that the transmitted symbols have equal a priori probability, $P_k = 1/K$ for $k = 1, 2, \ldots, K$, the decision can be based on the loglikelihood ratio

$$\Lambda(n|k) = \ln(m_k^n e^{-m_k}) = n \ln m_k - m_k \tag{5.206}$$

The receiver decides on the signal for which $\Lambda(n|k)$ is largest. This divides the sample space of N into a set of K decision regions. The boundary α_k between region k and $k + 1$ is determined by $\Lambda(n_{k+1}|n) = \Lambda(n_k|n)$. Substitution of (5.206) and solving for $n = \alpha_k$ yields

$$\alpha_k = \frac{m_{k+1} - m_k}{\ln m_{k+1} - \ln m_k} \tag{5.207}$$

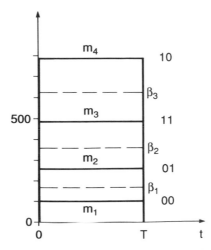

Figure 5.24   Multilevel amplitude shift keying (ASK) with four signal levels. The optimal levels m_i and the corresponding detector thresholds β_i are indicated.

which is the same relation as (5.26) for OOK. An integer-valued α_k is obtained in the same way as in the binary case (5.27).

The probability of correct detection $P_c = 1 - P_e$ is, cf. (5.24)

$$P_c = \frac{1}{4}\left[\sum_{n=0}^{\alpha_1} P(n|1) + \sum_{\alpha_1+1}^{\alpha_2} P(n|2) + \ldots + \sum_{\alpha_K+1}^{\infty} P(n|K)\right] \qquad (5.208)$$

The average number of received photoelectrons per symbol is

$$\hat{m} = \sum_{k=1}^{K} m_k \qquad (5.209)$$

and the average number per bit is $\hat{m}/\log_2 K$.

The variance of a Poisson variable grows with increased intensity and equally spaced intensity levels m_k are not optimal in terms of minimal error probability. An optimal set of intensity levels can in principle be obtained by maximizing (5.208) over the variables m_k with (5.209) as a constraint. A simple analytic solution, which is close to the optimal, can be obtained from the Gaussian approximation.

Substitution of $E_k = m_k$ and $\sigma_k = \sqrt{m_k}$ into (5.33) yields the Gaussian thresholds

$$\beta_k = \frac{\sigma_k E_{k+1} + \sigma_{k+1} E_k}{\sigma_k + \sigma_{k+1}} = \sqrt{m_k m_{k+1}} \qquad (5.210)$$

The Gaussian estimate of the error probability, when signal level k was transmitted, is analogous to (5.31)

$$P_e(k) = \Pr(N < \beta_{k-1}|m_k) + \Pr(N > \beta_k|m_k)$$
$$= Q\left(\frac{E_k - \beta_{k-1}}{\sigma_k}\right) + Q\left(\frac{\beta_k - E_k}{\sigma_k}\right) \qquad (5.211)$$
$$= Q(\sqrt{m_k} - \sqrt{m_{k-1}}) + Q(\sqrt{m_{k+1}} - \sqrt{m_k})$$

At the endpoints, $k = 1$ and $k = K$, the expression for P_e contains one term only.

From (5.211) it follows that the Gaussian approximation of the error probability for ASK modulation with K equally likely intensity levels is

$$P_e \approx \frac{2}{K}\sum_{k=1}^{K-1} Q(\rho_k) \qquad (5.212)$$

with

$$\rho_k = \sqrt{m_{k+1}} - \sqrt{m_k}$$

A set of intensity levels for which the terms in (5.212) are equal in magnitude minimizes the Gaussian approximation estimate of the error probability and they can be expected also to result in a low exact error probability.

If the first two levels m_1 and m_2 are specified, the level m_3 is determined by letting $\rho_1 = \rho_2$:

$$\sqrt{m_3} = 2\sqrt{m_2} - \sqrt{m_1} \tag{5.213}$$

In the same way m_4 is determined from m_3 and m_2, and so on. This generates a set of close-to-optimal intensity levels which, together with the maximum likelihood thresholds (5.207), specify a system with (almost) minimal error probability.

The multilevel ASK signals in Figure 5.24 illustrates intensity levels determined by the recursive procedure (5.213). The background noise level $m_1 = 100$ and the system is designed for an average symbol error probability of $P_e = 2 \times 10^{-9}$.

The symbol error probability of a multilevel system is not directly comparable to the bit error probability of a binary system. The relation between symbol and bit error depends on how the multi-valued symbols are coded in binary units. In Figure 5.24 Gray or reflected binary coding is indicated. The code has the property that it differs in one bit between adjacent signal levels. Decision errors to adjacent symbols are much more common than other errors, and with Gray coding the average bit error probability is approximately equal to half the symbol error probability.

The performance of a multilevel optical ASK system with $K = 4$ is illustrated in Figure 5.25. The diagram shows the bit error probability $P_b = P_e/2$ as a function of the average number of photoelectrons per bit $m_b = \hat{m}/2$. The background noise level is $m_1 = 100$. For comparison the error probability of signals with equally spaced intensity levels above m_1 is shown. A binary system operating with twice the speed, i.e. with symbol duration $T/2$, is included in the

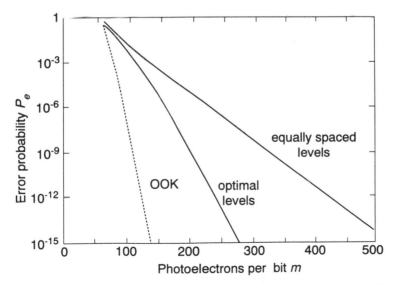

Figure 5.25 Performance of four level ASK with $m_1 = 100$. The bit error probability is shown as a function of the average number of photoelectrons per bit. The performance of on-off keying (OOK) is shown for comparison.

diagram. The background noise level is the same, i.e. the binary system has $m_0 = 50$. The improved Chernoff bounds (5.28) and (5.29) have been used to calculate the Poisson sums in the error probability (5.208).

It is clear that in optical systems it is more effective to transmit bits by on-off keying than by multilevel modulation.

5.12.2   Direct Detection FSK and PPM

In binary on-off keying the signals for 'zero' and 'one' appear in the same time interval. An alternative is to send the signals in separate channels. These are often orthogonal, which means that a signal in one channel will not interfere with a signal in another channel.

A widely used modulation format of this type is frequency-shift keying (FSK) where the optical field signals have the same shape but differ in frequency

$$\left.\begin{aligned} s_0(t) &= A(t)\cos(\omega_0 t + \phi) \\[2mm] s_1(t) &= A(t)\cos(\omega_1 t + \phi) \end{aligned}\right\} \quad 0 \le t \le T \qquad (5.214)$$

We consider wide band FSK, where the angular frequencies ω_1 and ω_0 are well separated making the signals orthogonal.

A direct detection optical FSK receiver consists of a beam splitter which directs the signal to two optical filters, each followed by a photodetector, see Figure 5.26. The filters have bandpass characteristics and are centered on wavelengths corresponding to ω_0 and ω_1, respectively. The outputs from the two detectors are lowpass filtered, typically by integrate-and-dump filters. The outputs are compared and the receiver decides in favour of the filter producing the largest output. The receiver in Figure 5.26 contains two independent branches, each identical to the on-off receiver in Figure 5.10, and it is easy to analyze using the results from Section 5.6.

The FSK signals are orthogonal due to their separation in optical wavelength or equivalently frequency. Another possibility is to separate them in time, which leads to pulse position modulation (PPM). In the binary case the PPM signals occupy separate halves of the signaling interval as shown in Figure 5.27a. The receiver, shown in Figure 5.27b, is simple. In principle it is an OOK

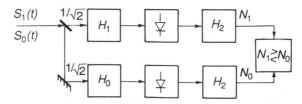

Figure 5.26   Receiver for direct detection frequency-shift keying (FSK).

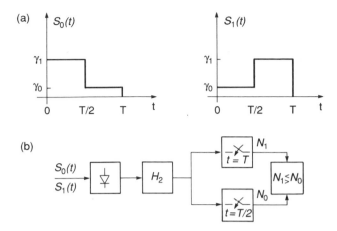

Figure 5.27 Binary pulse position modulation (PPM): (a) Examples of signals (b) Block diagram of receiver.

receiver where the integrate-and-dump filter is sampled twice during the time interval [0,T].

The decision variable N_0 produced at $t = T/2$ represents the number of received photoelectrons in the time interval $[0, T/2]$, and N_1 the number in $[T/2, T]$. The detector compares N_1 and N_0 and decides in favour of the signal alternative corresponding to the largest value. This is equivalent to a decision variable

$$N - N_1 \quad N_0$$

compared with a zero threshold.

The stochastic variable N is the difference between two independent Poisson distributed variables and its moment-generating function is the product of the MGFs for N_1 and N_0, see (5.60)

$$\Psi_N(s) = \exp[m_1(e^s - 1) + m_0(e^{-s} - 1)] \tag{5.215}$$

with $m_0 = \gamma_0 T/2$ and $m_1 = \gamma_1 T/2$.

Observe that N does not have a Poisson distribution. The sum of two Poisson variables is Poisson but the difference is not. The bit error probability of binary PPM is shown in Figure 5.28 as a function of the average number of received photoelectrons per bit $(m_1 + m_0)/2$.

The performance of an OOK system with the same γ_0 and m_b is indicated in the diagram. The two systems show practically identical performance. They differ at low background noise, however, and when $m_0 = 0$ on-off keying becomes 3 dB superior to binary PPM.

From the presentation above, it follows that the error probability of wide band FSK has the same mathematical form as that of binary PPM. However, the power splitter in the FSK receiver Figure 5.26 makes FSK inferior to PPM by a factor of 3 dB in sensitivity.

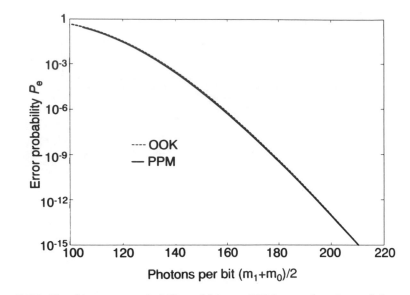

Figure 5.28 The bit error probability of binary PPM as a function of the average number of photons per bit. The background optical noise $m_0 = 100$. The performance of on-off keying (OOK) is shown for comparison.

5.12.3   *Direct Detection DPSK*

In the modulation methods above the intensity of the transmitted light is varied. This has the advantage that they are resistant to laser phase noise.

An alternative when a light source with a small spectral width is available is to modulate the phase of the optical field. A complication is that phase modulation requires a phase reference in the receiver. One possibility is differential phase-shift keying (DPSK) where the phase of the previous signal interval is used as a reference.

In DPSK the information is coded as phase shifts between successive signal intervals. Let ϕ_{-1} and ϕ_0 denote the phase of the optical field in the symbol intervals $[-T, 0]$ and $[0, T]$, respectively. The data symbol 'zero' is characterized by a phase shift $\phi_0 - \phi_{-1} = \pi$ and 'one' corresponds to $\phi_0 - \phi_{-1} = 0$.

A block diagram for the receiver is shown in Figure 5.29. The received signal is split into two parts and half the signal power is delayed by one symbol interval to be used as a phase reference for the next interval.

The optical field signal at the photodetector input is

$$B(t) = \frac{A(t)}{\sqrt{2}} \exp[j(\omega_0(t - T) + \phi_{-1})] + \frac{A(t)}{\sqrt{2}} \exp[j(\omega_0 t + \phi_0)] \qquad (5.216)$$

The intensity function of the shot noise current from the photodetector is proportional to the square envelope of the optical field. With the assumption

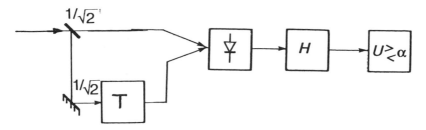

Figure 5.29 Reciver for direct detection of coherent differential phase-shift keying (DPSK).

that $\omega_0 T$ is a multiple of π

$$\Gamma(t) = \frac{1}{4}|B(t)|^2 = \frac{A^2(t)}{4}[1 + \cos(\phi_0 - \phi_{-1})] \tag{5.217}$$

The term $\cos(\phi_0 - \phi_{-1})$ is equal to -1 for the data symbol 'zero' which means that the optical fields cancel each other out and $\Gamma(t) = 0$. For 'one' $\cos(\phi_0 - \phi_{-1})$ is equal to $+1$ and the direct and delayed signals recombine to a signal of the same intensity as before the splitting.

From (5.217) it follows that ideal DPSK is equivalent to a direct modulation on-off system with $E_0 = 0$ and

$$E_1 = \int_0^T \Gamma_1(t)\, dt = \frac{1}{2}\int_0^T A^2\, dt = m \tag{5.218}$$

with m the average number of photons in the optical pulse $A(t)$.

In Chapter 5 it was shown that the optimum receiver for this situation is of the photoelectron-counting type and can be instrumented by an integrate-and-dump filter followed by a threshold device. The bit error probability is (5.16)

$$P_e = \frac{1}{2}e^{-m} \tag{5.219}$$

For $P_e = 10^{-9}$ the average number of photoelectrons per bit is $m = 20.0$ which is twice the quantum limit (5.17) of ideal on-off keying. The intensity of a DPSK signal is constant and equal to the peak intensity for OOK.

DPSK requires a stable and exact coherent signal for the cancellation when two signal of opposite phase are added. The performance of DPSK deteriorates rapidly in the presence of phase noise. See the analysis in Chapters 7 and 8.

5.13 Optical Information Theory

5.3.1 Introduction

Information theory is a science that is applicable to communication systems in general. Applied to optical systems it gives an insight into the theoretical capabilities of lightwave communication. For direct detection optical systems information theory not only reveals the ultimate limits of transmission capacity but it is also able to specify the modulation scheme that achieves the optimal performance.

A fundamental concept in information theory is channel capacity C which constitutes a limit on the amount of information that can be transmitted over a channel without deterioration. For data rates $R < C$ it is possible to communicate with arbitrarily low transmission error probability. When $R > C$ the error rate is bound to be larger than zero for any modulation and signal coding scheme.

The key process for achieving error free communication is error-correcting codes. In general long codes are needed for information rates approaching capacity, which imposes a delay T between transmitted and received message. The amount of coding necessary to achieve a specified bit error probability P_e can be estimated from the error exponent, $E(R)$ through the relation, see e.g. Gallager (1968) or Viterbi and Omura (1979).

$$P_e \leq e^{-TE(R)} \tag{5.220}$$

The theory proves, for sufficiently large T, the existence of an error correcting code such that (5.220) is satisfied for arbitrarily low P_e.

For most channels $E(R)$ is known only in terms of upper and lower bounds for part of its region of definition. For the optical channel, however, it can be determined exactly; an example is shown in Figure 5.30.

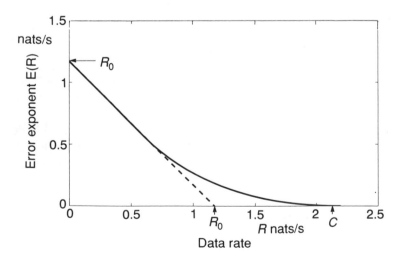

Figure 5.30 Error exponent for a direct detection optical channel. The received signal intensity is restricted within $\gamma_0 = 1$ and $\gamma_1 = 10$. The channel capacity $C = 2.19$ nats/s and error bound parameter $R_0 = 1.17$ nats/s are indicated in the diagram.

The error exponent $E(R)$ is larger than zero for $R < C$ and (5.220) shows that then P_e can be made arbitrarily small by increasing T, using codes of great length. An important quantity is the error bound parameter, R_0 indicated in the figure. For rates $R < R_0$ a simpler coding bound is

$$P_e \le e^{-T(R_0-R)} \tag{5.221}$$

The parameter R_0 is also called the cut-off rate since the sequential decoding algorithm for convolutional codes works only for $R < R_0$.

A digital communication channel is characterized by a finite set of input signals x_i, $i = 1, 2, \ldots, M$. Each transmitted signal represents a possible message and the transmission rate expressed in bits is

$$R_b = \log_2 M \text{ bits/transmission} \tag{5.222}$$

where $\log_2$ denotes the (binary) logarithm to the base 2. An alternative measure used in the mathematical derivations below is natural units (nats)

$$R = \ln M = R_b \ln 2 = 0.69 \times R_b \text{ nats/transmission} \tag{5.223}$$

The channel output is a sequence of noisy, in general continuous, multi-dimensional symbols y.

An important class of communication channels is those without memory which means that the noise affects the signals independently. Such a channel is characterized by a set of conditional probability densities between output and input symbols

$$\{p(y|x_i)\} \; ; \quad i = 1, 2, \ldots, M$$

An important concept in information theory is the mutual information $I(X; Y)$ between the input and output alphabets X and Y. For a continuous output, discrete input memoryless channel (CDMC)

$$I(X; Y) = \sum_{i=1}^{M} \int_{-\infty}^{\infty} p(x_i, y) \ln \left[\frac{p(y|x_i)}{p(y)} \right] dy \tag{5.224}$$

The channel capacity is defined as the maximum, or more exactly the supremum, of the mutual information maximized over all possible probability assignments of the input X.

$$C = \sup_{\{P(x_i)\}} I(X; Y) \tag{5.225}$$

The error bound parameter R_0 is often easier to determine than C. It is defined by the expression, see e.g. Gallager (1968).

$$R_0 = \sup_{P(x_i)} - \ln \left[\int_{-\infty}^{\infty} \left(\sum_{i=1}^{M} P(x_i) \sqrt{p(y|x_i)} \right)^2 dy \right] \tag{5.226}$$

The transmitted signals are assumed to have finite energy. Without such a restriction it would be possible for most channels to achieve an arbitrarily low transmission error probability by increasing the transmitted power. Light signals can have limited energy or limited intensity. It will be shown that these restrictions give different results with regard to channel capacity and the error bound parameter.

We assume that one of these signal constraints is imposed. It is also possible that both can be active simultaneously. For results under that condition see Wyner (1988).

Additional constraints, such as restrictions on the spectral bandwidth of the signals or maximal switching speed of the transmitting laser, are also possible. See Shlomo and Lapidoth (1993) and the references therein.

5.13.2 Optical Channel with Energy Limitation

The transmitter sends one of M possible optical signals, which at the receiver are characterized by optical intensity functions

$$\Gamma_i(t); \ 0 \geq t \geq T$$

each of time duration T. The signals have limited optical energy

$$\int_0^T \Gamma_i(t)\,dt \leq m \tag{5.227}$$

The channel is identical to the Poisson channel model introduced in Section 5.9. The receiver is assumed to be capable of detecting the arrival times of individual photons, resulting in the channel output

$$y = \mathbf{t}_n = (t_1, t_2, \ldots, t_n)$$

The number n of observed photons is a random variable having a Poisson distribution with a mean determined by the signal intensities

$$m_i = \int_0^T \Gamma_i(t)\,dt \ ; \quad i = 1, 2, \ldots, M \tag{5.228}$$

The arrival times t_k, considered individually without regard to their internal order, are statistically independent with a probability density (5.139)

$$p(t_k|i) = \frac{\Gamma_i(t_k)}{m_i} \tag{5.229}$$

The conditional probability density $p(y|x_i)$ is obtained from the results derived in Section 5.9. It is the product of a Poisson probability P_n and the multidimensional conditional density (5.140)

$$p(y|x_i) = p(\mathbf{t}_n, n|i) = P_n p(\mathbf{t}_n|i) = \frac{m_i^n \, e^{-m_i}}{n!} \prod_{k=1}^n \frac{\Gamma_i(t_k)}{m_i} = \frac{e^{-m_i}}{n!} \prod_{k=1}^n \Gamma_i(t_k) \tag{5.230}$$

We first determine the error bound parameter and the derivation will show that pulse position modulation, PPM is the best modulation method to maximize

R_0. We next calculate the channel capacity for that particular signal coding scheme and will find, somewhat surprisingly, that C is infinite, whereas R_0 is limited for an optical channel with energy limitation.

5.13.3   Error Bound Parameter

Expansion of the quadratic term in (5.226) gives with $Q_i = P(x_i)$

$$\int_{-\infty}^{\infty} \left(\sum_{i=1}^{M} P(x_i) \sqrt{p(y|x_i)} \right)^2 dy = \int_{-\infty}^{\infty} \sum_{i=1}^{M} \sum_{j=1}^{M} Q_i Q_j [p(y|x_i)p(y|x_j)]^{1/2} dy \quad (5.231)$$

Substitution of (5.230) yields

$$\int_{-\infty}^{\infty} Q_i Q_j [p(y|x_i)p(y|x_j)]^{1/2} dy$$

$$= Q_i Q_j \sum_{n=0}^{\infty} \frac{e^{-m_i/2}}{\sqrt{n!}} \frac{e^{-m_j/2}}{\sqrt{n!}} \prod_{k=1}^{n} \int_0^T \Gamma_i^{1/2}(t_k) \Gamma_j^{1/2}(t_k) \, dt_k \quad (5.232)$$

$$= Q_i Q_j \exp[-(m_i + m_j)/2] \sum_{n=0}^{\infty} \frac{c_{ij}^n}{n!}$$

where

$$c_{ij} = \int_0^T \Gamma_i^{1/2}(t) \Gamma_j^{1/2}(t) \, dt$$

The summation $\sum c_{ij}^n/n!$ is equal to $e^{c_{ij}}$ and (5.226) can be written as

$$R_0 = \sup_{Q_i} - \ln \sum_{i=1}^{M} \sum_{j=1}^{M} Q_i Q_j \exp\left\{ -\frac{1}{2} \int_0^T [\Gamma_i^{1/2}(t) - \Gamma_j^{1/2}(t)]^2 dt \right\} \quad (5.233)$$

Overlap between $\Gamma_i(t)$ and $\Gamma_j(t)$ cancel each other and do not contribute to R_0. This indicates that signals with a minimum of intersymbol interference, i.e. orthogonal signals, should be used. It is shown by Massey (1981) that (5.233) is maximal when the integral

$$\int_0^T [\Gamma_i^{1/2}(t) - \Gamma_j^{1/2}(t)]^2 dt \quad (5.234)$$

is equal for all indices i and j. The optimum input probability distribution is then uniform

$$Q_i = 1/M$$

A signal configuration that satisfies these requirements is pulse position modulation (PPM) which is defined as follows. The signaling interval $0 - T$ is divided into M timeslots. each of width T/M. The transmitter sends an optical

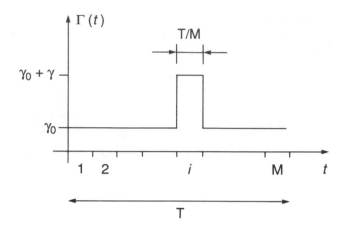

Figure 5.31 Pulse position modulation (PPM) with M time slots within the signal interval T.

pulse in one of the time slots. See Figure 5.31. The transmitted message is the position of the time slot.

The receiver is conceptually simple to implement. It can be realized as a photodetector followed by an integrate-and-dump filter sampled after each timeslot, producing a set of M photon (photoelectron) counts. See Section 5.6.

The signal set for PPM is

$$\Gamma_i(t) = \gamma_0 + \gamma(t - (i-1)T/M); \quad i = 1, 2, \ldots, M \tag{5.235}$$

where $\gamma(t)$ is the shape of the PPM-pulse of width T/M. See Figure 5.31.

Substitution of (5.235) into (5.233) gives

$$R_0 = -\ln\left[\frac{1}{M} + \frac{M-1}{M}\exp\left(-\int_0^{T/M}\left[\sqrt{\gamma(t) + \gamma_0} - \sqrt{\gamma_0}\right]^2 dt\right)\right] \tag{5.236}$$

The error bound parameter R_0 increases for decreasing background noise level γ_0. For the ideal case when $\gamma_0 = 0$ the integral in (5.236) is equal to m, the maximal optical signal energy (5.227), and

$$R_0 = -\ln[(1 - e^{-m})/M + e^{-m}] \tag{5.237}$$

The first term in the argument of the logarithm vanishes when M approaches infinity and the maximal value of R_0 is

$$R_0 = m \text{ nats/transmission} \tag{5.238}$$

Since m is the average number of photons in the received signals, the error bound parameter for the direct detection optical channel with energy limitation can be expressed as

$$R_0 = 1 \text{ nats/photon} \tag{5.239}$$

Note that for $\gamma_0 = 0$ the shape of the optical pulse $\gamma(t)$ can be arbitrary, so long as it is confined to the time slot.

In the general case with $\gamma_0 > 0$ the value of the integral in (5.236) does depend on the shape of $\gamma(t)$. The best signal form is a narrow pulse with area (5.227) equal to m. Substitution of an impulse function

$$\gamma(t) = m\delta(t)$$

into (5.236) gives an R_0 identical to (5.237) independent of the background noise level γ_0.

Next we consider the channel capacity of an optical direct detection system with signals restricted in energy. We have shown that PPM is the optimal modulation method to maximize the error bound parameter and it is natural to investigate the channel capacity for this channel. For simplicity we consider the ideal PPM channel with $\gamma_0 = 0$.

With this assumption, if there are any photons observed they will appear in the transmitted timeslot. The only situation when a transmission error can occur is when, due to the Poisson fluctuations, no photons are observed in the interval $0 - T$. The situation is the same as in the derivation of the quantum limit in Section 5.4. The ideal PPM channel is equivalent to an M-ary erasure channel, shown in Figure 5.32, where an erasure occurs if no photons are received. The probability for that event is, from (5.5)

$$P(n = 0) = \varepsilon = e^{-m} \tag{5.240}$$

The capacity for the M-ary erasure channel is, from Massey (1981)

$$C = (1 - e^{-m}) \ln M \text{ nats/transmission} \tag{5.241}$$

A verification of (5.241) is obtained by considering a system with a return channel between receiver and transmitter. When no photons are observed the receiver has no indication of what was transmitted and asks for a retransmission. A request for retransmission will occur on average for a

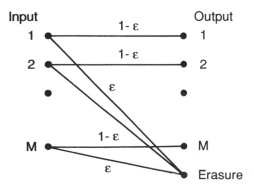

Figure 5.32 Channel transition diagram for an ideal optical PPM channel with $\gamma_0 = 0$. The channel is a M-ary erasure channel with erasure probability $\varepsilon = e^{-m}$.

fraction ε of the transmitted messages. A receiver which repeats the retransmission request until a non-erasure has occurred will never make an incorrect decision. Each successful transmission carries $\ln M$ nats of information and the scheme provides error-free transmission at an average rate $R = (1 - \varepsilon) \ln M$. A return channel would be a very simple method for operating with no errors at the channel capacity. A general theorem of information theory states that the presence of a return channel cannot increase the capacity beyond that of a system with the forward channel alone.

The capacity increases with M and grows without limit when M approaches infinity. This proves the strange fact that the capacity for the direct detection optical channel with energy limitation is infinite, but the error bound parameter is finite.

The capacity, expressed as the transmission rate per (average) received photon, is

$$C_m = C/m = \frac{1 - e^{-m}}{m} \ln M \text{ nats/photon} \tag{5.242}$$

The factor $(1 - e^{-m})/m$ approaches 1 when $m \to 0$ and the capacity per photon approaches infinity with increasing M even for an arbitrarily small transmitted optical energy.

For a channel with background noise the influence of $\gamma_0 > 0$ can be eliminated by letting the transmitted pulses be narrow in time in the same way as in the derivation of R_0 for $\gamma_0 > 0$. This means that the channel capacity approaches infinity for increasing M also when $\gamma_0 > 0$. A proof can be found in Davis (1980).

The result that the theoretical capacity of an optical channel is infinite is surprising. For a system with electrical signals disturbed by additive noise, such as a radio channel or cable system, the capacity is finite even if the system bandwidth is unlimited.

It may be speculated whether the result reflects a superiority of optical over electrical communication. That photons are better suited for transmission of information than electrons. However, an optical PPM system, even at moderate transmission rates, is difficult to implement.

Example 5.14   PPM system

Consider an optical direct detection PPM system with a signaling interval $T = 1$ s.

Assume that the received signal is weak and that on average one photon is received for each transmission, $m = 1$. The data rate of the system is $R_b = 1200 \text{ bit/s}$ which is modest compared with today's data transmission standards.

Determine the modulation parameters for a PPM system capable of operating at this rate.

Solution

The system needs a channel capacity that is at least equal to R_b. The capacity (5.241) expressed in bits per second is

$$C/T = \frac{(1 - e^{-m}) \log_2 M}{T}$$

The factor $1 - e^{-1} = 0.63$ and $C/T = 1200$ gives $\log_2 M = 1200T/0.63 = 1900$, which corresponds to

$$M = 2^{1900} = 10^{572}$$

The width of the timeslots is

$$T/M = 10^{-572} \text{ seconds}$$

which is impossible to realize in practice.   □

In the presentation above the quantities C and R_0 are measured in nats or bits per use of the channel. The capacity and error bound parameter in nats or bits per second are C/T and R_0/T, respectively. The symbol time T, however, is a free parameter in the derivation of R_0 and C and can in theory be made arbitrarily small. This means that even if R_0 (5.237) and (5.238) are finite R_0/T approaches infinity when T goes to zero.

5.13.4   Optical Channel with Intensity Limitation

Energy limitation produces a system with narrow optical pulses of extremely high intensity. A more realistic constraint is to restrict the signals to be limited in optical power.

Let the received signals be

$$\Gamma_i(t) = \gamma_0 + \gamma_i(t) \le \gamma_1; \ 0 \le t \le T \tag{5.243}$$

The parameter γ_0 represents background optical noise originating from dark current in the photo detector or zero level light from the transmitting laser.

The quantity $\gamma_i(t) \ge 0$ is the intensity of the received optical signal and it is clear from (5.243) that the signals also are limited in energy

$$\int_0^T \gamma_i(t) \, dt \le (\gamma_1 - \gamma_0)T \tag{5.244}$$

In practice binary on-off modulation, studied in Chapter 5, is used almost universally for optical digital communication. We determine the error bound parameter and the channel capacity for this type of modulation, which provides lower bounds on R_0 and C. However, upper bounds from the literature turn out to coincide with our results and the expressions derived here are the true capacity and error bound parameter for the intensity limited optical channel. An interesting conclusion is that channel capacity can be achieved by binary intensity modulation. The specific code that should be used is also, somewhat surprisingly, known. It is presented in Section 5.13.7.

5.13.5 Error Bound Parameter for Binary On-Off Modulation

Consider a binary system with on-off modulation using rectangular optical pulses. The parameter $M = 2$ and

$$\Gamma_i(t) = \begin{cases} \gamma_0 ; & i = 1 \\ \gamma_1 ; & i = 2 \end{cases} \tag{5.245}$$

The average number of received photons, when a 'zero' is transmitted is $m_0 = \gamma_0 T$, and $m_1 = \gamma_1 T$ when a 'one' is transmitted. The receiver contains a photon counter followed by a threshold (5.27)

$$\alpha = \text{int}\left(\frac{m_1 - m_0}{\ln m_1 - \ln m_0}\right) \tag{5.246}$$

The channel is equivalent to the unsymmetric memoryless binary channel, shown in Figure 5.33, with crossover probabilities from (5.24)

$$\left. \begin{aligned} P(y = 1|x = 0) = \varepsilon_0 &= \sum_{n=\alpha+1}^{\infty} \frac{m_0^n}{n!} e^{-m_0} \\ P(y = 0|x = 1) = \varepsilon_1 &= \sum_{n=0}^{\alpha} \frac{m_1^n}{n!} e^{-m_1} \end{aligned} \right\} \tag{5.247}$$

The error bound parameter is equal to the maximal value of the discrete output version of (5.226)

$$R_0 = \sup_{P(x)} -\ln\left[\sum_y \left(\sum_x P(x)\sqrt{P(y|x)}\right)^2\right] \tag{5.248}$$

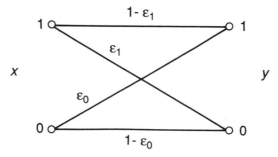

Figure 5.33 An asymmetric binary channel representing a direct detection optical system with on-off modulation.

For the binary channel (5.247) the double summation in (5.248) is equal to

$$[(1-Q)\sqrt{1-\varepsilon_0}+Q\sqrt{\varepsilon_1}]^2 + [(1-Q)\sqrt{\varepsilon_0}+Q\sqrt{1-\varepsilon_1}]^2$$
$$= (1-Q)^2 + Q^2 + 2Q(1-Q)(\sqrt{\varepsilon_1(1-\varepsilon_0)}\sqrt{\varepsilon_0(1-\varepsilon_1)}) \quad (5.249)$$

where $Q = P(x=1)$.

The error bound parameter expressed in nats per second is (5.248) divided by the signal interval T. It turns out that R_0/T assumes its greatest value when T approaches zero. The threshold then becomes $\alpha = 0$ and

$$\left.\begin{array}{l} \varepsilon_0 = 1 - e^{-\gamma_0 T} \\ \varepsilon_1 = e^{-\gamma_1 T} \end{array}\right\} \quad (5.250)$$

For $T \ll 1$ is $\varepsilon_0 \approx \gamma_0 T$ and $1 - \varepsilon_1 \approx \gamma_1 T$. Substitution into (5.249), neglecting terms of lower order, gives

$$(1-Q)^2 + Q^2 + 2Q(1-Q)[1 - (\gamma_0 + \gamma_1)T/2 + \sqrt{\gamma_0 T \gamma_1 T}]$$
$$= 1 - Q(1-Q)(\sqrt{\gamma_1} - \sqrt{\gamma_0})^2 T \quad (5.251)$$

The error bound parameter (5.248) is the logarithm of (5.251)

$$R_0/T \approx -\frac{1}{T}\ln[1 - Q(1-Q)(\sqrt{\gamma_1} - \sqrt{\gamma_0})^2 T] \quad (5.252)$$
$$\approx Q(1-Q)(\sqrt{\gamma_1} - \sqrt{\gamma_0})^2 \quad (5.253)$$

The optimum value for Q is $Q = 1/2$ resulting in

$$R_0/T = (\sqrt{\gamma_1} - \sqrt{\gamma_0})^2/4 \quad (5.254)$$

As an example let $\gamma_0 = 10^9$ and $\gamma_1 = 10^{10}$ photons/s which gives

$$R_0/T = (\sqrt{10} - \sqrt{1})^2 \times 10^9/4 = 1.169 \times 10^9 \text{ nats/s} = 1.686 \times \text{ Gbits/s} \quad (5.255)$$

The complete error exponent $E(R)$ for a channel with $\gamma_1/\gamma_0 = 10$ is shown in Figure 5.30.

For the ideal case $\gamma_0 = 0$ substitution into (5.254) gives

$$R_0/T = \gamma_1/4 \text{ nats/s} \quad (5.256)$$

The error bound parameter R_0 as a function of the received optical power is illustrated in Figure 5.34. The background optical noise is specified as a dark current i_d through the relation (5.18)

$$\gamma_0 = i_d/q \quad (5.257)$$

The maximal received intensity γ_1 is expressed as optical signal power P_o using the relation (5.2)

$$\gamma_1 - \gamma_0 = \frac{P_o}{hf} \quad (5.258)$$

The optical wavelength is assumed to be $\lambda = 1.55 \ \mu\text{m}$.

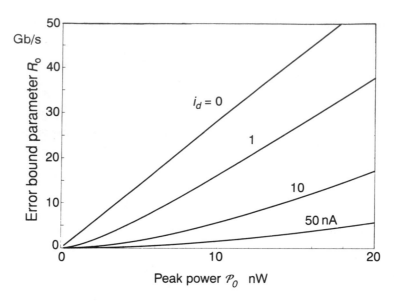

Figure 5.34   Error bound parameter R_0 in Gb/s as a function of the received peak signal power for an intensity limited optical channel. The background noise level γ_0 is specified as dark current i_d in the photodetector.

5.13.6   Capacity for Binary On-Off Modulation

Channel capacity in nats per second is equal to the maximal value of the mutual information

$$I(X;Y) = \sum_x \sum_y P(x,y) \ln\left[\frac{P(y|x)}{P(y)}\right] \qquad (5.259)$$

divided by the length T of the signal interval.

From the definitions of ε_0 and ε_1 follow

$$\left.\begin{array}{l} P(y=0) = Q\varepsilon_1 + (1-Q)(1-\varepsilon_0) \\[2mm] P(y=1) = Q(1-\varepsilon_1) + (1-Q)\varepsilon_0 \end{array}\right\} \qquad (5.260)$$

with $Q = P(x=1)$. The double summation in (5.259) contains four terms

$$\begin{aligned} I(X;Y) = &\,(1-Q)(1-\varepsilon_0)\ln[(1-\varepsilon_0)/P(y=0)] \\ &+ (1-Q)\varepsilon_0 \ln[\varepsilon_0/P(y=1)]] \\ &+ Q\varepsilon_1 \ln[\varepsilon_1/P(y=0)]] \\ &+ Q(1-\varepsilon_1)\ln[(1-\varepsilon_1)/P(y=1)]] \end{aligned} \qquad (5.261)$$

The capacity in nats per second is the maximum of $I(X;Y)/T$ which is achieved when T approaches zero. For $T \ll 1$, $\varepsilon_0 \approx \gamma_0 T$ and $1 - \varepsilon_1 \approx \gamma_1 T$. A series

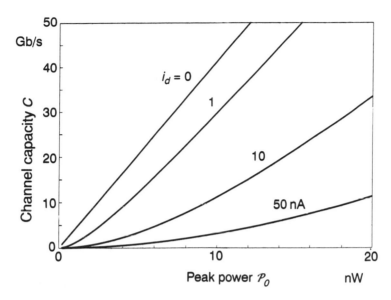

Figure 5.35 Channel capacity C in Gb/s as a function of the received peak signal power for an intensity limited optical channel. The background noise level γ_0 is specified as dark current i_d in the photodetector.

expansion of (5.261) in these variables, neglecting higher order terms, gives, after some algebraic manipulations

$$I(X;Y)/T = A\gamma_1 \ln(\gamma_1/\gamma_0) - \gamma_0 A \ln(A) \tag{5.262}$$

where

$$A = 1 + Q(\gamma_1 - \gamma_0)\gamma_0 \tag{5.263}$$

Letting the derivative of (5.262) with respect to Q be equal to zero yields the optimal value

$$Q = \frac{\gamma_0}{\gamma_1 - \gamma_0} \left[\frac{1}{e} \left(\frac{\gamma_1}{\gamma_0} \right)^{\gamma_1/(\gamma_1-\gamma_0)} - 1 \right] \tag{5.264}$$

Substitution of (5.264) into (5.262) determines the channel capacity

$$C = \max_q I(X;Y)/T = \frac{\gamma_0}{e} \left(\frac{\gamma_1}{\gamma_0} \right)^{\gamma_1/(\gamma_1-\gamma_0)} - \frac{\gamma_0\gamma_1}{\gamma_1 - \gamma_0} \ln \left(\frac{\gamma_1}{\gamma_0} \right) \tag{5.265}$$

As an example let $\gamma_0 = 10^9$ and $\gamma_1 = 10^{10}$ photons/s, which gives

$$C = \left[\frac{1}{e} \left(\frac{10}{1} \right)^{10/(10-1)} - \frac{1 \times 10}{10 \times 1} \ln \left(\frac{10}{1} \right) \right] \times 10^9 \tag{5.266}$$

$$= 2.193 \times 10^9 \text{ nats/s} = 3.164 \text{ Gbits/s}$$

The relation between $E(R)$, R_0 and C is illustrated in Figure 5.30.

When γ_0 approaches zero the first term in (5.265) vanishes and the remaining term becomes

$$C = \gamma_1/e; \quad \gamma_0 = 0 \qquad (5.267)$$

Comparison with (5.256) shows that for an intensity-limited Poisson channel, without any background noise, the capacity and the error bound parameter both grow linearly with the received optical signal power, maintaining a constant ratio $C/R_0 = 4/e = 1.47$ between these quantities.

5.13.7 Coding for Capacity

In information theory the results are often derived by a technique called random coding. The performance of a communication system is evaluated as a mean value over an ensemble of codes generated randomly. Since at least one member of the ensemble must fall below the mean this proves the existence of at least one code with the calculated performance. The procedure does not provide an explicit construction of a code. For the optical direct detection channel, however, Wyner (1988) has designed a code that asymptotically achieves channel capacity.

A code with M codewords of length L is specified in terms of a parameter k. It is constructed from a binary $M \times L$ matrix with

$$L = \binom{M}{k}$$

We illustrate with an example with $M = 4$ and $k = 2$.

$$
\begin{array}{cccccc}
1 & 1 & 1 & 0 & 0 & 0 \\
1 & 0 & 0 & 1 & 1 & 0 \\
0 & 1 & 0 & 1 & 0 & 1 \\
0 & 0 & 1 & 0 & 1 & 1
\end{array} \qquad (5.268)
$$

The columns of (5.268) are all binary words of length M containing exactly k ones. There are $L = 6$ such words and the codewords are the $M = 4$ rows of the matrix. The four optical signals produced by (5.268) are shown in Figure 5.36.

Channel capacity is achieved by letting the code length L approach infinity. This should be done in such a way that, according to Wyner (1988),

$$\frac{k}{M} = \frac{(1+s)^{1+s}}{s^s e} - s \qquad (5.269)$$

with $s = \gamma_0/\gamma_1$.

As an example, $\gamma_0/\gamma_1 = s = 0.1$ gives

$$k/M = 0.414$$

and for $\gamma_0 = 0$

$$k/M = e^{-1} = 0.368$$

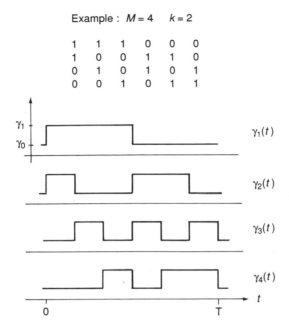

Example : $M = 4$ $k = 2$

$$
\begin{array}{cccccc}
1 & 1 & 1 & 0 & 0 & 0 \\
1 & 0 & 0 & 1 & 1 & 0 \\
0 & 1 & 0 & 1 & 0 & 1 \\
0 & 0 & 1 & 0 & 1 & 1
\end{array}
$$

Figure 5.36 Optimal coding for an intensity limited optical channel. The figure shows a short code with $M = 4$ codewords. To achieve channel capacity the number of codewords should approach infinity.

For $M - 1024$ each codeword represents $\log_2(1024) = 10$ bits of information. For $k/M = 0.414$ the length of the codewords is

$$
L = \left(\frac{1024}{424}\right) \approx 10^{300}
\tag{5.270}
$$

which is discouraging from an implementation point of view.

5.13.8 System Capability

Channel capacity determines the amount of information that it is possible to convey over a transmission system without any limitations on cost or complexity.

A way of characterizing the capability of a fiber optical communication system is in terms of transmission rate multiplied by the spacing between repeaters.

The complexity or cost of a long-distance system is in rough measures proportional to the number of repeaters $M_L = L_{tot}/L$ equal to the total length divided by the repeater spacing distance. An approximate measure of the

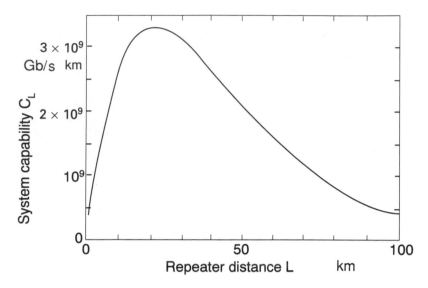

Figure 5.37 Example of system capability in Gb/s · km as a function of the repeater spacing. The transmitting laser produces an output optical power of 100 mW and the photodetector in the receiver has a dark current of 10 nA.

capacity per unit cost would be proportional to

$$\frac{C}{M_L} = \frac{L_{tot}}{CL}$$

For optical fiber systems

$$C_L = CL \tag{5.271}$$

is a therefore measure of system capability that to some extent takes cost and complexity into account. The product $R \times L$ of data rate and transmission distance is used frequently in the literature to evaluate different system solutions.

Figure 5.37 illustrates the variation of C_L as a function of L for a system with a fiber attenuation of $\mathbf{a} = 0.2$ dB/km. The diagram is obtained by assuming a fixed optical power of $\mathcal{P}_t = 100$ mW for the transmitting laser in the repeater. The receiver is assumed to be equipped with a photodetector with a dark current of $i_d = 10$ nA. The system is assume to operate at a wavelength $\lambda = 1.55\ \mu$m.

The signal intensities at the receiver input are

$$\gamma_0 = i_d/q \tag{5.272}$$

and

$$\gamma_1 = \frac{1}{hf} \mathcal{P}_t \times 10^{-\beta L/10} + \gamma_0 \tag{5.273}$$

Substitution of these parameters into (5.265) gives C and multiplication by L yields C_L.

The diagram shows that C_L attains its maximum at the rather short repeater distance of about $L = 20$ km.

6

Optical Receivers

6.1 Introduction

Chapter 5 dealt with the basic theory of optical detection under idealized conditions. In addition to the optical noise inherent in the signal itself, thermal noise was introduced as an independent additive term in the decision variable. In this chapter we investigate the origin of thermal noise and show how it is related to the characteristics of the photodiode and to the electrical circuitry of the receiver. The received signal is weak and amplification and also often equalization are needed before it can be fed into the decision circuit.

It turns out that the capacitance of the photodiode limits the data rate, and this has to be compensated for in the design of an optical receiver. A design philosophy based on this principle, suggested by Personick (1973), is presented in Section 6.5.

6.2 Thermal Noise

Thermal noise in electrical circuits is caused by random thermodynamic movements of the electrons responsible for the transport of electric energy. The resulting fluctuations in the electrical current or voltage can often be modeled as a stationary Gaussian random process with constant spectral density, i.e. white noise.

The thermal noise in an impedance Z can be represented by a random voltage generator e_n in series with Z, as shown in Figure 6.1a. The (two-sided) power spectral density of e_n is determined by the thermal equilibrium of the randomly moving electrical charges. It is equal to, see e.g. Bennett (1960).

$$S_E = 2k_B T_K R \quad [\text{V}^2/\text{Hz}] \qquad (6.1)$$

where $k_B = 1.38054 \times 10^{-23}\,\text{Ws/K}$ is Boltzmann's constant, T_K is the temperature in Kelvin, and $R = \text{Re}\{Z\}$ is the resistive (real) part of Z.

An alternative model giving the same results represents the noise by a current generator i_n in parallel with the impedance, as shown in Figure 6.1b.

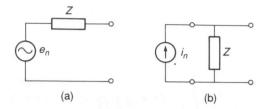

Figure 6.1　Equivalent circuits for thermal noise in an impedance Z: (a) voltage noise source e_n in series, (b) current noise source i_n in parallel.

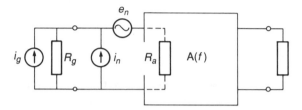

Figure 6.2　Equivalent circuit of a noisy, passive or active, linear two-port network.

The power spectral density of i_n is

$$S_I = \frac{2k_B T_K}{R} \quad [\text{A}^2/\text{Hz}] \tag{6.2}$$

It is easy to verify that both circuits produce the same noise at the output terminals.

The noise properties of a linear two-port device such as a filter or an amplifier can always be represented by two random sources at its input, followed by a noise-free device as described by Haus (1960). In general these sources may be nonwhite and they can be correlated.

The noise factor of a two-port device is defined as the noise power at its output divided by the output power that would result if the device was noiseless, in which case the output noise is caused solely by the internal impedance of the generator. The definition requires the input source to be at a standardized temperature $T_0 = 290$ K, equal to 17 C.

The power spectral density of the noise at the output of Figure 6.2 is, assuming that e_n and i_n are uncorrelated with spectral densities S_E and S_I, respectively

$$\left[(S_G + S_I) \left(\frac{R_g R_a}{R_g + R_a} \right)^2 + S_E \left(\frac{R_a}{R_g + R_a} \right)^2 \right] |A(f)|^2 \tag{6.3}$$

where $S_G = 2k_B T_0/R_g$ is the spectral density of the noise from the generator resistance R_g, R_a is the amplifier input resistance, and $A(f)$ is the frequency transfer function of the amplifier or filter. The part of the output spectral density that is due to R_g is

$$S_G \left(\frac{R_g R_a}{R_g + R_a} \right)^2 |A(f)|^2 \tag{6.4}$$

Dividing these expressions gives the noise factor or noise figure

$$F = 1 + \frac{S_I}{S_G} + \frac{S_E}{S_G R_g^2} \tag{6.5}$$

Note that the noise factor is independent of the input impedance R_a but it depends on the generator impedance R_g.

Under matched conditions with $R_g = R_a$ the output noise power spectral density, reduced to the input, becomes

$$R(f) = FS_G \left(\frac{R_g}{2} \right)^2 = \frac{Fk_B T_0 R_g}{2} \quad [\mathrm{V}^2/\mathrm{Hz}] \tag{6.6}$$

and the noise power measured in a frequency band B is

$$P = Fk_B T_0 R_g B \tag{6.7}$$

An alternative way of characterizing the noise of a two-port device is by a (effective or equivalent) noise temperature T_e. This is equal to the increase in temperature at the source that would produce the output noise with a noiseless network. The output noise power (6.7) is equal to

$$Fk_B T_0 R_g B = k_B (T_0 + T_e) R_g B$$

from which it follows that

$$T_e = (F - 1) T_0 \tag{6.8}$$

In our analysis we use the modified equivalent circuit shown in Figure 6.4. The input impedance R_a is considered as a part of the input circuitry and the random source e_n is placed to the right of R_a.

The current source i_n in Figure 6.4 represents the thermal noise in the amplifier input resistance R_a and the photodetector parallel resistance R_d. It is uncorrelated with e_n and its spectral density is given by (6.2)

$$S_I = \frac{2k_B T_K}{R} \quad [\mathrm{A}^2/\mathrm{Hz}] \tag{6.9}$$

with R equal to the resistance of R_d and R_a in parallel.

$$R = \frac{R_d R_a}{R_d + R_a} \tag{6.10}$$

In practice the internal noise in an amplifier is often dominated by the part produced by the input stage or by a pre-amplifier. A reasonable approximation is then to let e_n be white noise determined by the transconductance g_m of the input transistor. Its power spectral density is then

$$S_E = \frac{2k_B T_K}{g_m} \quad [\text{V}^2/\text{Hz}] \tag{6.11}$$

6.3   Ideal Photodetector and Rectangular Filter Amplifier

A realistic circuit model of a photodetector is shown in Figure 6.3. The resistance R_d represents the bias or load impedance of the diode. For a normal photodiode it is of the order of Mohm. The parallel capacitance C_d is the combined effect of the p-n junction, the leads and the packaging. Typical values for C_d are $1 - 10$ pF.

To show how thermal noise sources enter into the analysis of receiver performance, we first consider an idealized situation where the parallel capacitance of the photodiode is assumed to be zero and where the input capacitance of the front end amplifier can be neglected.

Combination of the circuits in Figure 6.2 and Figure 6.3 with C_d equal to zero gives the equivalent circuit diagram in Figure 6.4 of the complete receiver. The input impedance R_a of the amplifier is assumed to be resistive .
The output current from the photodetector is (5.50)

$$i_d(t) = \sum_k q\delta(t - t_k) \tag{6.12}$$

The input voltage to the amplifier is

$$y(t) = [i_d(t) + i_n(t)]R + e_n(t) \tag{6.13}$$

where R is the resistance of R_d and R_a in parallel.

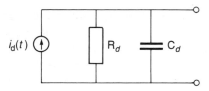

Figure 6.3   Realistic equivalent circuit for a photodetector.

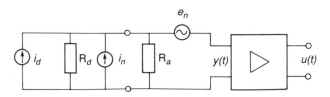

Figure 6.4   Equivalent circuit diagram of an optical receiver containing an ideal photo-detector (without capacitance) and a linear amplifier.

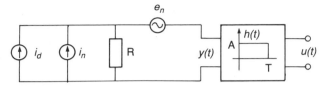

Figure 6.5 Equivalent circuit diagram of an ideal photodetector (without capacitance) followed by an integrate-and-dump filter with amplification factor A.

The amplifier is assumed to have the characteristics of a rectangular filter (5.49) with amplification factor A. The output at time $t = T$ is then (5.51)

$$u(t = T) = \int_0^T A\,y(t)\,dt = ARqN + A \int_0^T [R\,i_n(t) + e_n(t)]\,dt \qquad (6.14)$$

where N is the number of photoelectrons received in the signal interval.

Consider the normalized quantity, cf. (5.55)

$$\frac{u(t = T)}{ARq} = U = N + X \qquad (6.15)$$

with

$$X = \frac{1}{q} \int_0^T [i_n(t) + e_n(t)/R]\,dt \qquad (6.16)$$

The variance of a stochastic process with constant spectral density R_n is, after the passage of a rectangular filter with amplification A

$$R_n \int_{-\infty}^{\infty} h^2(t)\,dt = R_n \int_0^T A^2 dt = A^2 R_n T \qquad (6.17)$$

The noise signals $i_n(t)$ and $e_n(t)$ are uncorrelated stochastic processes and their combined power spectral density is

$$R_n = (S_I + S_E/R^2) \qquad (6.18)$$

The variance of X is from (6.17)

$$E\{X^2\} = \sigma_x^2 = \frac{T}{q^2}(S_I + S_E/R^2) \qquad (6.19)$$

The stochastic variable N has a Poisson distribution and X is a Gaussian zero mean variable.

The analysis of the receiver is now reduced to the situation dealt with in Section 5.6 of Chapter 5. The calculation of error probability is illustrated by two examples.

Example 6.1   System without ISI

A fiber optical system with on-off modulation operates at a bitrate of $B = 10$ Mb/s. The received optical pulse $p(t)$ is confined to the signaling interval $0 \leq t \leq T$ so that no interference between received pulses occur. The optical wavelength is $\lambda = 1.3$ μm. The photodiode has a quantum efficiency of $\eta = 0.7$ and a dark current of $i_d = 1$ nA. Its internal resistance is $R_d = 1.5$ Mohm. The front end amplifier is frequency-independent and has an input resistance $R_a = 3$ Mohm. The transconductance of the input transistor is $g_m = 5000$ μS. The amplifier is followed by an ideal rectangular filter.

Determine the received optical power necessary for a bit error probability of $P_e = 10^{-9}$.

Solution:

We use the Gaussian approximation to estimate the error probability. The mean of the decision variable U depends on the transmitted binary symbol. When 'zero' is transmitted it is

$$E_0 = m_0 = \gamma_d T$$

with $\gamma_d = i_d/q$ the dark current electron intensity When 'one' is transmitted

$$E_1 = E\{N\} = m_1 = \frac{\eta}{hf} \int_0^T p(t)\, dt + \gamma_d T = m + m_0$$

The variances are

$$\sigma_0^2 = m_0 + \sigma_x^2$$

and

$$\sigma_1^2 = m_1 + \sigma_x^2 = m + \sigma_0^2$$

respectively, with σ_x^2 equal to the variance of the thermal noise

$$\sigma_x^2 = \frac{2k_B T_K T}{q^2 R} \left(1 + \frac{1}{g_m R}\right) \tag{6.20}$$

with

$$R = \frac{R_a R_d}{R_a + R_d}$$

Substitution of the numerical values, assuming $T_K = 300$ K, gives

$$\sigma_x^2 = \frac{2 \times 1.38 \times 10^{-23} \times 300 \times 10^{-7}}{(1.602 \times 10^{-19})^2 10^6} \left(1 + \frac{1}{5 \times 10^{-3} \times 10^6}\right) = 3.23 \times 10^4$$

and

$$\sigma_0^2 = \frac{10^{-9} \times 10^{-7}}{1.602 \times 10^{-19}} + 3.23 \times 10^4 = 3.29 \times 10^4$$

The signal-to-noise ratio (5.35) in the Gaussian approximation is

$$\rho = \frac{E_1 - E_0}{\sigma_1 + \sigma_0} = \frac{m}{\sqrt{m + \sigma_0^2} + \sigma_0} \tag{6.21}$$

The desired error probability P_e is related to ρ ratio through the relation (5.32)

$$P_e = Q_1(\rho)$$

From (6.21) follows that

$$\rho\sqrt{m + \sigma_0^2} = m - \rho\sigma_0$$

Squaring both sides and solving for m yields

$$m = \rho^2 + 2\rho\sigma_0 \tag{6.22}$$

For $\rho = 6.0$, corresponding to $P_e = 10^{-9}$, it is

$$m = 2213.2$$

The optical noise has a standard deviation $\sigma_n = \sqrt{m} = 47.0$ which is about one-third of the thermal noise $\sigma_x = 179.7$. It is common to express the required optical signal strength in terms of the optical power P_1 averaged over the signaling interval. The quantity P_1 is related to m through

$$m = \frac{\eta}{hf} P_1 T \tag{6.23}$$

which gives

$$P_1 = \frac{hfm}{\eta T}$$

As an example, $\eta = 0.9$ and wavelength $\lambda = 1.3 \ \mu m$ give

$$P_1 = \frac{6.626 \times 10^{-34} \times 2.998 \times 10^8 \times 2213.2}{0.9 \times 10^{-7} 1.3 \times 10^{-6}} = 3.76 \times 10^{-9} \ W$$

The optical power P_1 refers to a signaling interval containing the symbol 'one'. The average P over a sequence of 'ones' and 'zeros', each occurring with probability one half, is $P = P_1/2$. The power P represents the sensitivity of the receiver. It is usually expressed in dBm, i.e. relative to a power level of 1 mW.

$$10 \log\left(\frac{P}{10^{-3}}\right) = 10 \log\left(\frac{3.76 \times 10^{-9}}{2 \times 10^{-3}}\right) = -57.3 \ dBm \qquad \Box$$

For an optical system without intersymbol interference the signal-to-noise ratio ρ (5.80) when thermal noise is present can be expressed as

$$\rho = \frac{m}{\sqrt{m + m_0 + \sigma_x^2} + \sqrt{m_0 + \sigma_x^2}} \tag{6.24}$$

where m is equal to the average number of photoelectrons in the optical pulse, see (5.22) and (6.23)

$$m = m_1 - m_0 = \frac{\eta}{hf} P_1 T \tag{6.25}$$

The numerator of (6.24) is proportional to T. Substitution of m from (5.20) and

σ_x^2 from (6.19) into (6.24) gives the denominator

$$\left[\frac{\eta PT}{hf} + \gamma_0 T + \gamma_d T + \frac{T}{q^2}(S_I + S_E/R^2)\right]^{1/2} + \left[\gamma_0 T + \gamma_d T + \frac{T}{q^2}(S_I + S_E/R^2)\right]^{1/2}$$

The denominator varies in proportion to the square root of the bit time T and the signal-to-noise ratio varies in proportion to the square root of the bit time T or inversely proportional to the square root of the transmission rate $B = 1/T$ of the system.

For an optical system with thermal noise the influence of P through the term m in the denominator of (6.24) can often be neglected which makes ρ approximately proportional to $P/\sqrt{B}$. It is then easy to calculate the change in received optical signal power for a change in transmission rate while maintaining the same system performance in terms of bit error rate. The relation is true for an arbitrary optical pulse $p(t)$ and filter $h(t)$ if the form of $p(t)$ and $h(t)$ are preserved when the transmission rate is increased.

Example 6.2   System with increased transmission rate

Determine the deterioration in transmission error probability that will result if the transmission rate in Example ex 7.3.1 is increased from 10 Mb/s. to 144 Mb/s. Calculate the increase in average received optical power needed to restore the error probability to $P_e = 10^{-9}$.

Solution:

Let ρ_1 denote the signal-to-noise ratio at transmission rate B_1. With constant average signal power the signal-to-noise ratio ρ_2 at rate B_2 is

$$\rho_2 = \rho_1 \sqrt{B_1/B_2} \tag{6.26}$$

With $B_1 = 10$ Mb/s and $\rho_1 = 6.0$ for $B_2 = 144$ Mb/s

$$\rho_2 = 6.0\sqrt{10/144} = 1.58$$

The bit error probability estimated by the Gaussian approximation is

$$P_{e2} = 7.23 \times 10^{-2}$$

If $\mathcal{P}_1$ is the optical signal power required at data rate B_1

$$P_2 \approx P_1 \sqrt{B_2/B_1} \tag{6.27}$$

is required to maintain the signal-to-noise ratio and thereby the transmission error probability at rate B_2. Expressed in dBm this is

$$\mathcal{P}_2 \text{ [dBm]} \approx \mathcal{P}_1 \text{ [dBm]} + 5\log(B_2/B_1)$$

For $B_1 = 10$ Mb/s and $\mathcal{P}_1/2 = -57.3$ dBm an increased transmission rate $B_2 = 144$ Mb/s gives

$$\mathcal{P} = \mathcal{P}_2/2 \approx -57.3 + 5\log(144/10) = -51.5 \text{ dBm}$$

The result is based on the approximative relation (6.27). An exact expression is

$$P_2 = \frac{m_2 B_2}{m_1 B_1} \tag{6.28}$$

where m_1 and m_2 denotes the parameter m at rate B_1 and B_2 respectively. Expressing m_1 ad m_2 in the form (6.22) yields

$$P_2 = \frac{\rho^2 \sqrt{B_2/B_1} + 2\rho\sigma_{01}}{\rho^2 + 2\rho\sigma_{01}} P_1 \sqrt{\frac{B_2}{B_1}} \tag{6.29}$$

The variance $\sigma_{01}^2 = 3.29 \cdot 10^4$ is calculated in Example 6.1 and substitution into (6.29) gives for the exact value of P

$$P = -51.3 \text{ dBm}$$

which is close to the approximate result. □

A system with intersymbol interference is easily analyzed using the methods of Section (5.5). The mean values (5.47) and (5.46) are

$$m_0 = m_{d0} + \delta m \tag{6.30}$$

and

$$m_1 = m_{d0} + (1 - \delta)m \tag{6.31}$$

with $m_{d0} = \gamma_0 T + \gamma_d T$ and where δ is the ISI parameter (5.45)

$$1 - \delta = \int_0^T \gamma(t) \, dt \bigg/ \int_{-\infty}^{\infty} \gamma(t) \, dt \tag{6.32}$$

The signal-to-noise ratio ρ (6.24) for a system with intersymbol interference is

$$\rho = \frac{m(1 - 2\delta)}{\sqrt{m_{d0} + (1 - \delta)m + \sigma_x^2} + \sqrt{m_{d0} + \delta m + \sigma_x^2}} \tag{6.33}$$

The first square root in the denominator contains the term $-\delta m$ and the second square root contains the term $+\delta m$. These two terms have a tendency to cancel each other, with the result that the denominator can be considered approximately to be a constant, independent of δ, equal to its value at $\delta = 0$, corresponding to no ISI. A comparison with a system without intersymbol interference (6.21) shows that ISI approximately reduces the signal-to-noise ratio by the factor $1 - 2\delta$ in the numerator which is called the ISI power penalty.

The power penalty is usually expressed in dB as a function of the bandwidth-dispersion product $B\sigma_t$. As an example consider received pulses of exponential form

$$p(t) = \exp(-a|t|) \tag{6.34}$$

The parameter δ is

$$\delta = \exp(-aT/2) \tag{6.35}$$

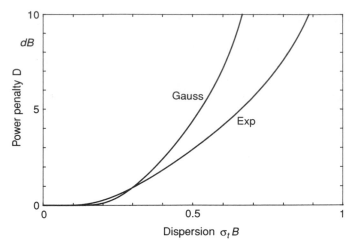

Figure 6.6   Power penalty due to intersymbol interference for optical pulses of Gaussian and exponential form.

The mean square pulse width (dispersion) is from (4.9)

$$\sigma_t^2 = \int_{-\infty}^{\infty} t^2 p(t) \, dt \Big/ \int_{-\infty}^{\infty} p(t) \, dt = 2/a^2 \tag{6.36}$$

This gives $a = \sqrt{2}/\sigma_t$, and since $T = 1/B$ is

$$\delta = \exp(-1/\sqrt{2}B\sigma_t) \tag{6.37}$$

the power penalty in dB becomes

$$D = 10 \log(1 - 2\delta)$$

This is shown in Figure 6.6 as a function of $B\sigma_t$.

For a Gaussian pulse, the ISI parameter can be expressed in terms of the Q-function (5.32)

$$\delta = 2Q(1/2B\sigma_t) \tag{6.38}$$

The corresponding power penalty is shown in Figure 6.6.

6.4   Real Photodetector and Equalizing Receiver

In the preceding section the photodetector was ideal, having no internal or external capacitance. A real photodiode has a small capacitance and in this section it is shown how this needs to be taken into account in the design of the receiver, especially for high data rates.

Consider the receiver model shown in Figure 6.7a. The resistance R is the internal resistance of the photodetector and the input resistance of the amplifier

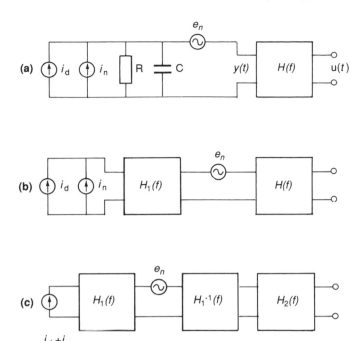

Figure 6.7 Equivalent circuit diagrams of an optical receiver with a real photodiode. (a) Original diagram. (b) The R and C represented as a filter $H_1(f)$. (c) The filter $H(f)$ replaced by an inverse filter $H_1^{-1}(f)$ and $H_2(f)$ in cascade.

in parallel, and the capacitance C is the sum of the the detector capacitance and the input capacitance of the amplifier.

The R and the C form a linear filter with a current-to-voltage impulse response

$$h_1(t) = (1/C)e^{-t/RC}u(t) \tag{6.39}$$

where $u(t)$ is the unit step function.

The frequency function is

$$H_1(f) = \frac{R}{1 + j2\pi fRC} \tag{6.40}$$

The circuit model in Figure 6.7a is equivalent to the block diagram in Figure 6.7b. The effect of the capacitance C is illustrated by the following example.

Example 6.3   System without ISI

Investigate the effect of a parallel capacitance $C = 10$ pF on a system operating at a bit rate of $B = 10$ Mb/s. The amplifier input resistance is $R = 1$ Mohm.

Solution:

For simplicity, assume that the received pulses are rectangular.

$$p(t) = \mathcal{P} \, \text{rect}(t/T) = \begin{cases} \mathcal{P}; & |t| < T/2 \\ 0; & |t| > T/2 \end{cases}. \tag{6.41}$$

A single received pulse generates a signal $p_1(t)$ at the amplifier input that is the convolution of a rectangular pulse of width T and the impulse response $h_1(t)$. See Figure 6.8.

$$p_1(t) = \mathcal{P} \int_0^T h_1(t-s) \, ds = \begin{cases} \mathcal{P}R\left(1 - e^{-t/RC}\right); & 0 \le t \le T \\ \mathcal{P}R\left(e^{T/RC} - 1\right)e^{-t/RC}; & t \ge T \end{cases} \tag{6.42}$$

The pulse $p_1(t)$ is extended in time, which will cause intersymbol interference. To investigate the amount of ISI we calculate the time it takes for $p_1(t)$ to decay to 1% of its peak value. See Figure 6.8b.

$$e^{-t_d/RC} = 0.01$$

which gives

$$t_d/T = (RC/T) \ln(100) = (10^6 \times 10^{-11}/10^{-7}) \times 4.61 = 461$$

It will take 461 signal intervals before the pulse have decayed to a reasonably small value and it is clear that the intersymbol interference will be extremely large making the system worthless for communication, at least at the transmission rate 10 Mb/s. The problem is that the time constant RC is too large compared with the symbol interval T. A simple way of reducing RC is to decrease R, but unfortunately this has a negative effect of the system

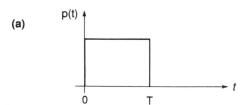

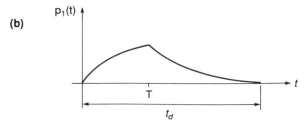

Figure 6.8 Input optical signal $p(t)$ and output electrical signal $p_1(t)$ from a photodiode with resistance R and capacitance C.

performance. For the pulse $p_1(t)$ to decay to 1% of its peak value within one
signaling interval, i.e. $t_d = T$, the resistance R should be

$$R = T/C\ln(100) = 10^{-7}/(10^{-11} \times 4.61) = 2200 \text{ ohm}$$

for $C = 10\text{pF}$.

Unfortunately a low value of R increases the thermal noise. From (6.20) it
follows that the variance of the thermal noise varies approximately inversely
to R. For a reduction of R from $R_1 = 1\text{Mohm}$ to $R_2 = 2200$ ohm the thermal
noise will dominate over the optical noise, and the signal-to-noise ratio (6.21)
is reduced by

$$10\log(\rho_1/\rho_2) = 10\log(\sqrt{R_1/R_2}) = 5\log(10^6/2200) = 13.3 \text{ dB}$$

In addition to the decline in signal-to-noise ratio there will also be a power
penalty due to the ISI, and it is clear that the reduction of R is not a good idea.

□

To eliminate the problems caused by the input capacitance C the receiver has to
contain an equalizing function. The effect of $H_1(f)$ can, at least theoretically, be
cancelled by an inverse filter

$$H_1^{-1}(f) = (1 + j2\pi fRC)/R$$

The receiver filter $H(f)$ now consists of $H_1^{-1}(f)$ followed by $H_2(f)$, which still has
to be determined. See Figure 6.7c.

One possible choice would be to let $H_2(f)$ be the rectangular (integrate-and-
dump) filter (5.49). Since the influence of C is eliminated by $H_1^{-1}(f)$ the result
would be the idealized situation analyzed in the preceding chapter.

This would give

$$H(f) = H_1^{-1}(f) \times H_2(f)$$

with

$$H_2(f) = A\frac{\sin(\pi fT)}{\pi fT}e^{-j\pi fT} \tag{6.43}$$

The noise source e_n in Figure 6.7 is filtered through the filter $H(f)$ and its
variance at the filter output is

$$\sigma_e^2 = \int_{-\infty}^{\infty} S_E|H(f)|^2 df = \int_{-\infty}^{\infty} S_E|H_{-1}(f)|^2|H_2(f)|^2 df$$

$$= \frac{2k_BT_K}{g_m}A^2\int_{-\infty}^{\infty} \frac{1+(2\pi fRC)^2}{R^2}\left(\frac{\sin(\pi fT)}{\pi fT}\right)^2 df \tag{6.44}$$

The integral in (eq 7.23) does not converge, which means that the noise e_n has
unlimited variance at the filter output.

The filter $H_1^{-1}(f)$ amplifies high frequencies and it has to be followed by a
filter $H_2(f)$ which has a larger attenuation at high frequencies than the
rectangular filter. A choice that works is to let $H_2(f)$ be the strictly bandlimited

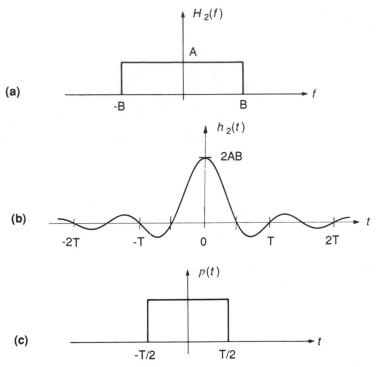

Figure 6.9 Ideal band-limited receiver filter: (a) frequency function $H_2(f)$, (b) impulse response $h_2(t)$, (c) optical signal $p(t)$ assumed in the analysis in the text.

filter shown in Figure 6.9.

$$H_2(f) = \begin{cases} A; & |f| < B \\ 0; & |f| > B \end{cases} \tag{6.45}$$

with bandwidth $B = 1/T$. The impulse response of the filter is

$$h_2(t) = 2AB \frac{\sin(2\pi Bt)}{2\pi Bt} \tag{6.46}$$

The filter H_2 is not causal but can be implemented with an arbitrarily small deviation by a realizable filter with delay. Its impulse response is symmetrical, which means that the decision time for symbol b_0 is chosen to be $t = 0$.

The output signal consists of three parts

$$u(t) = u_d(t) + u_i(t) + u_e(t)$$

where $u_d(t)$ is due to the optical signal and the other two terms are produced by the thermal noise sources $i_n(t)$ and $e_n(t)$, respectively. Their variances are

$$\mathrm{Var}\{u_i^2\} = \int_{-\infty}^{\infty} S_I |H_2(f)|^2 df$$

$$= \int_{-B}^{B} S_I A^2 df = S_I A^2 2B \tag{6.47}$$

and

$$\mathrm{Var}\{u_e^2\} = \int_{-\infty}^{\infty} S_I |H(f)|^2 df$$

$$= \int_{-\infty}^{\infty} S_E |H_1^{-1}(f)|^2 |H_2(f)|^2 \, df$$

$$= \int_{-B}^{B} S_E A^2 [1 + (2\pi f RC)^2] \, df \tag{6.48}$$

$$= S_E A^2 2B(1 + (2\pi RCB)^2/3)$$

The signal $u_d(t)$ is shot noise, produced by the photodetector current source $i_d(t)$, and filtered through the filter H_2. With the decision time $t = 0$ the function $v(t)$ in Campbell's theorem is equal to $v(t) = qh_2(-t)$. Substitution into (5.131) gives the mean and the variance of $u_d(0)$

$$E\{u_d(0)\} = q \int_{-\infty}^{\infty} \Gamma(t) h_2(-t) \, dt \tag{6.49}$$

and

$$\mathrm{Var}\{u_d(0)\} = q^2 \int_{-\infty}^{\infty} \Gamma(t) h_2^2(-t) \, dt \tag{6.50}$$

respectively.

The possibility to solve these integrals analytically depends on the particular form of $\Gamma(t)$ and $h_2(t)$. A simple case, that is easy to deal with, is when the optical pulse is rectangular with duration T, which corresponds to straightforward on-off modulation.

Let the received optical pulse have the rectangular form

$$\gamma(t) = \begin{cases} \gamma; & -T/2 < t < T/2 \\ 0; & \text{otherwise} \end{cases} \tag{6.51}$$

shown in Figure 6.9c. The symmetrical form is motivated by the choice of decision time $t = 0$.

The intensity function $\Gamma(t)$ is, cf. (5.42)

$$\Gamma(t) = \gamma_0 + \frac{\eta}{hf} \sum_k b_k \gamma(t - kT) \tag{6.52}$$

The pulse $\gamma(t)$ is confined to the symbol interval $-T/2 \le t \le T/2$ and exhibits no intersymbol interference. The filtered signal $u_d(t)$, however, has a longer time duration and ISI does occur at the decision point.

When a symbol 'one' is transmitted, i.e. when $b_0 = 1$, the worst situation with regard to intersymbol interference occurs when all transmitted symbols are 'one', i.e. $b_k = 1$; $\forall k$. The function $\Gamma(t)$ (6.52) is then a constant

$$\Gamma_1(t) = \gamma_0 + \gamma = \Gamma_1; \ \forall t \tag{6.53}$$

Substitution into (6.49) gives

$$E_1 = q\Gamma_1 \int_{-\infty}^{\infty} h_2(-t)\, dt = q\Gamma_1 H_2(0) = q\Gamma_1 A \tag{6.54}$$

The variance of $u_d(0)$ is from (6.50)

$$\sigma_1^2 = q^2\Gamma_1 \int_{-\infty}^{\infty} h_2^2(-t)\, dt$$

$$= q^2\Gamma_1 \int_{-\infty}^{\infty} |H_2(f)|^2 df = q^2\Gamma_1 A^2 2B \tag{6.55}$$

When the thermal noise is dominating and the optical intersymbol interference can be neglected, the worst situation when a 'zero' is transmitted occurs when $b_k = 0;\ \forall k$. The intensity function $\Gamma(t)$ is then

$$\Gamma(t) = \gamma_0;\ \forall t \tag{6.56}$$

The mean value obtained from (6.49) is

$$E_0 = q\gamma_0 A \tag{6.57}$$

The variance obtained from (6.50) and (6.55) is

$$\sigma_0^2 = q^2\gamma_0 A^2 2B \tag{6.58}$$

The thermal noise is given by (6.47) and (6.48)

$$\sigma_t^2 = E\{u_i^2\} + E\{u_e^2\} = S_I A^2 2B + S_E A^2 2B \left(\frac{1}{R^2} + \frac{(2\pi CB)^2}{3} \right)$$

The signal-to-noise ratio for the Gaussian approximation is

$$\rho = \frac{E_1 - E_0}{\sqrt{\sigma_0^2 + \sigma_t^2} + \sqrt{\sigma_1^2 + \sigma_t^2}}$$

$$= \frac{\gamma}{\sqrt{2B}\left\{ [\gamma_0 + W^2]^{1/2} + [\gamma_0 + \gamma + W^2]^{1/2} \right\}} \tag{6.59}$$

where

$$W^2 = \frac{2k_B T_K}{q^2 R} \left[1 + \frac{1}{g_m R} \left(1 + (2\pi RCB)^2/3 \right) \right] \tag{6.60}$$

The performance of the receiver is illustrated by an Example.

Example 6.4   Optical receiver with bandlimited filter H_2

Consider an optical system with transmission rate $B = 10$ Mb/s and having $R = 1$ Mohm, $C = 10$ pF and $g_m = 5000\ \mu$S. The photodetector has quantum efficiency $\eta = 0.9$ and a dark current of 8 pA. The optical wavelength is $\lambda = 1.3\ \mu$m.

Solution:

For $T_K = 300$ K is

$$\frac{2k_B T_K}{q^2 R} = \frac{2 \times 1.38 \times 10^{-23} \times 300}{(1.60 \times 10^{-19})^2 \times 10^6} = 3.23 \times 10^{11}$$

The thermal noise (6.60) becomes

$$W^2 = 3.23 \times 10^{11} \left[1 + \frac{1}{5 \times 10^{-3} \times 10^6} \left(1 + \frac{(2\pi \times 10^6 \times 10^{-11} \times 10^7)^2}{3} \right) \right]$$

$$= 8.83 \times 10^{12}$$

If the Gaussian approximation is used to estimate the error probability the signal-to-noise ratio has the form (6.59) with the parameter $\gamma = \eta P_1/hf$. A certain transmission error probability corresponds to a specific value of γ and the necessary optical power P_1 is determined by solving (6.59) for γ with ρ fixed. In many situations the thermal noise W^2 dominates over the optical noise, i.e. $W^2 \gg i_0/q + \gamma$ and ρ is approximately equal to

$$\rho \approx \frac{\gamma}{2W\sqrt{2B}} \tag{6.61}$$

Solving for γ gives

$$\gamma \approx \rho 2W\sqrt{2B} \tag{6.62}$$

It is easy to verify that the approximation is valid in the present case. An error probability of $P_e - 10^{-9}$ requires a signal-to-noise ratio $\rho = 6.0$ and the corresponding value of γ obtained from (6.62) is

$$\gamma = 6.0 \times 2 \times 2.97 \times 10^6 \times \sqrt{2 \times 10^7} = 1.59 \times 10^{11}$$

which gives the received optical power

$$P_1 = \frac{hf}{\eta} \gamma = \frac{6.626 \times 10^{-34} \times 3 \times 10^8}{0.9 \times 1.3 \times 10^{-6}} \, 1.59 \times 10^{11} = 2.70 \times 10^{-8}$$

The average optical power level expressed in dBm is

$$10 \log \left(\frac{P_1/2}{10^{-3}} \right) = 10 \log \left(\frac{2.70 \times 10^{-8}}{2 \times 10^{-3}} \right) = -48.7 \, \text{dBm} \qquad \square$$

The receiver filter used in Example 6.4 is not the best one. An optimum filter should compensate for the input capacitance C and reduce thermal noise while causing little intersymbol interference. These requirements are contradictory since narrowing the receiver filter would decrease the thermal noise but increase the ISI. The problem is complicated by the fact that, in contrast to an additive noise channel, the ISI of an optical channel affects both the signal strength and the noise variance.

6.5   Design of Optical Receivers

The theoretical analysis of a receiver with an equalizer compensating for the RC-characteristics of the photodetector is in general difficult. The effects caused by intersymbol interference are complicated, and an optimum receiver in terms of minimum error probability is hard to specify.

The analysis of the MSE receiver in Chapter 5 showed that a zero-forcing equalizer is close in performance to the slightly better mean-square-error equalizer. A zero-forcing equalizer has the advantage that intersymbol interference is eliminated, which greatly facilitates the analysis.

The bandwidth of the receiver is an important design parameter. If it is small the thermal noise is reduced, but so is the data signal. A narrow filter will also introduce intersymbol interference. When the bandwidth is large the received signal power is preserved but the noise is apt to be large. The design of an efficient receiver is a balance between these two conditions.

Personick (1973) has suggested a design procedure that constitutes an approximate solution to receiver optimization. The receiver filter $H(f)$ of Figure 6.7 is specified such that the signal $u(t)$ at the decision point is a so-called Nyquist function with zero crossings at the decision times $t = kT$. Such a signal has no ISI in the signal amplitude and the ISI effect on the signal variance is easy to calculate. The error probability is estimated by the Gaussian approximation for a worst-case ISI situation. The bandwidth of the receiver filter is varied to obtain a maximized signal-to-noise ratio ρ and thereby a minimum of the error probability estimate. Since the optimization is performed over an one-dimensional system parameter it is computationally easy. The presentation below is based on a technical report by Andersson (1990).

The receiver is linear and a constant factor of amplification is of no importance. By normalizing the signals in amplitude and with respect to the data symbol time T the design procedure is adapted to optical systems of arbitrary transmission rate.

We start with the received optical intensity (5.42)

$$\Gamma(t) = \gamma_0 + \sum_{k=-\infty}^{\infty} b_k \gamma(t - kT) \tag{6.63}$$

It is convenient to introduce the normalized signal pulse

$$s(t) = \gamma(t)/\gamma T \tag{6.64}$$

where

$$\gamma = \frac{1}{T} \int_{-\infty}^{\infty} \gamma(t)\, dt \tag{6.65}$$

is the average optical intensity of the optical pulse $\gamma(t)$ normalized to the symbol interval T.

It is clear from the definition that

$$\int_{-\infty}^{\infty} s(t)\, dt = 1 \tag{6.66}$$

We will study systems with two different received signals $s(t)$ of rectangular and Gaussian form, shown in Figure 6.10 and Figure 6.11, respectively. The rectangular pulse is an example of a received signal confined to the signaling interval and therefore without intersymbol interference.

$$s(t) = \begin{cases} 1/aT; & -aT/2 < t < aT/2 \\ 0; & \text{otherwise} \end{cases} \tag{6.67}$$

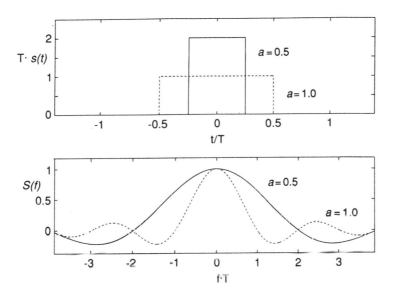

Figure 6.10 Rectangular received optical pulse $s(t)$ and its frequency function $S(f)$.

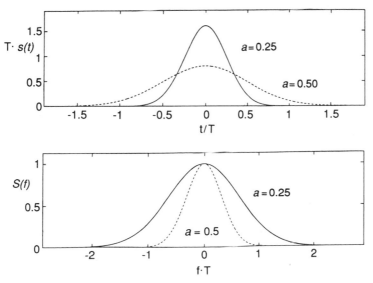

Figure 6.11 Gaussian received optical pulse $s(t)$ and its frequency function $S(f)$.

Its frequency function is

$$S(f) = \frac{\sin(\pi a f T)}{\pi a f T} \tag{6.68}$$

The Gaussian signal is not limited in time and it represents a case where dispersion has broadened the received pulse.

$$s(t) = \frac{1}{aT\sqrt{2\pi}} \exp[-t^2/2(aT)^2] \tag{6.69}$$

Its frequency function is

$$S(f) = \exp[-(2\pi a f T)^2/2] \tag{6.70}$$

The parameter a is a measure of the pulse width.

The photodetector is an avalanche photodiode with average gain M and quantum efficiency η. Its dark current is composed of a primary bulk dark current I_{bd}, which is multiplied by the avalanche mechanism, and a surface dark current I_{sd}, which is unaffected by the avalanche gain. I_{bd} and I_{sd} are modeled as Poisson processes with electron intensities $\gamma_{bd} = I_{bd}/q$ and $\gamma_{sd} = I_{sd}/q$, respectively.

The intensity function for the primary shot noise current at the photodetector output can be written as

$$\Gamma_d(t) = \Gamma_a(t) + \Gamma_b(t) \tag{6.71}$$

with

$$\Gamma_a(t) = \gamma_0 + \gamma_{bd} + \gamma T \sum_{k=-\infty}^{\infty} b_k s(t - kT) \tag{6.72}$$

representing the part that is multiplied by the avalanche gain and

$$\Gamma_b(t) = \gamma_{sd} \tag{6.73}$$

the part that is unaffected of the gain.

The receiver makes decisions on the transmitted symbols based on the output signal $u(t)$. It can be divided into three parts

$$u(t) = u_d(t) + u_i(t) + u_e(t) \tag{6.74}$$

with

$$u_d(t) = \int_{-\infty}^{\infty} h_2(t - s)i_d(s)\, ds \tag{6.75}$$

where $h_2(t)$ is the impulse response (current to voltage) of the filter denoted $H_2(f)$ in Figure 6.7. The current $i_d(t)$ is shot noise with intensity $\Gamma_d(t)$.

The signals $u_i(t)$ and $u_e(t)$ are produced by thermal noise with $u_i(t)$ equal to

$$u_i(t) = \int_{-\infty}^{\infty} h_2(t - s)i_n(s)\, ds \tag{6.76}$$

with $i_n(t)$ a zero mean white Gaussian noise stochastic process with spectral density S_I. The signal $u_e(t)$ is white Gaussian noise with spectral density S_E filtered through the filter $H(f)$ of Figure 6.7.

$$u_e(t) = \int_{-\infty}^{\infty} h(t - s)i_e(s)\, ds \tag{6.77}$$

The receiver samples $u(t)$ at the end of each symbol interval and compares this value with a threshold to decide on the transmitted symbols. For notational simplicity the time scale is normalized in such a way that the decision time is equal to $t = 0$. The decision variable is

$$U = u(t = 0) \tag{6.78}$$

The received optical pulse $s(t)$, (6.67) and (6.69), is defined as being symmetrical around $t = 0$. The specification of the decision time $t = 0$ requires the receiver filter to be non causal. This puts no restrictions on generality, however. Realizable filters are obtained by introducing a suitable delay in the receiver.

To estimate the bit error probability by the Gaussian approximation, the mean and variance of U are needed. We start with the mean.

The thermal noise signals $i_n(t)$ and $e_n(t)$ have zero mean which makes $E\{u_i(kT)\} = 0$ and $E\{u_e(kT)\} = 0$. The mean $E\{U\}$ is then equal to $E\{u_d(kT)\}$. Since $u_d(t)$ is a shot noise process its mean is obtained from Campbell's Theorem (5.131) with $v(t) = qh_2(-t)$ and $E\{a_k\} = M$

$$E\{U\} = Mq \int_{-\infty}^{\infty} \Gamma_a(t)h_2(-t)\, dt + q \int_{-\infty}^{\infty} \Gamma_b(t)h_2(-t)\, dt \tag{6.79}$$

Substitution of (6.72) and (6.73) yields

$$E\{U\} = Mq\gamma T \int_{-\infty}^{\infty} \sum_{k=-\infty}^{\infty} b_k s(t - kT)h_2(-t)\, dt$$
$$+ q(M\gamma_0 + M\gamma_{bd} + \gamma_{sd}) \int_{-\infty}^{\infty} h_2(-t)\, dt \tag{6.80}$$

The first part of (6.80) can be expressed as a convolution

$$f(t) = Mq \int_{-\infty}^{\infty} s(t - s)h_2(s)\, ds \tag{6.81}$$

The integral in the second term is equal to

$$\int_{-\infty}^{\infty} h_2(s)\, ds = H_2(0) \tag{6.82}$$

Substitution of (6.81) and (6.82) into (6.80) gives

$$E\{U\} = \gamma T \sum_{k=-\infty}^{\infty} b_k f(-kT) + q(M\gamma_0 + M\gamma_{bd} + \gamma_{sd})H_2(0) \tag{6.83}$$

We next put the restriction on $f(t)$ to be a normalized Nyquist function, see Nyquist (1928)

$$f(t) = \frac{\sin(\pi t/T)\cos(\pi bt/T)}{(\pi t/T)[1 - (2bt/T)^2]} \tag{6.84}$$

The appearance of $f(t)$ is shown in Figure 6.12 together with its frequency function $F(f)$.

$$F(f) = \begin{cases} T; & -(1-b)/2T \le |f| \le (1-b)/2T \\ (T/2)[1 - \sin(\pi/b)(|f|T - 1/2)]; & (1-b)/2T \le |f| \le (1+b)/2T \\ 0; & \text{otherwise} \end{cases} \tag{6.85}$$

The parameter b varies between $b = 0$ and $b = 1$ and determines the bandwidth of the filter.

The function $f(t)$ is the overall impulse response of the transmission system. Nyquist pulses do not interfere at the sampling times $t = kT$ which means that the mean of the decision variable is affected only by the symbol b_k to be decided, which in turn simplifies the analysis. However, the system is not free from intersymbol interference. As will be shown later the variance of the decision variable depends on the complete sequence b_k of data symbols.

Note that (6.84) is not necessarily the best choice for min P_e, but it can be expected to be close.

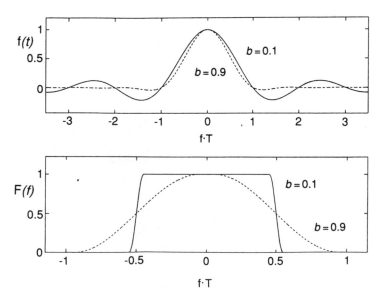

Figure 6.12 The overall impulse response $f(t)$ and the corresponding frequency function $F(f)$ of an optical system with zero-forcing equalization.

An $f(t)$ according to (6.84) yields

$$E\{U\} = \gamma Tb_k + q(M\gamma_0 + M\gamma_{bd} + \gamma_{sd})H_2(0) \tag{6.86}$$

From (6.81) the following relation in terms of frequency functions is derived

$$F(f) = MqS(f)H_2(f) \tag{6.87}$$

Substitution of $H_2(f)$ from (6.87) into (6.86) gives the means of the decision variable under the hypothesis that $b_k = 1$ and $b_k = 0$, respectively

$$E_1 = \gamma T + \left(\gamma_0 + \gamma_{bd} + \frac{\gamma_{sd}}{M}\right)\frac{F(0)}{S(0)} \tag{6.88}$$

and

$$E_0 = \left(\gamma_0 + \gamma_{bd} + \frac{\gamma_{sd}}{M}\right)\frac{F(0)}{S(0)} \tag{6.89}$$

To evaluate the variance we notice that the three components of (6.74) are statistically independent. The variance of $u_d(t)$ at the decision time $t = 0$ is with the use of Campbell's Theorem (C.31)

$$\sigma_d^2 = M^2 F_a(M)q^2 \int_{-\infty}^{\infty} \Gamma_a(t)h_2^2(-t)\,dt + q^2 \int_{-\infty}^{\infty} \Gamma_b(t)h_2^2(-t)\,dt \tag{6.90}$$

After considerable manipulations, utilizing Fourier transform properties, (6.90) can be written as

$$\sigma_d^2 - F_a(M)\gamma T \int_{-\infty}^{\infty} \left(\sum_{k=-\infty}^{\infty} b_k e^{-j2\pi fkT}\right) S(f)\left(\frac{F(f)}{S(f)} * \frac{\Gamma(f)}{S(f)}\right) df$$

$$+ [F_a(M)(\gamma_0 T + \gamma_{bd}) + \gamma_{sd}] \int_{-\infty}^{\infty} \left|\frac{F(f)}{S(f)}\right|^2 df \tag{6.91}$$

The output signals $u_i(t)$ and $u_e(t)$ are filtered white noise stochastic processes. Their variances are

$$\sigma_i^2 = \int_{-\infty}^{\infty} S_I|H_2(f)|^2 dt = (Mq)^{-2} S_I \int_{-\infty}^{\infty} \left|\frac{F(f)}{S(f)}\right|^2 dt \tag{6.92}$$

and analogously to (6.44)

$$\sigma_e^2 = \int_{-\infty}^{\infty} S_E|H(f)|^2 dt = (Mq)^{-2}\frac{S_E}{R^2} \int_{-\infty}^{\infty} \left|\frac{F(f)}{S(f)}\right|^2 [1 + (2\pi RCf)^2]\,dt \tag{6.93}$$

respectively. The variance of the decision variable $U = u(0)$ is

$$\sigma^2 = \sigma_d^2 + \sigma_i^2 + \sigma_e^2 \tag{6.94}$$

Note that σ, in contrast with $E\{U\}$, depends on the complete sequence b_k of transmitted data symbols. This means that intersymbol interference is present in spite of the choice of $f(t)$ as a Nyquist function. This is characteristic for optical transmission where the noise is signal-dependent.

One way of handling the situation is to consider the worst cases, as was done in the analysis of ISI in Chapter 5. The least favourable data sequences are those maximizing σ. It turns out that for both $b_k = 0$ and $b_k = 1$

$$b_j = 1; \ \forall \, j \neq k$$

is the worst case.

It is convenient to normalize the frequency functions in the integrals with respect to the symbol interval T. Let

$$S'(\nu) = S(\nu/T) \tag{6.95}$$

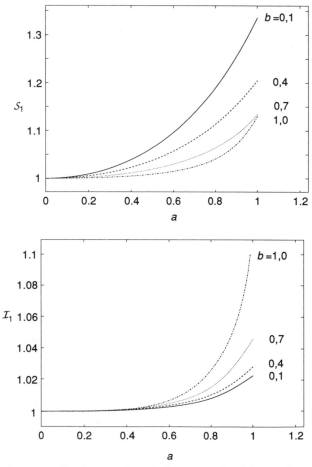

Figure 6.13 The normalized sum S_1 and the normalized integral $\mathcal{I}_1$ as functions of relative width a of the received pulse. The curves are plotted with the relative system bandwidth b as a parameter. The received pulse $s(t)$ is rectangular, as shown in Figure 6.10.

and

$$F'(\nu) = \frac{1}{T}F(\nu/T) \tag{6.96}$$

where ν is normalized frequency

$$\nu = fT \tag{6.97}$$

This gives normalized integrals which are independent of T and apply to systems of arbitrary transmission rate.

$$\mathcal{S}_1 = \sum_{k=-\infty}^{\infty} S'(k)\left(\frac{F'(k)}{S'(k)} * \frac{F'(k)}{S'(k)}\right) \tag{6.98}$$

$$\mathcal{I}_1 = \int_{-\infty}^{\infty} S'(\nu)\left(\frac{F'(\nu)}{S'(\nu)} * \frac{F'(\nu)}{S'(\nu)}\right) d\nu \tag{6.99}$$

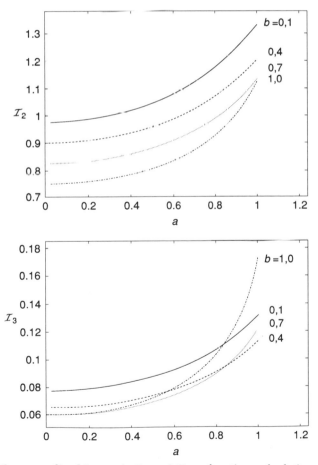

Figure 6.14 The normalized integrals $\mathcal{I}_2$ and $\mathcal{I}_3$ as functions of relative width a of the received pulse. The curves are plotted with the relative system bandwidth b as a parameter. The received pulse $s(t)$ is rectangular, as shown in Figure 6.10.

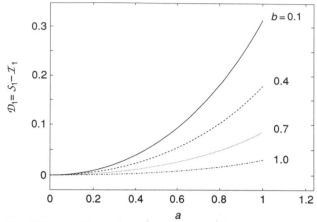

Figure 6.15 The difference $\mathcal{D}_1 = \mathcal{S}_1 - \mathcal{I}_1$ as a function of relative width a of the received pulse. The curves are plotted with the relative system bandwidth b as a parameter. The received pulse $s(t)$ is rectangular shown in Figure 6.10.

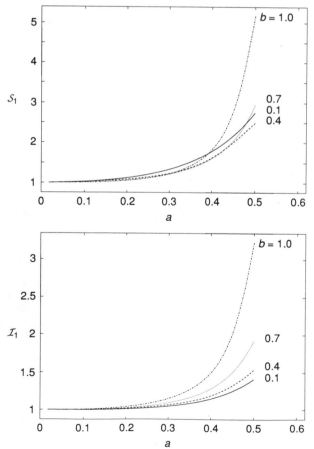

Figure 6.16 The normalized sum $\mathcal{S}_1$ and the normalized integral $\mathcal{I}_1$ as functions of relative width a of the received pulse. The curves are plotted with the relative system bandwidth b as a parameter. The received pulse $s(t)$ is Gaussian, as shown in Figure 6.11.

$$\mathcal{I}_2 = \int_{-\infty}^{\infty} \left| \frac{F'(\nu)}{S'(\nu)} \right|^2 d\nu \tag{6.100}$$

$$\mathcal{I}_3 = \int_{-\infty}^{\infty} \left| \frac{F'(\nu)}{S'(\nu)} \right|^2 \nu^2 \, d\nu \tag{6.101}$$

The quantities (6.98) – (6.101) are calculated as functions of the parameters a and b for the rectangular received signal Figures 6.13–6.15 and for the Gaussian signal in Figures 6.16–6.18.

The variance (6.94) is, when a data symbol 'one' was transmitted ($b_k = 1$), expressed in terms of the normalized quantities (6.98) – (6.101)

$$\sigma_1^2 = F_a(M)T[\gamma S_1 + \gamma_0 \mathcal{I}_2] +$$
$$(Mq)^{-2}T[F_a(M)M^2q^2\gamma_{bd} + q^2\gamma_{sd} + S_I + S_E/R^2]\mathcal{I}_2 + \tag{6.102}$$
$$(Mq)^{-2}(1/T)(2\pi C)^2 S_E \mathcal{I}_3$$

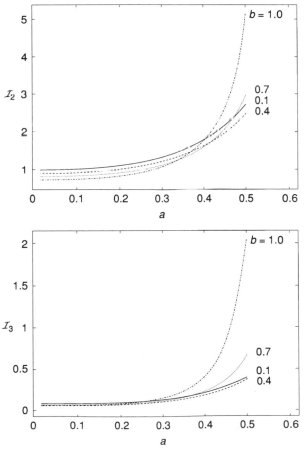

Figure 6.17 The normalized integrals $\mathcal{I}_2$ and $\mathcal{I}_3$ as functions of relative width a of the received pulse. The curves are plotted with the relative system bandwidth b as a parameter. The received pulse $s(t)$ is Gaussian, as shown in Figure 6.11.

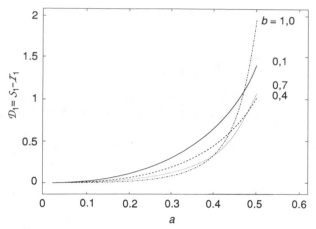

Figure 6.18 The difference $\mathcal{D}_1 = \mathcal{S}_1 - \mathcal{I}_1$ as a function of relative width a of the received pulse. The curves are plotted with the relative system bandwidth b as a parameter. The received pulse $s(t)$ is Gaussian, as shown in Figure 6.11.

For the data symbol 'zero' ($b_k = 0$)

$$
\begin{aligned}
\sigma_0^2 = {} & F_a(M)T[\gamma(\mathcal{S}_1 - \mathcal{I}_1) + \gamma_0\mathcal{I}_2] + \\
& (Mq)^{-2}T[F_a(M)M^2q^2\gamma_{bd} + q^2\gamma_{sd} + S_I + S_E/R^2]\mathcal{I}_2 + \\
& (Mq)^{-2}(1/T)(2\pi C)^2 S_E\mathcal{I}_3
\end{aligned}
\tag{6.103}
$$

The formulas derived assume an avalanche photodiode (APD). Results for an ordinary (PIN) photodiode are obtained by letting $M = 1$ and $F_a(M) = 1$. The error probability follows from the Gaussian approximation (5.34).

Example 6.5   Receiver design

Use the Personick design procedure to determine the receiver sensitivity of a repeater for an undersea system with a line rate $B = 296$ Mb/s operating at a wavelength $\lambda = 1.3$ μm. The laser at the transmitter has an extinction ratio of 15 dB. The received optical pulses have Gaussian shape and a relative dispersion of $\sigma_t B = 0.25$. The front end amplifier has an input impedance $R_a = 2$ Mohm and and the additional thermal noise can be represented by a transconductance $g_m = 8$ mS

Determine the relative bandwidth parameter b for an equalizing filter and calculate the bit error probability. Determine the receiver sensitivity defined as the input power needed to achieve $P_e = 10^{-9}$

(a) The photodetector is a PIN diode with quantum efficiency $\eta = 0.75$ and dark current $i_d = 10$ nA. Its internal impedance is $R_d = 1.5$ Mohm and shunt capacitance $C = 2.5$ pF.

(b) For an avalanche photodiode having an excess noise factor $F(M) = M^x$ with $x = 0.4$. The quantum efficiency of the diod is $\eta = 0.8$ and the dark current $i_d = 100$ nA. Its internal impedance $R_d = 1$ Mohm and shunt capacitance $C = 4$ pF. Determine the best value for the avalanche gain M. Use the parameter b obtained from (a).

Solution:

(a) The extinction ratio $10\log(\gamma_1/\gamma_0) = 15$ dB which gives $\gamma_0 = 10^{-15/10}\gamma_1 = 0.032\gamma_1$ and $\gamma = \gamma_1 - \gamma_0 = 0.97\gamma_1$.

The signal-to-noise ratio $\rho = (E_1 - E_0)/(\sigma_1 + \sigma_0)$ is obtained by substitution into (6.88), (6.89), (6.102) and (6.103) with the weight factors S_1, $\mathcal{I}_1$, etc calculated for a Gaussian pulse with $a = 0.25$. The SNR is a function of the received optical power $\mathcal{P}_1 = \gamma_1 hf/\eta$ and of the filter parameter b.

The estimated error probability $P_e = Q_1(\rho)$ for input levels $10\log(\mathcal{P}_1/10^{-3}) = -34$ dBm and -35 dBm as a function of b is shown in Figure 6.19. The best performance is obtained for a filter with a bandwidth corresponding to $b = 0.55$ and interpolation in the diagram determines the receiver sensitivity for $P_e = 10^{-9}$ as -33.9 dBm. In terms of received photoelectrons per bit this corresponds to $m_1 = \gamma_1/B = 6755$.

(b) The same procedure for the APD-receiver results in Figure 6.20 which shows the error probability as a function of the avalanche gain M. The value $b = 0.55$ is used in the calculation of the curves. The diagram shows that error probability is lowest at $M = 18$. For higher values the performance is deteriorated by the increased excess noise. The receiver sensitivity for $P_e = 10^{-9}$ is -41.1 dBm, corresponding to $m_1 = 1373$ photoelectrons per bit.

□

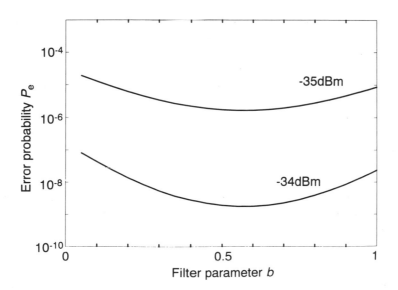

Figure 6.19   Example 6.5a. The bit error probability as a function of the equalizing filter bandwidth parameter b for two received signal levels.

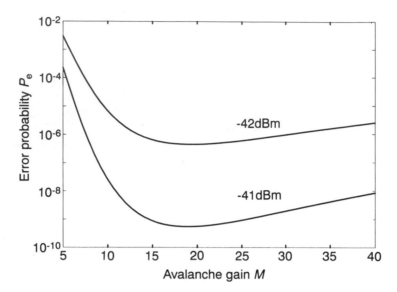

Figure 6.20 Example 6.5b. The bit error probability as a function of the photodiode avalance gain for two received signal levels. The equalizing filter bandwidth parameter $b = 0.55$.

6.6 Simplified System Analysis

In this section we present a simple system analysis, which does not result in an exact evaluation of system performance, but rather gives an overview of the capability of fiber optical system using present-day components.

We consider a fiber optical system with standard on-off modulation. The system error probability is restricted to be lower than some given design criterion which, in the examples below, we let be equal to $P_e = 10^{-12}$. The goal is, for a given information rate B, to design a system with as large transmission distance L as possible. The result depends on the attenuation and dispersion of the optical fiber together with the intensity of the light source and the sensitivity of the receiver.

We first consider an idealized situation where the attenuation of the fiber is the only transmission impairment. More realistic systems with dispersion are studied at the end of the section.

The presentation is based on ideas introduced by P. S. Henry (1985). The results depend on the system parameters assumed, and the numerical values obtained should be considered as examples and not as exact guidelines.

Extended studies, with a more detailed analysis, can be found in Kimura (1988) and Heidemann *et al.* (1993).

6.6.1 Ideal Systems

To investigate the effect of fiber attenuation on transmission distance we consider systems with an ideal optical receiver where the effect of thermal noise is neglected.

The error probability is estimated by the Gaussian approximation (5.41)

$$P_e \approx Q_1(\rho) = Q_1(\sqrt{m_1} - \sqrt{m_0}) \tag{6.104}$$

where $m_1 = \gamma_1 T$ and $m_0 = \gamma_0 T$ denotes the number of photoelectrons at the receiver and Q_1 is the function (5.40). The signal-to-noise ratio ρ can be expressed in terms of the photoelectron intensities

$$\rho = (\sqrt{\gamma_1} - \sqrt{\gamma_0})/\sqrt{B} \tag{6.105}$$

where $B = 1/T$ is the data rate which also represents the system bandwidth.

The signal-to-noise ratio for which (6.104) gives a bit error probability of $P_e = 10^{-12}$ is $\rho = 7.0$

The received optical power is related to the transmitted intensity γ^* by the transmittance factor of the optical fiber

$$\mathcal{A} = 10^{-aL/10} \tag{6.106}$$

where a is the fiber attenuation in dB/km and L is the distance in km between transmitter and receiver. The received intensity is

$$\gamma = \mathcal{A}\eta\gamma^* \tag{6.107}$$

with η the quantum efficiency of the photodetector.

Consider a system that uses a single-mode fiber with attenuation $a = 0.20$ dB/km and operates at an optical wavelength $\lambda = 1.55$ μm. The receiver is of the photoelectron counting type, with a photodetector having quantum efficiency $\eta = 0.9$. The peak output power of the laser is assumed to be $P_T = 1$ mW.

The quantity γ_0 represents background noise and the analysis differs if it is generated at the transmitter or at the receiver. We first consider a situation where it is caused by a light source with nonzero extinction ratio

$$\epsilon = \frac{\gamma_0^*}{\gamma_1^*} \tag{6.108}$$

with $\gamma_1^* = \gamma^* + \gamma_0^*$ the laser peak intensity. The square of the signal-to-noise ratio (6.105) can be expressed as

$$\rho^2 = \left(\sqrt{\mathcal{A}\eta\gamma_1^*} - \sqrt{\mathcal{A}\eta\gamma_0^*}\right)^2/B$$
$$= \mathcal{A}\eta\gamma^*(1 - \sqrt{\epsilon})^2/B \tag{6.109}$$

Substitution of $\mathcal{A}$ (6.106) and solving for L yields

$$L = \frac{10}{a}\log\left(\frac{\eta\gamma_1^*(1 - \sqrt{\epsilon})^2}{\rho^2 B}\right) \tag{6.110}$$

A transmitted optical power of $\mathcal{P}_T = 1$ mW at $\lambda = 1.55$ μm corresponds to

$$\gamma^* = \frac{\mathcal{P}_T \lambda}{hc} = \frac{1 \times 10^{-3} \times 1.55 \times 10^{-6}}{6.626 \times 10^{34} \times 2.998 \times 10^8} = 7.83 \times 10^{15} \text{ photons/s} \qquad (6.111)$$

For a fiber attenuation of $\mathbf{a} = 0.2$ dB/km, with $\epsilon = 0.1$ and $\eta = 0.9$ the relation (6.110) for $\rho = 7.0$ becomes

$$L = 391.4 - 50.0 \log B \qquad (6.112)$$

where L is in km and B is the data rate in Mb/s. It is shown in Figure 6.21.

A system for which L is determined by an expression like (6.112) is called attenuation limited.

Another source of background noise is dark current in the photodetector. Consider a situation where the background noise is caused by a dark current $i_d = 10$ nA.

The signal-to-noise ratio (6.105) is then

$$\rho = \left(\sqrt{A\eta\gamma^* + \gamma_0} - \sqrt{\gamma_0}\right)/\sqrt{B} \qquad (6.113)$$

where γ_0 is the electron intensity of the dark current (5.18)

$$\gamma_d = \frac{i_d}{q} \qquad (6.114)$$

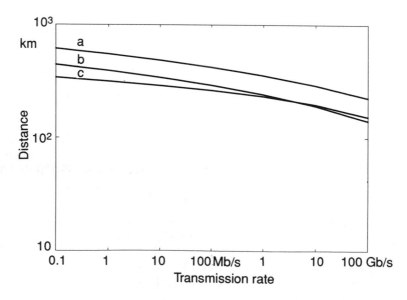

Figure 6.21   Repeater spacing as a function of the transmission rate for an ideal direct detection systems without dispersion: (a) quantum limit. (b) a laser with extinction ratio 0.1 in the transmitter. (c) a photodetector with 10 nA dark current in the receiver.

Rewrite equation (6.113) as

$$\rho\sqrt{B} + \sqrt{\gamma_0} = \sqrt{A\eta\gamma^* + \gamma_0} \tag{6.115}$$

Squaring both sides yields

$$A\eta\gamma^* = \rho^2 B + 2\rho\sqrt{B\gamma_0} \tag{6.116}$$

Substitution of (6.106) and solving for L gives the relation

$$L = \frac{10}{a}\left[\log\left(\frac{\eta\gamma^*}{\rho^2}\right) - \log\left(B + \frac{2\sqrt{\gamma_0}}{\rho}\sqrt{B}\right)\right] \tag{6.117}$$

For the same system parameters as before

$$L = 407.9 - 50.0\log(B + 71.4\sqrt{B}) \tag{6.118}$$

with L in km and B in Mb/s. The relation (6.118) is shown in Figure 6.21.

The quantum limit for optical detection implies the further idealization that no background noise is present at the transmitter or receiver and that the photodetector has a quantum efficiency equal to unity.

As shown in Chapter 5 the Gaussian approximation is not accurate in this case. The correct error probability is given by (5.16)

$$P_e = \frac{1}{2} e^{-m_1} \tag{6.119}$$

where $m_1 = \gamma_1 T$ is the average number of received photons for the binary symbol 'one'. For $P_e = 10^{-12}$ it is

$$m_1 = -\ln 2P_e = -\ln 2 \times 10^{-12} - 26.9 \text{ photons} \tag{6.120}$$

The quantity m_1 is related to the transmitted optical intensity γ_1^* by

$$m_1 = A\gamma_1^* T \tag{6.121}$$

Substitution of A from (6.106) and $B = 1/T$ and solving for L gives

$$L = \frac{10}{a}\left[\log\frac{\gamma_1^*}{m_1} - \log B\right] \tag{6.122}$$

Let the optical wavelength be $\lambda = 1.55\ \mu$m and let the fiber attenuation $a = 0.154$ dB/km which is close to the minimal value for silica fibers, see Section 4.9. For a transmitted power $\mathcal{P}_T = 1$ mW the relation (6.122) become

$$L = 549.6 - 64.9\log B \tag{6.123}$$

with L in km and B in Mb/s. The relation (6.123) is shown in Figure 6.21. For a system with transmission data rate $B = 10$ Gb/s the maximal repeater distance calculated by (6.123) is $L = 290$ km.

For systems with direct detection the curves of Figure 6.21 are optimistic in two respects. Thermal noise in the receiver is neglected and the effect of fiber dispersion is not included.

The distance L is determined by the optical power of the transmitted light, the fiber attenuation and the receiver sensitivity. The fiber attenuation cannot be improved much, compared to $a = 0.154$ dB/km, as long as silica glass is used. In Chapter 8 it will shown that phase-shift keying (PSK) with coherent optical signals results in a receiver sensitivity that is close to what can be achieved by direct detection with an ideal photon counting receiver. The heterodyning in coherent systems provides amplification of the received signal, and thermal noise need not be a limiting factor in such systems.

A larger distance L than given by (6.123) can be obtained by increasing the output optical power $\mathcal{P}_T$, which can be achieved by optical amplification. A receiver with an optical preamplifier, however, has a lower theoretical sensitivity than a pure photon counting receiver. See Chapter 7 where optical fiber systems with optical amplifiers are studied.

The results obtained from the quantum limit for direct detection do not represent the ultimate theoretical capacity of optical communication. In Section 5.13 limits on transmission capability are derived by applying information theory to the optical fiber channel.

6.6.2   Realistic Direct Detection Receiver

The sensitivity for an optical fiber receiver is usually expressed as the optical power $\mathcal{P}_R$ required to result in an acceptable bit error probability, say $P_e = 10^{-12}$. For a given $\mathcal{P}_R$ the transmission distance L in km is determined by the relation

$$L\mathbf{a} = 10 \log \mathcal{P}_T/\mathcal{P}_R - b \tag{6.124}$$

where $\mathbf{a}$ is the fiber attenuation in dB/km and $\mathcal{P}_T$ is the output optical power from the light source. The parameter b is a system design margin introduced to account for ageing and fluctuations in system components. In practice a value of $3-6$ dB is common.

The analysis in Section 5.6 and Example 6.5 indicates that, in the presence of thermal noise, the received pulses should contain approximately 6000 photons for the error probability to be $P_e \leq 10^{-12}$ and this value is used in the examples below.

For high transmission rates the dispersion effects of the fiber becomes important. Dispersion causes the pulses to overlap and the resulting intersymbol interference increases the error probability, as illustrated in Section 5.5. In Example 5.2 received pulses with an RMS width $\sigma_t = T/2$ are causing inter symbol interference, resulting in an unacceptably high error probability.

To avoid the negative effects of dispersion we use the relation

$$\sigma_t \leq T/4 \tag{6.125}$$

as a design condition to ensure that intersymbol interference will not degrade performance too much. The dispersion σ_t grows in proportion to L, and the relation (6.125) is a straight line with slope -1 in the performance diagram Figure 6.22. For high rates B it determines the distance L in which case the system is dispersion-limited.

Consider a system operating at $\lambda = 0.85 \ \mu m$, using a multimode fiber. The fiber attenuation is assumed to be $\mathbf{a} = 1.6 \ dB/km$. This is a slightly higher than the value Figure 4.18 to allow for the splicing losses that occur in practical systems. A light-emitting diode producing $\mathcal{P}_T = 0.1 \ mW$ is assumed at the transmitter.

The required number of received photons $m = 6000$ corresponds to an optical signal energy

$$E_R = \frac{mhc}{\lambda} = \frac{6000 \times 6.625 \times 10^{-34} \times 2.998 \times 10^8}{0.85 \times 10^{-6}} = 1.40 \times 10^{-15} Ws$$

The received optical power is $\mathcal{P}_T = E_R/T = E_R B$, and at $B = 10 \ Mb/s$ the receiver sensitivity expressed in dBm is

$$S = 10 \log \frac{E_R B}{10^{-3}} = -48.5 \ dBm \tag{6.126}$$

Substitution into (6.124) with a system margin $b = 6 \ dB$ gives

$$L = 28.5 - 6.25 \log B \tag{6.127}$$

with L in km and B in Mb/s.

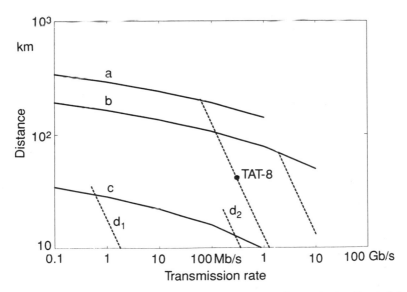

Figure 6.22 Repeater spacing as a function of bit rate for realistic direct detection systems. The solid lines indicate attenuation limitation and the dashed lines dispersion limitation: (a) single-mode fiber at $\lambda = 1.55 \ \mu m$. (b) dispersion-compensated fiber at $\lambda = 1.3 \ \mu m$. (c) multimode fiber at $\lambda = 0.85 \ \mu m$, step-index (d_1) and graded-index (d_2).

The dispersion of a step index fiber is from (4.57)

$$\sigma_t = \frac{\Delta}{2\sqrt{3}} \frac{N_1 L}{c}$$

For a silica fiber with $N_1 = 1.46$ and $\Delta = 0.01$ the relation (6.125) gives

$$BL \leq \frac{2\sqrt{3}c}{4\Delta N_1} = \frac{2 \times \sqrt{3} \times 2.998 \times 10^8}{4 \times 1.46 \times 0.01} = 17.8 \text{ Mb/s} \cdot \text{km} \qquad (6.128)$$

For a graded index fiber of the same material from (4.60)

$$\sigma_t = \frac{\Delta^2}{4\sqrt{3}} \frac{N_1 L}{c}$$

and

$$BL \leq 3.56 \text{ Gb/s} \cdot \text{km} \qquad (6.129)$$

The relations (6.128) and (6.129) are shown in Figure 6.22. From the diagram it is evident that multimode transmission over a step index fiber is practical for short distances and low transmission rates only.

The dispersion limit is relaxed if single-mode fibers are used. Chromatic dispersion is then dominating and from (4.83) it follows that

$$\sigma_t = \sigma_\lambda DL \qquad (6.130)$$

where σ_λ is the spectral linewidth of the light source and D is the combined waveguide and material dispersion.

Consider a system operation at $\lambda = 1.55 \ \mu\text{m}$ where the fiber attenuation has its minimum assumed to be $\mathbf{a} = 0.20$ dB/km. The dispersion indicated in Figure 4.10 is

$$D = 20 \text{ ps/nm} \cdot \text{km}$$

For a transmitting laser with spectral width $\sigma_\lambda = 1$ nm the dispersion limit (6.125) is

$$BL \leq \frac{1}{4\sigma_\lambda D} = 12.5 \text{ Gb/s} \cdot \text{km}$$

The attenuation limitation for a transmitted power $P_T = 1$ mW with a receiver sensitivity (6.126) and a system margin of 3 dB is, from (6.124)

$$L = 290.7 - 50.0 \log B \qquad (6.131)$$

A system with a dispersion-compensated fiber operating at $\lambda = 1.3 \ \mu\text{m}$ has a low dispersion, see Figure 4.10. The dispersion calculated in Example 4.3 is

$$D = 1.9 \text{ ps/nm} \cdot \text{km}$$

The dispersion limit becomes

$$BL \leq 131.6 \text{ Gb/s} \cdot \text{km}$$

The attenuation at $\lambda = 1.3$ μm, however, is higher than at $\lambda = 1.55$ μm. A value of $\mathbf{a} = 0.35$ dB/km can be estimated from Figure 4.18 resulting in

$$L = 163.9 - 28.6 \log B \tag{6.123}$$

The performance limits derived are shown in Figure 6.22. The undersea system TAT-8 presented in Chapter 1 is indicated in the diagram.

The system analysis presented above is based on a series of simplifying assumptions and cannot replace an analysis based on a more complete theory of fiber optical systems and components.

6.7 Analog Optical Transmission

The main application of optical fiber systems is the transmission of voice and digital data in the telephone network. It is a standard practice to convert analog information to digital form, usually by pulse code modulation (PCM), to facilitate the transmission and switching functions. Binary on-off modulation (OOK) is used almost exclusively in present-day systems.

It is sometimes advantageous to transmit analog signals as speech, radio or television signals in analog form, see e.g. Wagner and Menendez (1989). There is then no need for analog-to-digital (A/D) conversion at the transmitter and digital-to-analog reconstruction (D/A) at the receiver, which often results in simple and inexpensive system solutions. Analog optical fiber transmission has found use primarily in cable television (CATV) systems. The transmission distance of analog optical fiber systems is limited and they are suitable for local networks only.

The analog signal modulates the intensity of the light source, which can be a laser diode (LD) or a light-emitting diode (LED). The requirements in terms of bandwidth, linearity and dynamic range make LD the choice except for simple indoor systems.

The information signal, which can contain many separate channels, is usually modulated on a carrier before being submitted to the laser. Amplitude modulation (AM), which is simple and bandwidth-effective, is one alternative; another is frequency modulation (FM), which requires larger bandwidth, but is less sensitive to noise. In the presentation below we consider AM, but the same basic principles of analysis apply to FM. First single-channel systems are studied and then a multi-user system.

There are many practical design issues in connection with analog multi-channel optical transmission. We present the basic theory that illustrates the system aspects of this type of optical fiber communication.

6.7.1 Single-Channel Transmission

The information signal is usually modulated on a carrier by amplitude, frequency or phase modulation producing a bandpass signal $X(t)$. The signal

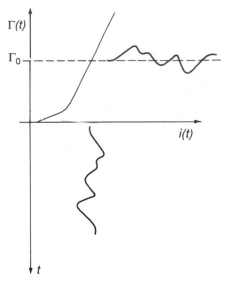

Figure 6.23 Analog amplitude modulation of light. The information carrying signal modulates the drive current of a laser diode.

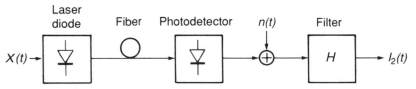

Figure 6.24 Analog optical transmission system. The transmitter is a laser diode and the receiver contains a photodiode followed by a bandpass filter.

$X(t)$ is imposed on the injection current of a laser, and it controls the intensity of the transmitted light, see Figure 6.23.

Consider amplitude modulation and let $\Gamma(t)$ denote the optical intensity at the receiver after transmission through the fiber

$$\Gamma(t) = [1 + AX(t)]\Gamma_0 \qquad (6.133)$$

From Figure 6.23 it is clear that the amplitude of $X(t)$ must be limited to avoid distortion and we assume that it is normalized such that $|X(t)| \leq 1$. The parameter A is the modulation index. The modulation is performed on the linear part of the laser characteristics, and in practice A falls between 0.3 and 0.6.

A block diagram of the system is shown in Figure 6.24. The receiver is an uncomplicated device, consisting of a photodetector followed by an electrical amplifier and a bandpass filter. For simplicity we let the filter be an ideal bandpass filter with frequency function $H(f)$ shown in Figure 6.25a.

The received optical signal illuminates the the photodetector which produces a shot noise process $I(t)$ with intensity proportional to (6.133). The receiver

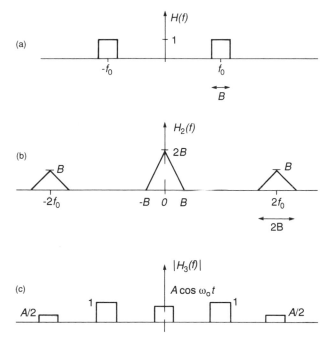

Figure 6.25 Filter transfer functions: (a) the receiver bandpass filter $H(f)$. (b) the filter $H_2(f) = H(f) * H(f)$ determining the variance of the output signal. (c) the filter function $|H_3(f)|$ determining the variance of the intensity noise.

output signal $I_2(t)$ is this process filtered by the filter H. Its basic statistical properties can be calculated from the Campbell Theorem, see Appendix C.

The mean value of the output signal at time t is from (C.31)

$$E\{I_2(t)\} = \eta q \int_{-\infty}^{\infty} \Gamma(s)h(t-s)\, ds \qquad (6.134)$$

where q is the electron charge, η the quantum efficiency of the photo detector and $h(t)$ the impulse response of the receiver filter, i.e. the inverse Fourier transform of $H(f)$. The variance of $I_2(t)$ is from (C.31)

$$\mathrm{Var}\{I_2(t)\} = \eta q^2 \int_{-\infty}^{\infty} \Gamma(s)h^2(t-s)\, ds \qquad (6.135)$$

Note that $E\{I_2(t)\}$ and $\mathrm{Var}\{I_2(t)\}$ are time-varying quantities which are determined by linear filtering operations on $\Gamma(t)$.

If the bandwidth of $X(t)$ is less than or at most equal to the bandwidth B of the receiver filter it will pass the filter unchanged. The 'direct current' term Γ_0, however, will be blocked by the filter and

$$E\{I_2(t)\} = \eta q A \Gamma_0 X(t) \qquad (6.136)$$

The variance (6.135) is equal to the signal $\Gamma(t)$ filtered through a filter with impulse response $h_2(t) = h^2(t)$. The frequency function of $h_2(t)$, shown in Figure

6.25b, is equal to the convolution of $H(f)$ with itself. It is evident that only the term Γ_0 of (6.133) will pass the filter and

$$\text{Var}\{I_2(t)\} = \sigma_o^2 = 2\eta q^2 \Gamma_0 B \tag{6.137}$$

The instantaneous signal-to-noise ratio at the receiver is

$$\rho(t) = \frac{[\text{E}\{I_2(t)\}]^2}{\text{Var}\{I_2(t)\}} = \frac{\eta A^2 \Gamma_0 X^2(t)}{2B} \tag{6.138}$$

An average signal-to-noise ratio is defined as

$$\rho = \text{E}\{\rho(t)\} = \frac{\eta A^2 \Gamma_0 \{X^2(t)\}}{2B} = \frac{\eta A^2 \Gamma_0 P_s}{2B} \tag{6.139}$$

with $P_s = \{X^2(t)\}$ the average power of the transmitted signal $X(t)$.

It is common to characterize analog systems by a carrier-to-noise ratio (CNR). Let the signal be an unmodulated carrier

$$X(t) = \cos 2\pi f_0 t \tag{6.140}$$

The average power of such a signal is

$$P_s = \{X^2(t)\} = 1/2 \tag{6.141}$$

The CNR is obtained by substitution of (6.141) into (6.139)

$$\rho_c = \frac{\eta A^2 \Gamma_0}{4B} \tag{6.142}$$

A laser does not produce light that is absolutely stable in intensity. The basic physical mechanism of a laser is amplification by stimulated emission, which is random in nature. The intensity exhibits small random fluctuations which appear as noise in the demodulated signal. This effect is called intensity noise. The intensity noise is a random component in $\Gamma(t)$ and (6.133) is modified to

$$\Gamma(t) = [1 + AX(t)][\Gamma_0 + \Gamma_i(t)] \tag{6.143}$$

where $\Gamma_i(t)$ is a zero mean stochastic process. Its power spectral density $R_i(f)$ depends on the design of the laser and of its pumping conditions. At frequencies of interest for analog optical transmission the (double-sided) spectral density can be considered to be constant: $R_i(f) = R_i$.

It is common to express laser intensity noise as a logarithmic measure relative to the mean intensity Γ_0, relative intensity noise (RIN)

$$10 \log R_i = 10 \log(2R_i/\Gamma_0^2) \text{ dB/Hz} \tag{6.144}$$

The quantity R_i decreases with an increase in the laser output power, as shown in Figure 6.26, illustrating a typical relationship.

The primary shot noise process $I(t)$ is a doubly stochastic Poisson process in the presence of intensity noise. The result of filtering such a process is studied in Appendix C. The mean and variance of the output process are determined by

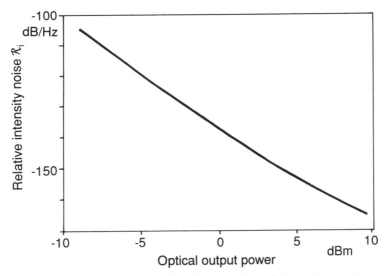

Figure 6.26   Relative intensity noise. The diagram shows $10 \log \mathcal{R}_i$ as a function of the optical output power in dBm for a typical laser diode.

the generalized Campbell Theorem presented in the appendix, equations (C.33) and (C.36).

The average of $I_2(t)$

$$E\{I_2(t)\} = E\left\{\eta q \int_{-\infty}^{\infty} [1 + AX(\tau)][\Gamma_0 + \Gamma_i(\tau)]h(t - \tau)\,d\tau\right\} \tag{6.145}$$

is equal to (6.134) since $\Gamma_i(t)$ is a zero mean process.

The variance is obtained from (C.36)

$$\mathrm{Var}\{I_2(t)\} = E\left\{\eta q^2 \int_{-\infty}^{\infty} \Gamma(\tau)h^2(t - \tau)\,d\tau\right\} + \mathrm{Var}\left\{\eta q \int_{-\infty}^{\infty} \Gamma(\tau)h(t - \tau)\,d\tau\right\}$$

$$\tag{6.146}$$

Owing to the zero mean of $\Gamma_i(t)$ the first term in (6.146) is equal to (6.135). For $X(t) = \cos 2\pi f_0 t$ the second term becomes

$$\mathrm{Var}\left\{\eta q \int_{-\infty}^{\infty} [1 + A\cos 2\pi f_0\tau][\Gamma_0 + \Gamma_i(\tau)]h(t - \tau)\,d\tau\right\}$$

$$= E\left\{\left(\eta q \int_{-\infty}^{\infty} [1 + A\cos 2\pi f_0\tau]\Gamma_i(\tau)h(t - \tau)\,d\tau\right)^2\right\}$$

$$= E\left\{\left(\eta q \int_{-\infty}^{\infty} \Gamma_i(\tau)h_3(t - \tau)\,d\tau\right)^2\right\} \tag{6.147}$$

The frequency function $H_3(f)$ of the 'filter' $h_3(\tau) = 1 + A\cos[2\pi f_0(t - \tau)]h(\tau)$ is shown in Figure 6.25c.

The expression (6.147) is equal to

$$\eta^2 q^2 \int_{-\infty}^{\infty} R_i(f)|H_3(f)|^2 \, df = 2\eta^2 q^2 R_i[1 + A^2(1/2 + \cos 2\pi f_0 t)/2]B \qquad (6.148)$$

The variance due to the intensity noise (6.148) is a time-varying quantity with an average value

$$\sigma_i^2 = 2\eta^2 q^2 R_i(1 + A^2/2)B \qquad (6.149)$$

The factor $(1 + A^2/2)$ is equal to the average square of the factor $[1 + A \cos 2\pi f_0 t]$ in (6.143).

An additional noise source is a dark current i_d in the photodetector which generates a photoelectron intensity $\gamma_d = i_d/q$. Its contribution to the variance (6.137) is

$$\sigma_d^2 = 2qi_d B \qquad (6.150)$$

The resistance in the photodiode and the amplifier following it adds thermal noise to $I_2(t)$. The origin and properties of thermal noise are studied in Chapter 6. The double-sided power spectral density of the equivalent current noise source representing thermal noise is (6.2) multiplied by the amplifier noise factor F:

$$R(f) = \frac{2k_B T_0}{R_L} F \qquad (6.151)$$

with R_L the photodetector load resistance.

The thermal noise is filtered through the filter $h(t)$ and the variance after the filter is from (6.19)

$$\sigma_t^2 = \int_{-\infty}^{\infty} R(f)|H(f)|^2 \, df = 4Fk_B T_0 B/R_L \qquad (6.152)$$

The complete average carrier-to-noise ratio is, after substitution of $\mathcal{P}_s = 1/2$,

$$\rho_c = \frac{(\eta q A\Gamma_0)^2/2}{\sigma_0^2 + \sigma_d^2 + \sigma_i^2 + \sigma_t^2} \qquad (6.153)$$

The average current in the receiver produced by light of constant intensity Γ_0 is $I_0 = \eta q \Gamma_0$. The intensity noise density R_i is related to Γ_0 by the RIN parameter (6.144) and the carrier-to-noise ratio of analog optical bandpass signal transmission can be expressed as

$$\rho_c = \frac{A^2 I_0^2}{4q(I_0 + i_d)B + (2 + A^2)I_0^2 R_i B + 8Fk_B T_0 B/R_L} \qquad (6.154)$$

The requirement on the signal-to-carrier ratio to provide satisfactory quality for standard TV is $\rho_c = 40 - 60$ dB.

Example 6.6   Single-Channel Analog System

An analog optical system with a bandwidth $B = 10$ mHz is operating at a wavelength of $\lambda = 1.3$ μm. The transmitted optical power is $\mathcal{P}_T = 1$ mW and the modulation index $A = 0.45$. The photodiode in the receiver has a quantum efficiency $\eta = 0.7$ and a dark current $i_d = 6$ nA. Its load resistance is $R_L = 600$ ohm. The amplifier in the receiver has a noise figure of $10 \log F = 5$ dB.

Determine the carrier-to-noise ratio at the receiver as a function of the received optical power.

Solution

The average output current from the photodiode is related to the received optical power $\mathcal{P}_0$ by the photodetector responsivity $\mathbf{R}_0$:

$$I_0 = \mathbf{R}_0 \mathcal{P}_0 = \eta q \Gamma_0 = \eta q \frac{\lambda}{hc} \mathcal{P}_0 = 0.84 \, \mathcal{P}_0 \qquad (6.155)$$

The relative intensity noise at $\mathcal{P}_T = 1$ mW is, from Figure 6.26, equal to $10 \log \mathcal{R}_i = -138$ dB. Substitution of the data into (6.154) results in the SCR shown in Figure 6.27. In the diagram ρ_c is a function of the optical fiber attenuation between transmitter and receiver. With a fiber attenuation of 0.4 dB/km the maximal distance for $\rho_c \geq 50$ dB is $L = 13.1/0.4 = 33$ km. If the transmitted power is reduced to $\mathcal{P}_T = 0.5$ mW, the intensity noise according to Figure 6.26 increases to $10 \log \mathcal{R}_i = -127$ dB, and the carrier-to-noise ratio at the receiver is reduced, as shown in Figure 6.27. The SCR now falls below 50 dB. □

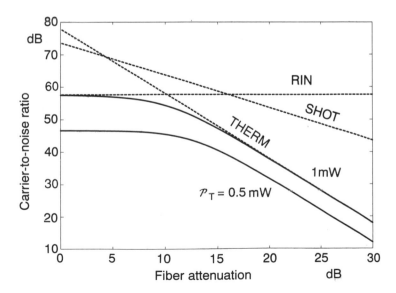

Figure 6.27 Example 6.6. Carrier-to-noise ratio CNR as a function of the fiber attenuation for an analog optical system with bandwidth $B = 10$ MHz. The parameter $\mathcal{P}_T$ is the output optical power of the transmitting laser. The dashed lines indicate the contributions of the different noise sources.

The contributions from the different noise sources are indicated in the diagram. The carrier-to-noise ratio of the thermal noise and the photodetector dark current increase with 2 dB per dB increase in signal power. For the shot noise the increase is 1 dB per dB while the SCR of the intensity noise ρ_c is constant, constituting an upper limit to the transmission quality.

6.7.2 Multi-Channel Transmission

The large bandwidth of an optical fiber makes it possible to transfer wide-band signals containing a large number of multiplexed channels. One application is distribution of broad-band services to subscribers, Olshansky *et al.* (1989). An example is cable television (CATV) where a number of TV channels are multiplexed and transmitted over a single optical fiber. Another application is subscriber loop networks, Winston (1989).

In subcarrier multiplexing (SCM) the basic channels are modulated on separate electrical carriers and combined into a wide-band frequency division multiplex (FDM) signal which modulates the intensity of the light source. Figure 6.28 shows an SCM system with vestigial sideband modulation.

An additional distortion appearing in multi-channel systems is the interchannel interference caused by the inevitable deviation from exact linearity of the input-output characteristics of the transmitting light source. Intermodulation distortion is a well-known effect in coaxial cable telephone systems and is analysed, for example, in the handbook of Bell Telephone Laboratories (1971).

The relation between the laser drive current $i(t)$ and the intensity $\gamma(t)$ of the transmitted light is in practice not exactly linear, but it contains a small amount of nonlinearity.

$$\gamma(t) = a_1 i(t) + a_2 i^2(t) + a_3 i^3(t) \tag{6.156}$$

To facilitate the analysis we model the compound input signal $i(t)$ as a bandpass Gaussian stochastic process. The effect of the nonlinarities can then be determined by straightforward calculations as described by Davenport and Root (1958). Nonlinearities will also affect single-channel systems by causing signal distortion. However, it less harmful than in multi-channel systems where intermodulation has the character of additive noise.

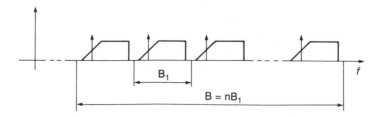

Figure 6.28   Vestigial sideband multichannel subcarrier modulation.

Let the system contain n channels, each of bandwidth B_1 and carrying the same load. The power spectrum $R_s(f)$ of $i(t)$ can then, to simplify the calculations, be considered to be constant over the system bandwidth B, see Figure 6.29a

$$R_s(f) = R_s; \quad t_0 - B/2 < |f| < t_0 + B/2 \tag{6.157}$$

The power spectral density of $\gamma(t)$ is determined by its autocorrelation function

$$r_\gamma(i_1, i_2) = \{[a_1 i_1 + a_2 i_1^2 + a_3 i_1^3][a_1 i_2 + a_2 i_2^2 + a_3 i_2^3]\} \tag{6.158}$$

where the notation $i_1 = i(t_1)$ and $i_2 = i(t_2)$ is introduced.

For a Gaussian stochastic process the two-dimensional moment-generating function (MGF) of i_1 and i_2 is

$$\Psi(s_1, s_2) = \{\exp(s_1 i_1 + s_2 i_2)\} = \exp(\sigma_1^2 s_1^2/2 + r_s(t_1, t_2) + \sigma_2^2 s_2^2/2) \tag{6.159}$$

where σ_1^2 and σ_2^2 are the variances of i_1 and i_2, respectively, and $r_s(t_1, t_2)$ is the autocorrelation function of $i(t)$

$$r_s(t_1, t_2) = E\{i(t_1)i(t_2)\} \tag{6.160}$$

Evaluation of (6.158) requires calculation of moments $\{i_1^\nu i_2^\mu\}$ which are related to the MGF (6.159) through

$$\{i_1^\nu i_2^\mu\} = \left. \frac{\partial^{\nu+\mu} \Psi(s_1, s_2)}{\partial s_1^\nu \partial s_2^\mu} \right|_{s_1=s_2=0} \tag{6.161}$$

Moments with $\nu + \mu$ an odd integer are equal to zero due to the symmetry of the Gaussian distribution. The even moments are obtained from (6.161) and (6.159) by straightforward calculations

$$\left. \begin{array}{l} E\{i_1 i_2^3\} = 3\sigma_2^2 r_s(t_1, t_2) \\ E\{i_1^3 i_2\} = 3\sigma_1^2 r_s(t_1, t_2) \\ E\{i_1^2 i_2^2\} = \sigma_1^2 \sigma_2^2 + 2r_s^2(t_1, t_2) \\ E\{i_1^3 i_2^3\} = 9\sigma_1^2 \sigma_2^2 r_s(t_1, t_2) + 6r_s^3(t_1, t_2) \end{array} \right\} \tag{6.162}$$

The input signal $i(t)$ is stationary, which makes the variances equal $\sigma_1^2 = \sigma_2^2 = \sigma^2$, and $r_s(t_1, t_2)$ a function of $\tau = t_2 - t_1$.

Substitution of (6.162) into the expanded form of (6.158) yields

$$r_\gamma(\tau) = a_2^2 \sigma^4 + [a_1^2 + 6a_1 a_3 \sigma^2 + 9a_3^2 \sigma^4] r_s(\tau) + 2a_2^2 r_s^2(\tau) + 6a_3^2 r_s^3(\tau) \tag{6.163}$$

The power spectral density of $\gamma(t)$ is the Fourier transform of $r_\gamma(\tau)$. For a transfer function with small nonlinearities, $a_2/a_1 \ll 1$ and $a_3/a_1 \ll 1$

$$R_\gamma(f) = a_2^2 \sigma^4 \delta(f) + a_1^2 R_s(f) + 2a_2^2 R_s^{(2)}(f) + 6a_3^2 R_s^{(3)}(f) \tag{6.164}$$

where $R_s^{(2)}(f)$ is the convolution of $R_s(f)$ with itself and $R_s^{(3)}(f)$ is the convolution of $R_s(f)$ with $R_s^{(2)}(f)$, see Figure 6.29.

To determine the shape of the spectral density $R_\gamma(f)$ it is convenient to express $R_s(f)$ in terms of its equivalent lowpass spectrum $\tilde{R}(f)$

$$R_s(f) = R_s[\tilde{R}(f - f_0) + \tilde{R}^*(-f - f_0)] \tag{6.165}$$

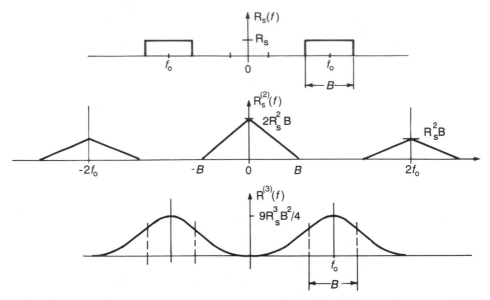

Figure 6.29 Frequency functions determining the intermodulation distortion: (a) the transmitted signal spectrum $R_s(f)$. (b) the convolution $R_s^{(2)}(f) = R_s(f) * R_s(f)$. (c) the second-order convolution $R_s^{(3)}(f) = R_s(f) * R_s(f) * R_s(f)$.

with $\tilde{R}(f)$ a normalized rectangular baseband function

$$\tilde{R}(f) = \begin{cases} 1; & |f| < B/2 \\ 0; & \text{otherwise} \end{cases} \tag{6.166}$$

The first-order convolution generates a triangular spectrum of bandwidth B

$$\tilde{R}^{(2)}(f) = \int_{-\infty}^{\infty} \tilde{R}(f-s)\tilde{R}(s)\, ds = B(1 - |f|/B); \quad |f| \le B \tag{6.167}$$

The second-order convolution $\tilde{R}^{(3)}(f) = \tilde{R}^{(2)}(f) * \tilde{R}(f)$ is

$$\tilde{R}^{(3)}(f) = \begin{cases} B^2[3/4 - (f/B)^2]; & 0 \le |f| \le B/2 \\ B^2[3/2 - |f|/B]^2/2; & B/2 \le |f| \le 3B/2 \end{cases} \tag{6.168}$$

The spectral densities in (6.164) are related to (6.167) and (6.168) by

$$R_s^{(2)}(f) = R_s^2[2\tilde{R}^{(2)}(f) + \tilde{R}^{(2)}(f - 2f_0) + \tilde{R}^{(2)}(f + 2f_0)] \tag{6.169}$$

and neglecting parts outside the frequency bands of interest

$$R_s^{(3)}(f) = 3R_s^3[\tilde{R}^{(3)}(f - f_0) + \tilde{R}^{(3)}(f + f_0)] \tag{6.170}$$

The spectral density $R_s^{(2)}(f)$ represents second-order intermodulation products. It is evident from Figure 6.29 that they are not harmful if $f_0 > 3B/2$, since they then fall outside the primary signal bandwidth.

The third-order products, however, fall into the signal band and create intermodulation noise. The noise power is largest in a channel located at the center of the system frequency band:

$$\sigma_3^2(0) = 12a_3^2 \int_{f_0-B_1/2}^{f_0+B_1/2} R_s^{(3)}(f) \, df \approx 36a_3^2 R_s^3 B_1 \tilde{R}^{(3)}(0) = 27a_3^2 R_s^3 B_1 B^2 \qquad (6.171)$$

At the edge of the band intermodulation noise takes its smallest value

$$\sigma_3^2(B/2) = 36a_3^2 R_s^3 B_1 \tilde{R}^{(3)}(B/2) = 18a_3^2 R_s^3 B_1 B^2 \qquad (6.172)$$

which is 2/3 of (6.171). The ratio of signal to intermodulation noise power for a central channel is

$$\rho_3 = \frac{2a_1^2 R_s B_1}{\sigma_3^2(0)} = \frac{8}{27(a_3/a_1)^2 \mathcal{P}_T^2} \qquad (6.173)$$

where $\mathcal{P}_T = 2R_s B$ is the total transmitted optical power.

The ratio a_3/a_1 can be determined as follows. Let the input signal $i(t) = A \cos(2\pi f_0/3)$ be a sinusoid of frequency $f_0/3$. The output contains the input frequency and a third order harmonic

$$\gamma(t) = \left(a_1 A + \frac{3a_3 A^3}{4} \right) \cos(2\pi f_0/3) + \frac{a_3 A^3}{4} \cos(2\pi f_0) \qquad (6.174)$$

The ratio in power between the harmonic and the fundamental tone is the third-order intermodulation coefficient

$$K_3 = \frac{(a_3 A^3/4)^2/2}{[a_1 A + (3a_3 A^3)/4]^2/2} \approx \frac{a_3^2 A^4}{16a_1^2} - \frac{a_3^2 \mathcal{P}_3^2}{4a_1^2} \qquad (6.175)$$

where $\mathcal{P}_3 - A^2/2$ is the power of the input signal.

The coefficient K_3 depends on the signal level employed in its measurement. An increase in $\mathcal{P}_3$ of 1 dB makes the measured value of K_3 greater by 2 dB.

The measurement of the nonlinearity can also be made with two input frequencies f_1 and f_2 in the vicinity of f_0. The nonlinearity parameter K_3 is related to the amplitude of and $2f_1 - f_2$ by elementary trigonometric identities. This procedure has the advantage that all frequencies in the measurement operation fall within the transmission band.

Substitution of a_3/a_1 from (6.175) into (6.173) yields

$$\rho_3 = \frac{2\mathcal{P}_3^2}{27K_3 \mathcal{P}_T^2} \qquad (6.176)$$

which shows that the intermodulation SNR decreases as the square of the power ratio $\mathcal{P}_T/\mathcal{P}_3$.

Example 6.7   Multi-Channel Analog System

Consider an analog optical system for TV transmission with $n = 100$ channels, each of bandwidth $B_1 = 6$ mHz. The optical wavelength $\lambda = 1.3$ μm and the modulation index $A = 0.45$. The output power of the transmitting laser is

$P_T = 3.2$ mW $= 5$ dBm. The laser has a third-order intermodulation coefficient $10 \log(1/K_3) = 85$ dB measured with $P_3 = 0.5$ mW $= -3$ dBm. The receiver has the same basic data as that of Example 6.6 with a photodiode having a quantum efficiency $\eta = 0.7$ and a dark current $i_d = 6$ nA. The amplifier in the receiver has a noise figure of $10 \log F = 8$ dB and an input impedance $R_{in} = 600$ ohm.

Determine the intermodulation distortion and estimate the complete signal-to-noise ratio of the individual TV channels.

Solution

The intermodulation SNR (6.176) expressed in dB is

$$10 \log \rho_3 = 20 \log P_3 - 10 \log K_3 - 20 \log P_T - 10 \log (27/2) \qquad (6.177)$$

Substitution of $P_T = 5.0$ dBm and $P_3 = -3.0$ dBm yields

$$10 \log \rho_3 = 2 \times (-3.0) + 85 - 2 \times 5.0 - 11.3 = 57.7 \text{ dB} \qquad (6.178)$$

For a channel at the edge of the transmission band ρ_3 is $10 \log 3/2 = 1.8$ dB larger.

The intermodulation distortion is generated at the transmitter and depends only on the transmitted power level P_T in contrast to thermal and shot noise which are functions of the signal level at the receiver.

At $P_T = 5.0$ dBm the relative intensity noise of Figure 6.26 is $R_i = -152$ dB/Hz. Substitution into (6.154) with system bandwidth

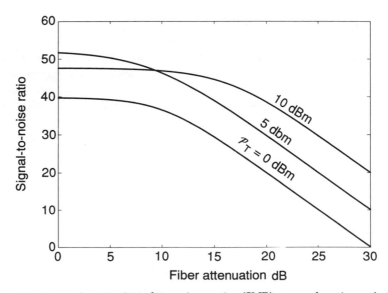

Figure 6.30 Example 6.7. Signal-to-noise ratio (SNR) as a function of the fiber attenuation for an analog optical system with 100 channels, each of bandwidth $B_1 = 6$ MHz. The parameter P_T is the output optical power of the transmitting laser.

$B = 600$ MHz gives a carrier-to-noise ratio ρ_c at zero fiber attenuation, ie. $\mathcal{P}_0 = \mathcal{P}_T$

$$10 \log \rho_c = 53.0 \text{ dB}$$

An estimate of the total signal-to-noise ratio is

$$\rho_{tot} = 1/(1/\rho_c + 1/\rho_3) \tag{6.179}$$

which for zero fiber attenuation is

$$\rho_{tot} = 51.7 \text{ dB}$$

The total distortion ρ_{tot} is shown as a function of the transmission attenuation in Figure 6.30. Also shown in the diagram is ρ_{tot} for an output powers $\mathcal{P}_T = 10$ dBm and $\mathcal{P}_T = 0$ dBm . Increased optical power will not always improve system performance since intermodulation distortion puts an upper limit on the SNR.   □

Subcarrier modulation is also used for digital signals. It requires a larger bandwidth than analog AM or FM but is less sensitive to internal and external noise.

7
Optical Amplifiers

7.1 Introduction

Optical receivers have two functions: they can be part of the terminal equipment of a communication link, or they can act as repeaters to extend the distance between terminals, as in the long-distance submarine optical fiber systems discussed in Chapter 1.

A terminal receiver detects the incoming optical signal and converts it to an electrical signal which is distributed to its destination, often after switching and further processing, e.g. digital-to-analog conversion.

The common way of regenerating an optical signal is to detect it by a photodetector receiver, as described in Chapter 5, amplify the resulting electrical signal and let it control an optical transmitter, producing a reconstructed replica of the original transmitted data signal; see Section 1.3. An alternative, which has been developed more recently, is to amplify the signal optically without any conversion to an electrical signal. An optical amplifier has a more straightforward implementation than an electro-optical repeater and also the advantage that it is transparent. It can amplify optical data signals of different information rate or even different modulation methods and also deal simultaneously with several signals at different optical wavelengths, i.e. multiplex signals. A repeater has to be designed for a specific modulation format and the data rate is usually fixed by the synchronization circuitry.

There are several types of optical amplifier. For semiconductor laser amplifiers see, for instance, O'Mahony (1988) and for optical fiber amplifiers Bjarklev (1993). The theory presented here applies to both types of amplifier.

Amplification of an optical signal is based on the principle of stimulated emission described in any standard text on photonics, such as Saleh and Teich (1991). The incoming photons interact with an atomic medium, causing the production of secondary photons, which can be done in a laser cavity or in a glass fiber doped with a rare earth metal such as erbium. The atoms are transposed to an upper energy level by an auxiliary physical process, called pumping. In a laser amplifier the pumping is usually done electrically and in an erbium-doped fiber amplifier (EDFA) optically by an external light source. The

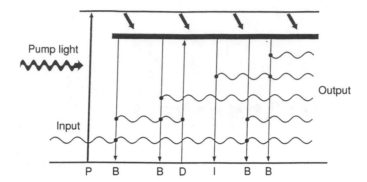

Figure 7.1 Schematic diagram of optical amplification. Atoms in the amplifying medium are shifted into an excited state by an external field (P = pumping). The incoming weak optical signal stimulates atoms to return to the ground state thereby emitting photons (B = birth). Photons can be emitted spontaneously (I = immigration), which acts as an unwanted disturbance. They may also be absorbed (D = death) in the process.

effect of the pumping process is to establish an inverted population of energy states, compared with thermal equilibrium.

The incoming photons cause atoms at the upper energy level to undergo transitions to a lower level and thereby emit new photons. These will, in turn, stimulate the emission of additional photons, which results in a large number of output photons, i.e. amplification. Figure 7.1 illustrates the amplification process and Figure 7.2 shows the appearance of an erbium-doped optical amplifier.

Photons occur randomly according to the laws of quantum mechanics. The strength of an optical signal fluctuates stochastically, acting as an inherent noise source causing transmission impairments.

There are essentially two mathematical statistical models of optical amplification. One characterizes the generation of amplified photons as a birth process, which is presented in Sections 7.2 and 7.3. In the other, studied in Section 7.4, the amplifier is modeled as a frequency-independent constant gain device, together with an additive noise term representing the imperfections of the amplifying process.

To present these models and analyse their properties we study the performance of an optical amplifier as a preamplifier in an optical terminal receiver. Similar applications, which can be analysed with the same theory, are when the amplifier is used as a booster to enhance the transmitted light fed to the optical fiber and when it is a line amplifier compensating for fiber loss.

7.2   Counting Processes

The fundamental principle for optical detection is photon counting. It is therefore natural to apply the theory of stochastic counting processes to analyze optical amplifiers.

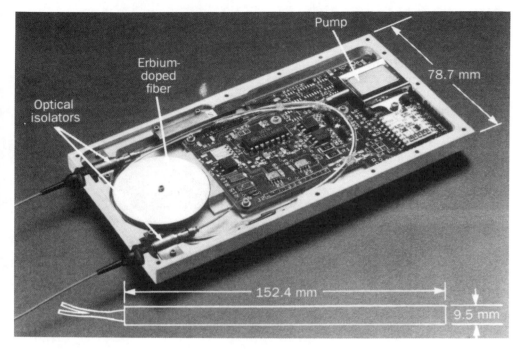

Figure 7.2 An optical amplifier module (EDFA). The amplifying medium is an optically pumped erbium-doped fiber. Copyright © 1992 AT&T. All rights reserved. Reproduced by permission.

A counting process is a continuous-time, integer-valued stochastic process $N(t)$ representing the number of events that have occurred by some random mechanism in a time interval $[0, t[$, see for instance Parzen (1962). For the present application it is convenient to let the interval be semi-closed, closed at 0 and open at t. The value of $N(t)$ can change one unit up or down at a time, and a counting process can be considered as a Markov process where transitions can occur to neighbouring states only. The process can represent the evolution of a population of individuals or, as in our case of particles, photons. The occurrence of a new member in the population is called a 'birth'. In an optical amplifier it represents stimulated emission of photons. The elimination of photons by absorption is, with the same terminology, called a 'death' of a member. A fundamental assumption is that the probability of occurrence of births or deaths is proportional to the present size of the population.

New members can also be introduced by a statistical mechanism that is independent of the size of the population. This is called 'immigration', and in an optical amplifier it represents spontaneous emission.

In this section we present the basic theory of counting processes with birth, death, and immigration which we call a BDI process. As a starting point the simplest such process, one with immigration only, is studied. It turns out that the resulting process is the Poisson process dealt with extensively in Chapter 5.

Consider a spontaneous emission (immigration) process starting at time $t = 0$, representing the occurrence of photons or photoelectrons. It is assumed that the events occurring in an arbitrary time interval $[t, t + h[$ are independent of what has happened in the interval $[0,t[$. Let the process have constant intensity γ, which means that the probability that an immigration, i.e. the spontaneous emission of a photon, occurs during the time interval $[t, t + h[$ is equal to $\gamma h + o(h)$. It is also assumed that the probability that more than one event occurs is equal to $o(h)$, i.e. it goes to zero more rapidly than h.

Let $P_n(t) = \Pr\{N = n\}$ denote the probability that exactly n events occur during the interval $[0, t[$. The probability that a total number of n events are observed in the interval $[0, t + h[$ is

$$P_n(t + h) = P_n(t)(1 - \gamma h) + P_{n-1}(t)\gamma h + o(h) \tag{7.1}$$

Rearranging the terms and dividing by h yields

$$\frac{P_n(t + h) - P_n(t)}{h} = -\gamma P_n(t) + \gamma P_{n-1}(t) + \frac{o(h)}{h} \tag{7.2}$$

Letting $h \to 0$, a set of differential equations are obtained

$$\frac{dP_n(t)}{dt} = -\gamma P_n(t) + \gamma P_{n-1}(t); \quad n \geq 1 \tag{7.3}$$

and for $n = 0$

$$\frac{dP_0(t)}{dt} = -\gamma P_0(t); \quad n = 0 \tag{7.4}$$

The equation (7.4) can be solved directly and (7.3) recursively with the solution of (7.4) as starting point. An alternative, which produces the distribution for N in one step, is to use a transform method. It is convenient to use the z-transform which in mathematical statistics is called the (probability) generating function

$$\Phi(z, t) = E\{z^N\} = \sum_{n=0}^{\infty} z^n P_n(t) \tag{7.5}$$

Comparison with the definition (5.58) of the moment-generating function shows that it is obtained from the generating function by the substitution $z = e^s$

$$\Psi(s) = \Phi(e^s) \tag{7.6}$$

Multiplying both sides of (7.1) by z^n and summing over n from $n = 1$ to infinity yields

$$\sum_{n=1}^{\infty} z^n P_n(t + h) = (1 - \gamma h)\sum_{n=1}^{\infty} z^n P_n(t) + \gamma h \sum_{n=1}^{\infty} z^n P_{n-1}(t) + o(h)\sum_{n=1}^{\infty} z^n \tag{7.7}$$

from the definition (7.5) it follows that this can be expressed as

$$\Phi(z, t + h) - P_0(t + h) = (1 - \gamma h)[\Phi(z, t) - P_0(t)] + \gamma h z \Phi(z, t) + o(h)\frac{z}{1 - z} \tag{7.8}$$

Rearranging the terms, dividing by h, and letting h approach zero gives

$$\frac{d\Phi(z,t)}{dt} - \frac{dP_0(t)}{dt} = \gamma(z-1)\Phi(z,t) + \gamma P_0(t) \tag{7.9}$$

Using (7.4) the differential equation for $\Phi(z,t)$ becomes

$$\frac{d\Phi(z,t)}{dt} = \gamma(z-1)\Phi(z,t) \tag{7.10}$$

This can be solved directly:

$$\Phi_0(z,t) = \exp[\gamma t(z-1)] \tag{7.11}$$

under the assumption that the process starts at $t=0$ with $n=0$, i.e. the distribution at $t=0$ is $P_n(0) = \delta_{n0}$ and consequently $\Phi(z,0) = 1$.

The mean number of events in the interval $[0,t[$ is $m = \gamma t$ and (7.11) with substitution of $z = e^s$ is the moment-generating function (5.60) for the Poisson distribution.

The result is readily generalized to a process with time varying intensity $\gamma(t)$, in which case γt in (7.11) is replaced by, cf. (5.4)

$$m = \int_0^t \gamma(s)\,ds$$

A pure immigration process cannot produce optical amplification since the events occur spontaneously and the generation of new particles does not depend on any input signal to the system.

To obtain amplification stimulated emission is needed. We illustrate the amplification mechanism of a counting process by considering a pure birth process with linear birth rate, which means that the probability of the occurrence of a new member of the population is directly proportional to the size of the present population. Let μ denote the birth rate, i.e. the probability per unit time of a birth from a single member of the population. The probability that a total number of n events are observed in the interval $[0, t+h[$ is

$$P_n(t+h) = P_n(t)(1 - n\mu h) + P_{n-1}(t)(n-1)\mu h + o(h) \tag{7.12}$$

Multiplying both sides of (7.1) by z^n and summing over n from $n=1$ to infinity yields

$$\sum_{n=1}^{\infty} z^n P_n(t+h) = \sum_{n=1}^{\infty} z^n P_n(t) - \mu h \sum_{n=1}^{\infty} nz^n P_n(t) +$$
$$\mu h \sum_{n=1}^{\infty}(n-1)z^n P_{n-1}(t) + o(h) \sum_{n=1}^{\infty} z^n \tag{7.13}$$

The first two sums can be expressed in terms of $\Phi(z, t+h)$ and $\Phi(z,t)$ and the succeeding two sums can be evaluated with the aid of the relation

$$\frac{\partial \Phi(z,t)}{\partial z} = \sum_{n=1}^{\infty} nz^{n-1} P_n(t) = z^{-1} \sum_{n=1}^{\infty} nz^n P_n(t) \tag{7.14}$$

Rearranging the terms, dividing by h and letting h approach zero gives

$$\frac{\partial \Phi(z,t)}{\partial t} - \frac{\partial P_0(t)}{\partial t} = \mu z(z-1)\frac{\partial \Phi(z,t)}{\partial z} \tag{7.15}$$

From the original equation (7.12) follows that for $n=0$

$$P_0(t+h) = P_0'(t)$$

and the derivative $\partial P_0(t)/\partial t$ in (eq 60.7.8.3) is equal to zero. The generating function of a pure birth process is thus the solution to the partial differential equation

$$\frac{\partial \Phi(z,t)}{\partial t} = \mu z(z-1)\frac{\partial \Phi(z,t)}{\partial z} \tag{7.16}$$

The birth process needs a nonzero starting condition since, if $n=0$ at $t=0$ the generating of any new members is zero. Let the initial condition be a single photon at $t=0$, i.e. $P_n(0) = \delta_{n1}$ and $\Phi(z,0) = z$. The solution of (7.16) is then

$$\Phi_1(z,t) = \frac{z}{G(t) - [G(t)-1]z} \tag{7.17}$$

with $G(t) = e^{\mu t}$. It is the generating function of the geometric distribution in Feller (1957) with his parameter $p = 1/G(t)$. In physics it is referred to as the Bose-Einstein distribution, see e.g. Saleh and Teich (1991).

The inverse of (7.17) is obtained from a table of elementary z-transforms, and the probability that the size of the population has grown to a total of n members at time t is

$$P_n(t) = \begin{cases} [1/G(t)][1 - 1/G(t)]^{n-1}; & n \geq 1 \\ 0; & n = 0 \end{cases} \tag{7.18}$$

The mean of the size of the population at time t is obtained from the derivative of $\Phi_1(z,t)$ with respect to z at $z=1$, see (7.14)

$$E\{N(t)\} = \left.\frac{\partial \Phi_1(z,t)}{\partial z}\right|_{z=1} = G(t) \tag{7.19}$$

The second derivative of $\Phi_1(z,t)$ with respect to z is

$$\frac{\partial^2 \Phi_1(z,t)}{\partial z^2} = \sum_{n=2}^{\infty} n(n-1)z^{n-2}P_n(t) \tag{7.20}$$

From (7.20) it follows that

$$\mathrm{Var}\{N(t)\} = \Phi_1''(1,t) + \Phi_1'(1,t) - [\Phi_1'(1,t)]^2 = G(t)[G(t)-1] \tag{7.21}$$

The population of a birth process grows exponentially in time. This is characteristic for optical amplification in an atomic medium with an inverted population. In a laser amplifier of the travelling wave type and in a doped fiber amplifier, the optical signal travels through the amplifying medium in a time T_L equal to the length of the medium divided by the group velocity of the signal. In what follows, we fix the time parameter to $t = T_L$ and drop the variable t in the notation.

A fundamental parameter of an optical amplifier is the (average) gain

$$G = G(T_L) = e^{\mu T_L} \tag{7.22}$$

The gain of an optical amplifier varies randomly and (7.21) shows that the variance grows almost quadratic with the gain.

In the derivation of $\Phi_1(z)$ it was assumed that the initial condition was a single photon. In practice the input signal is an optical pulse containing a random number of photons. If the time duration of the input optical pulse is short, the output distribution is obtained by solving (7.16) with a specified initial distribution $\{P_n(0)\}$.

First consider the case when the initial condition is exactly k photons. For a birth process with linear birth rate, the output is the same as the sum from k independent processes each starting with $k = 1$, which means that the generating function is equal to $\Phi_1^k(z)$. When the input is a random number of photons the generating function for the number of photons at the output is obtained by forming the average of $\Phi_1^k(z)$ over the variable k

$$\Phi(z) = \sum_{k=0}^{\infty} \Phi_1^k(z) P_k(0) = \Phi_S[\Phi_1(z)] \tag{7.23}$$

with $\Phi_S[z]$ the generating function of the distribution at the start of the process.

When the input light is in a coherent state, $\{P_n(0)\}$ is a Poisson distribution and $\Phi_S(z)$ is given by (7.11). Substitution $\Phi_1(z)$ from (7.17) gives

$$\Phi(z) = \exp[m(\Phi_1(z) - 1)] = \exp\left(\frac{mG(z-1)}{G - (G-1)z}\right) \tag{7.24}$$

with m equal to the average number of photons in the input distribution.

There appears to be no closed-form expression for the output probability distribution $P_n = P_n(T_L)$ of a pure birth process. An example of the appearance of P_n is shown in Figure 7.3a. It is obtained by numerical inversion of (7.24) by saddlepoint integration. A comparison with the Poisson distribution Figure 7.3b with the same mean shows that the amplification process causes an increase in the spread of the number of photons.

The mean number of photon at the amplifier output is

$$E\{N\} = \Phi'(1) = mG \tag{7.25}$$

and the variance is in the same way as (7.21)

$$\text{Var}\{N\} = \Phi''(1) + \Phi'(1) - [\Phi'(1)]^2 = mG(2G - 1) \tag{7.26}$$

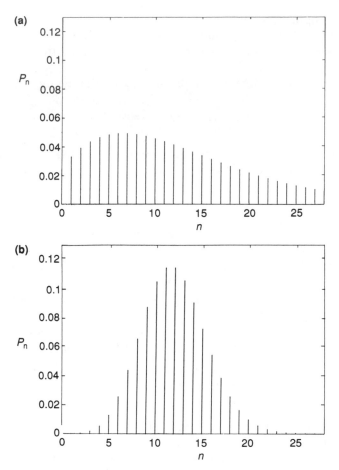

Figure 7.3 Probability P_n for the number of photons. (a) The photon distribution for the output of a pure birth process with a Poisson input distribution. (b) The photon distribution for a Poisson process. The distributions (a) and (b) have the same mean $m = 12$.

7.3 BDI Model

To obtain a complete model of an optical amplifier, based on counting process statistics, stimulated emission and absorption together with spontaneous emission have to be included. These mechanisms are represented as birth, death and immigration, respectively, which motivates the notation BDI model.

The application of BDI processes to optical amplification was suggested by Shimoda *et al.* (1957) and has, more recently, been further developed by Li and Teich (1991, 1993) and others.

The basic equation analogous to (7.11) and (7.12) is now

$$P_n(t + h) = P_n(t)[1 - n(\mu + \nu)h - \gamma h] +$$
$$P_{n-1}(t)[(n - 1)\mu h + \gamma h] + P_{n+1}(t)(n + 1)\nu h + o(h) \tag{7.27}$$

where μ and ν denote the birth and death rate, respectively, i.e. the probability per photon per unit time of a stimulated emission of a new photon or absorption of an old one. The parameter γ is the probability per unit time of a spontaneous emission which does not depend on the size of the photon population.

A partial differential equation for the generating function for the counting process that represents the number of photons in the amplifying medium at time t, is derived from (7.27) in the same way as (7.16)

$$\frac{\partial \Phi(z, t)}{\partial t} = (\mu z - \nu)(z - 1)\frac{\partial \Phi(z, t)}{\partial z} + \gamma(z - 1)\Phi(z, t) \qquad (7.28)$$

The solution for $t = T_L$ with a single photon at $t = 0$ as initial condition is

$$\Phi_1(z) = [1 + (G - K)(z - 1)][1 - K(z - 1)]^{-(1+\gamma/\mu)} \qquad (7.29)$$

with the gain G

$$G = e^{(\mu-\nu)T_L} \qquad (7.30)$$

To obtain amplification the stimulated emission must be greater than the absorption, i.e. $\mu > \nu$. The parameter

$$K = n_{sp}(G - 1) \qquad (7.31)$$

where

$$n_{sp} = \frac{\mu}{\mu - \nu} \qquad (7.32)$$

is the population inversion factor or spontaneous emission parameter which among other things depends on the geometry and pumping mechanism of the amplifying medium. It is greater than, or in ideal cases equal to, one.

The generating function (7.29) can be considered as representing the sum of two independent stochastic variables. It is the product of

$$\Phi_{BD}(z) = \frac{1 + (G - K)(z - 1)}{1 - K(z - 1)} \qquad (7.33)$$

which is the generating function of a birth and death process without immigration i.e. $\gamma = 0$, and

$$\Phi_I(z) = [1 - K(z - 1)]^{-\gamma/\mu} \qquad (7.34)$$

For an initial distribution characterized by $\Phi_S(z)$ the generating function for the output becomes, cf. (7.23)

$$\Phi(z) = \Phi_S[\Phi_{BD}(z)]\Phi_I(z) \qquad (7.35)$$

For a Poisson-distributed initial input, $\Phi_S(z)$ is equal to (7.11) and

$$\Phi(z) = \frac{1}{[1 - K(z - 1)]^M} \exp\left[\frac{mG(z - 1)}{1 - K(z - 1)}\right] \qquad (7.36)$$

where m is the average number of photons in the input signal and the parameter

$$M = \frac{\gamma}{\mu} \tag{7.37}$$

represents the number of modes excited by the spontaneous emission. In practice M is reduced by an optical filter at the amplifier output.

The probability of the number N of photons in an optical pulse after the passage of an optical amplifier is the inverse z-transform of (7.36).

$$P_n = \frac{K^n}{(1+K)^{n+M}} \exp\left(-\frac{mG}{1+K}\right) L_n^{(M-1)}\left(-\frac{mG}{K[1+K]}\right) \tag{7.38}$$

where $L_n^{(M-1)}(\ )$ is a generalized Laguerre polynomial defined, for example, in Abramowitz and Stegun (1968):

$$L_n^{(\mu)}[x] = \sum_{k=0}^n (-1)^k \binom{n+\mu}{n-k} \frac{1}{k!} x^k ; \quad \mu \geq 0 \tag{7.39}$$

The distribution (7.38) is called a noncentral negative-binomial (NNB) distribution by Li and Teich (1991), or Laguerre distribution by Gagliardi and Karp (1976).

Laguerre polynomials obey recurrence relations, see for instance Abramowitz and Stegun (1968), which makes evaluating them much more simple than by the series (7.39).

The characteristics of P_n are illustrated in Figure 7.4 at low and medium gain G for three different values of the parameter M.

For an on-off system where binary 'zero' is represented by the absence of a received pulse, $m_0 = 0$, the generating function for the decision variable is (7.36) with $m = 0$

$$\Phi(z) = [1 - K(z-1)]^{-M} \tag{7.40}$$

The probability distribution is

$$P_n = \binom{M+n-1}{n} \frac{K^n}{(1+K)^{n+M}} \tag{7.41}$$

which is the negative binomial (NB) distribution in Feller (1957) with his parameter $p = 1/(1+K)$.

Let $P_0(\alpha)$ denote the probability that N falls above an integer valued decision threshold α when a 'zero' was transmitted, and $P_1(\alpha)$ the probability that it falls below or on the threshold when a 'one' was transmitted.

With equal a priori probabilities for 'zero' and 'one', the error probability is, compare with (5.24)

$$P_e = \frac{1}{2} P_0(\alpha) + \frac{1}{2} P_1(\alpha) = \frac{1}{2} \sum_{n=\alpha+1}^{\infty} P_{n0} + \frac{1}{2} \sum_{n=0}^{\alpha} P_{n1} \tag{7.42}$$

where P_{n0} and P_{n1} are (7.38) with $m = m_0$ and $m = m_1$, respectively, and where

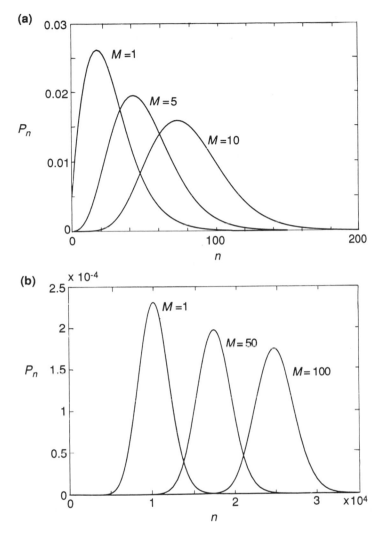

Figure 7.4 Probability P_n for the number of output photons from an optical amplifier with $n_{sp} = 1.5$ at various values of the parameter M: (a) Low amplifier gain $G = 5$. The average number of input photons $m = 4$. (b) Amplifier gain $G = 100$. The average number of input photons $m = 100$.

m_0 and m_1 are the average number of photons in the received optical pulse at the optical amplifier input for 'zero' and 'one', respectively.

The shape of the statistical distribution functions $P_0(\alpha)$ and $P_1(\alpha)$ is shown in Figure 7.5 for a system with gain $G = 316$ (25 dB) and spontaneous emission parameter $n_{sp} = 1.4$. The parameter $M = 30$ and the input optical signals have strength $m_0 = 150$ and $m_1 = 440$. The error probability (7.42) achieves its lowest value for a threshold close to the crossing of the curves. The figure shows that a decision threshold $\alpha/G = 318.1$ yields the minimum error probability $P_e = 2.3 \times 10^{-7}$.

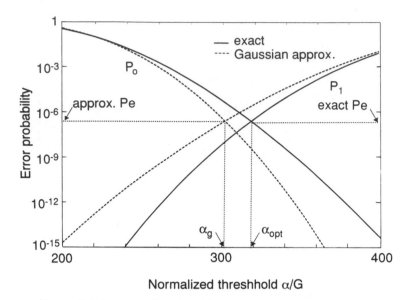

Figure 7.5 The probabilities $P_0(\alpha)$ and $P_1(\alpha)$ for the number of output photons from an optical amplifier with $G = 25$ dB, $n_{sp} = 1.4$ and $M = 30$. The input signal levels are $m_0 = 150$ and $m_1 = 440$. The dashed curves represent Gaussian distributions with the same mean and variance as the correct distributions.

Figure 7.6 illustrates how the bit error probability varies as a function of the average number of photons per bit for various values of m_0 and M. It will be shown below that the performance of an optical amplifier receiver approaches an asymptotic value at high gain G, and the results in Figure 7.6 refer to amplifiers with gain larger than about 10 dB.

The results presented in in Figure 7.5 and Figure 7.6 are calculated using a discrete saddlepoint approximation based on the generating function (7.36).

The Gaussian approximation provides a computationally simple estimate of the error probability with an accuracy that is often sufficient in practice. It requires the mean and variance of the decision variable which are readily obtained from the first and second derivatives of (7.35), see (7.25) and (7.26)

$$E\{N\} = \Phi'(1) = mG + MK \tag{7.43}$$

and

$$\mathrm{Var}\{N\} = \Phi''(1) + \Phi'(1) - [\Phi'(1)]^2 = mG(1 + 2K) + MK(1 + K) \tag{7.44}$$

The Gaussian approximation is based on the signal-to-noise ratio (5.35)

$$\rho = \frac{E_1 - E_0}{\sigma_1 + \sigma_0} \tag{7.45}$$

where the indices 0 and 1 denotes transmitted 'zero' and 'one', respectively. From (7.43) and (7.43)

$$\rho = \frac{(m_1 - m_0)G}{\sigma_1 + \sigma_0} \tag{7.46}$$

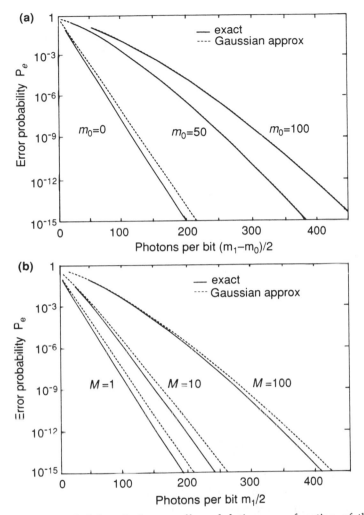

Figure 7.6 Error probability P_e for on-off modulation as a function of the average number of received photons per bit $(m_1 - m_0)/2$. The amplifier gain is 30 dB: (a) For optical filter bandwidth $M = 1$. The parameter m_0 is the average number of received photons for symbol 'zero'. (b) For background noise $m_0 = 0$ with M as a parameter. The dashed curves are obtained by the Gaussian approximation.

with

$$\sigma_i = \sqrt{m_i G(1 + 2K) + MK(1 + K)}; \quad i = 0, 1 \tag{7.47}$$

where m_0 and m_1 are the average numbers of received photons at the amplifier input for 'zero' and 'one', respectively. Substitution of data from Figure 7.5 gives $\rho = 5.0$ and with the aid of (5.40):

$$P_e \approx Q_1(\rho) = \frac{1}{\rho\sqrt{2\pi}} \exp(-\rho^2/2) = 2.7 \times 10^{-7} \tag{7.48}$$

which is close to the exact value $P_e = 2.3 \times 10^{-7}$.

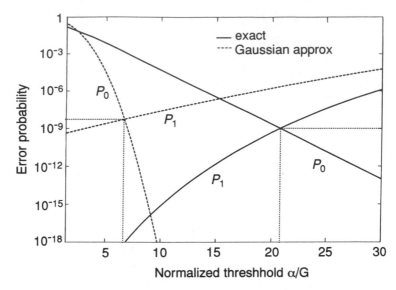

Figure 7.7 The probabilities $P_0(\alpha)$ and $P_1(\alpha)$ for an on-off system with an optical preamplifier at the quantum limit $P_e = 10^{-9}$. The dashed curves represent Gaussian distributions with the same mean and variance as the correct distributions.

An ideal optical amplifier has a minimal population inversion ratio $n_{sp} = 1$, which means that absorption is neglected. For best performance the output filter should eliminate all modes but the one occupied by the signal which corresponds to $M = 1$. For $n_{sp} = 1$ the parameter $K = G - 1$ and an ideal amplifier is characterized by its gain G only.

The most efficient transmission occurs when $m_0 = 0$ and the quantum limit for an optical preamplifier receiver with large gain $G \gg 1$ is a function of m_1 only. The probability functions $P_0(\alpha)$ and $P_1(\alpha)$ are shown in Figure 7.7 for a system with gain $G = 5 \times 10^3$ and $m_1 = 76.8$. The resulting error probability is shown to be equal to $P_e = 10^{-9}$ which is the usual reference for quantum limits. On average half the signal intervals contain optical pulses and the average number of photons per transmitted bit of information is

$$m_1/2 = 38.4 \text{ photons/bit} \tag{7.49}$$

which is the quantum limit for an optical preamplifier receiver in a system with on-off modulation.

It is interesting to compare the exact result with the estimate obtained from the Gaussian approximation. Substitution of $K = G - 1$ and $M = 1$ together with $m_0 = 0$ into (7.47) and (7.46) gives a SNR

$$\rho = \frac{m_1\sqrt{G}}{\sqrt{m_1(2G-1) + G - 1} + \sqrt{G-1}} \tag{7.50}$$

which for large gain $G \gg 1$ approaches

$$\rho = \frac{m_1}{\sqrt{2m_1 + 1} + 1} \tag{7.51}$$

Solving for m_1 yields

$$m_1 = 2\rho(\rho + 1) \tag{7.52}$$

The Gaussian approximation (7.48) for $P_e = 10^{-9}$ requires $\rho = 6.0$, which substituted into (7.52) gives $m_1 = 84$. On average, half the time intervals contain optical pulses, and the average number of photons per transmitted bit of information is

$$m_1/2 = 42.0 \text{ photons/bit} \tag{7.53}$$

which is the quantum limit for an optical preamplifier receiver calculated by the Gaussian approximation.

It is evident from Figure 7.7 that the Gaussian distribution differs markedly from the correct distribution at the quantum limit. The crossing of the curves, however, occurs at about the same level and the Gaussian approximation produces a reasonable estimate of P_e.

For $m_1 = 76.8$ the signal-to-noise ratio (7.51) $\rho = 5.79$ and the approximate error probability

$$P_e \approx Q_1(\rho) = 3.62 \times 10^{-9}$$

to be compared with the correct value $P_e = 10^{-9}$. The Gaussian approximation works well in this case. The difference in signal-to-noise ratio is only

$$10 \log \left(\frac{42.0}{38.4} \right) = 0.4 \text{ dB}$$

Note that the Gaussian approximation cannot be used to determine the decision threshold in the receiver. The fact that the distributions differ so much makes the Gaussian threshold much smaller than the correct one

The parameter K represents the noise caused by the spontaneous emission. It increases almost linearly with the gain G and, as illustrated by (7.51), the signal-to-noise ratio for an optical amplifier approaches a constant value when G is increased. The performance of an amplifier with low gain, however, is sensitive to the exact value of G. Figure 7.8 shows how the quantum limit for on-off modulation varies as a function of G. The lowest value $m = 10$ photons per bit is obtained when $G = 1$. Such an amplifier has no amplification and the situation is identical to ordinary direct detection without amplifier. It is, however, hardly possible in practice to operate at this limit. The very low signal levels in the receiver will be seriously perturbed by thermal nose. A preamplifier with a substantial G generates an output signal much larger than the thermal noise, which makes the quantum limit (7.49) of 38.4 photons per bit achievable in practice.

In Figure 7.8 the spontaneous emission parameter n_{sp} is kept constant and equal to unity while the gain G is decreased. In the limit when $G = 1$ the

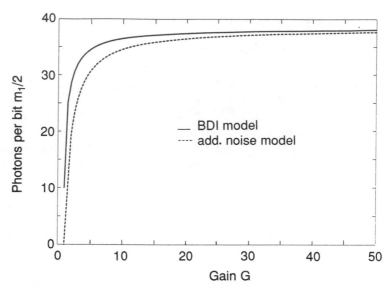

Figure 7.8   The variation of the quantum limit for an optical amplifier as a function of the gain G. The ordinate shows the average number of received photon needed to achieve a bit error probability $P_e = 10^{-9}$.

spontaneous emission noise parameter K is then equal to zero. A receiver with $G = 1$ can also be obtained by letting $\mu - \nu$ approach zero. Substitution of (7.30) and (7.32) into (7.31) gives

$$K = n_{sp}(G - 1) = \frac{\mu}{\mu - \nu}(e^{(\mu - \nu)T_L} - 1) \tag{7.54}$$

For $\mu - \nu \ll 1$ the last term can be expanded in a power series and

$$K = \frac{\mu}{\mu - \nu}[1 + (\mu - \nu)T_L - 1) = \mu T_L \tag{7.55}$$

An optical amplifier with $\nu = \mu$ has unity gain $G = 1$ but nonzero spontaneous emission noise.

7.4   Additive Noise Model

An alternate way of characterizing an optical amplifier, often used in the literature, is as an optical field amplifier with frequency-independent gain G in combination with additive noise representing spontaneous emission.

The input optical signal, assumed to be time-limited $[0, T]$, can be expressed as the real part of a complex field function

$$s(t) = \text{Re}\{A(t)\exp(j\omega_0 t)\} \tag{7.56}$$

where $\omega_0 = 2\pi f_0$ with f_0 the optical frequency. The function $A(t)$ is the (complex) envelope of $s(t)$ which we, without any loss of generality, assume to be real. It is convenient to normalize $A(t)$ in such a way that $|A(t)|^2$ represents the photon intensity and

$$m = \frac{1}{hf} \int_0^T \mathcal{P}(t)\, dt = \frac{1}{2} \int_0^T |A(t)|^2 dt \qquad (7.57)$$

is the average number of photons in $s(t)$. The factor $1/2$ follows from the complex notation.

The optical field, in complex signal representation, at the output of an amplifier with power gain G is

$$Y(t) = \sqrt{G}A(t) + X(t) \qquad (7.58)$$

where $X(t)$ is a white Gaussian stochastic process representing spontaneous emission noise. Its power spectral density is determined by general properties of optical amplification of light in an atomic medium; see Yariv (1985) and Henry (1986). To be compatible with the normalization of $A(t)$ the density of $X(t)$ should be expressed in photons per second and it is then equal to

$$\mathcal{N}_0 = n_{sp}(G - 1) \qquad (7.59)$$

with $n_{sp} \geq 1$ the spontaneous emission parameter (7.32).

A receiver for on-off modulation with an optical preamplifier specified by the additive noise model is shown in the block diagram Figure 7.9a. The effect of the noise $X(t)$ is reduced by a narrow bandpass optical filter H_1. The photodetector is modeled as a memoryless square law device with the filtered received optical field signal as input. This is an approximation since it does not include shot noise, which will be dealt with later. The output from the square law device is filtered in a postdetector filter H_2. To simplify the analysis we assume that that it is an integrate-and-dump filter with impulse response (5.49)

$$h_2(t) = \begin{cases} 1; & 0 < t < T \\ 0; & \text{otherwise} \end{cases} \qquad (7.60)$$

The communication situation shown in Figure 7.9 is a well-studied problem in electrical communication theory. It has been pointed out by Henry (1989) and also by Humblet and Azizoglu (1991) that it is equivalent to the detection of known binary signals in additive white Gaussian noise using a square law detector for noncoherent reception.

The spectral properties and the signal-to-noise ratio of the signal after the quadratic detector are easily calculated for an arbitrary filter from its frequency function $H_1(f)$, see e.g. Chapter 12 of Davenport and Root (1958). For evaluation of the transmission error probability, however, it is convenient to let H_1 be a special type of filter specified by a Fourier series expansion, as suggested by Marcuse (1991) and also by Humblet and Azizoglu (1991).

A bandpass filter operating at optical frequencies is extremely narrowband and the impulse response $h_1(t)$ of H_1 can be characterized by an equivalent complex valued lowpass function $f(t)$; see e.g. pp. 148–157 of Proakis (1989).

$$h_1(t) = 2\, \mathrm{Re}\{f(t)e^{j\omega_0 t}\} \qquad (7.61)$$

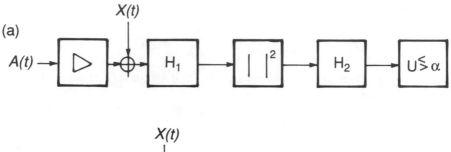

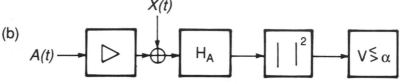

Figure 7.9 Receivers with enveloped detector in the approximate additive noise model: (a) Receiver with an optical bandpass filter H_1 and a postdetector filter H_2 integrating the signal over the data symbol interval [0,T]. (b) An optimal receiver with H_A is a filter matched to the input signal $A(t)$. No postdetector filter is needed in this case.

The signal $A(t)$ is confined in the bit interval and the impulse response $f(t)$ is assumed to be limited to the same time interval and can therefore be expanded in a Fourier series

$$f(t) = \sum_{k=-\infty}^{\infty} f_k e^{j2\pi kt/T}; \quad 0 \le t \le T \tag{7.62}$$

A bandpass filter of bandwidth $B_o = 2(L+1)/T$ with approximately equal attenuation in the pass band is obtained by letting $f_k = 0$ for $|k| > L$ and $f_k = 1$ for $|k| \le L$

$$f_1(t) = \sum_{k=-L}^{L} e^{j2\pi kt/T}; \quad 0 \le t \le T \tag{7.63}$$

The transfer function of such a filter with $L = 5$, i.e. the Fourier transform of (7.63), is shown in Figure 7.10.

The bandwidth B_o of the filter is assumed to be such that the optical signal (7.56) can pass it unaltered, which means that $A(t)$ can be expressed as the Fourier series expansion

$$A(t) = \sum_{k=-L}^{L} a_k e^{j2\pi kt/T}; \quad 0 \le t \le T \tag{7.64}$$

This is true if L is large enough but not necessarily so for small values. Due to the orthogonality of the base functions $e^{j2\pi kt/T}$

$$\int_0^T A^2(t) \, dt = T \sum_{k=-L}^{L} a_k^2 \tag{7.65}$$

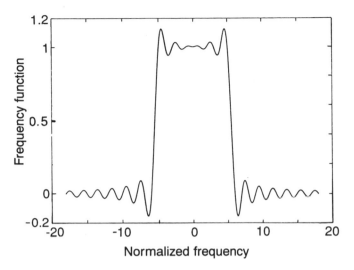

Figure 7.10 Equivalent baseband transfer function for an optical filter with bandwidth parameter $L = 5$. The figure shows the Fourier transform of the filter impulse response.

Expand the equivalent baseband noise $X(t)$ in the time interval $[0,T]$ in a Fourier series

$$X(t) = \sum_{k=-\infty}^{\infty} X_k e^{j2\pi kt/T}; \quad 0 \le t \le T \tag{7.66}$$

The filter (7.63) eliminates all Fourier components with indices $|k| > L$ and the noise signal at the filter output is

$$X_1(t) = \sum_{k=-L}^{L} X_k e^{j2\pi kt/T}; \quad 0 \le t \le T \tag{7.67}$$

where X_k are independent, equally distributed, complex-valued Gaussian variables

$$X_k = \frac{1}{T} \int_0^T X(t) e^{-j2\pi kt/T} dt \tag{7.68}$$

The real and imaginary parts of $X_k = X_{ck} + jX_{sk}$, the quadrature components, are independent and have equal variances for all $-L \le k \le L$

$$\text{Var}\{X_{ck}\} = \text{Var}\{X_{sk}\} = \mathcal{N}_0/T \tag{7.69}$$

with $\mathcal{N}_0$ the power density (7.59) of $X(t)$.

The quadratic device representing the photodetector forms the square envelope of the filtered received signal, and the decision variable U is the

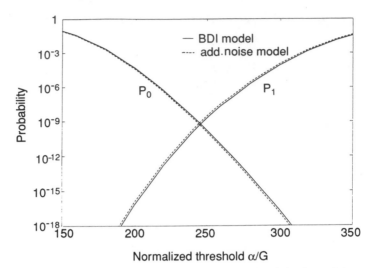

Figure 7.11   Probability distributed functions $P_0(\alpha)$ and $P_1(\alpha)$. The solid curves refer to the BDI-model and the dahsed curves to the additive noise model. The system parameters are $G = 180$, $n_{sp} = 1$, $m_0 = 129$ and $m_1 = 398$. $m_0 = 129$ and $m_1 = 398$.

output at time $t = T$ from the integrate-and-dump filter (7.60)

$$U = \frac{1}{2}\int_0^T \left|\sqrt{G}A(t) + X_1(t)\right|^2 dt \tag{7.70}$$

Substitution of (7.64) and (7.67) into (7.70) yields, analogously to (7.65)

$$U = \frac{T}{2}\sum_{k=-L}^{L} |a_k\sqrt{G} + X_k|^2 = \frac{T}{2}\left(\sum_{k=-L}^{L}[a_k\sqrt{G} + X_{ck}]^2 + X_{sk}^2\right) \tag{7.71}$$

The decision variable (7.71) is the sum of squares of $2(2L + 1)$ independent Gaussian variables with equal variance

$$\sigma^2 = \mathrm{Var}\{X_{ck}\sqrt{T}/\sqrt{2}\} = \mathrm{Var}\{X_{sk}\sqrt{T}/\sqrt{2}\} = \mathcal{N}_0/2 \tag{7.72}$$

obtained from (7.69).

The first $2L + 1$ terms in (7.71) have nonzero means $a_k\sqrt{G}$. From (7.65) and (7.57) follows that the sum of the squares of the means is

$$\frac{T}{2}\sum_{-L}^{L} a_k^2 G = mG \tag{7.73}$$

The decision variable U therefore has a noncentral chi-square distribution with a moment-generating function; see e.g. Proakis (1989)

$$\Psi_1(s) = \frac{1}{(1 - \mathcal{N}_0 s)^{2L+1}}\exp\left(\frac{mGs}{1 - \mathcal{N}_0 s}\right) \tag{7.74}$$

The probability density function can be found in Proakis (1989):

$$p_1(u) = \frac{1}{N_0} \left(\frac{u}{mG}\right)^L \exp\left(-\frac{u+mG}{N_0}\right) I_{2L}\left(-\frac{2\sqrt{mGu}}{N_0}\right) \tag{7.75}$$

where $I_{2L}(\ )$ is the modified Bessel function of order $2L$.

Consider a system with on-off modulation. Let m_0 and m_1 denote the average number of received photons for 'zero' and 'one', respectively. The bit error probability is, analogously to (7.42)

$$P_e = \frac{1}{2}P_0(\alpha) + \frac{1}{2}P_1(\alpha) = \frac{1}{2}\int_\alpha^\infty p_0(u)\,du + \frac{1}{2}\int_0^\alpha p_1(u)\,du \tag{7.76}$$

where $p_0(u)$ and $p_1(u)$ are equal to (7.75) with $m = m_0$ and $m = m_1$, respectively.

The integrals in (7.76) can be expressed in terms of the Marcum Q-function, see e.g. Appendix A of Schwartz *et al.* (1966):

$$\int_0^\alpha p(u)\,du = 1 - Q_M(\sqrt{4m/N_0}, \sqrt{2\alpha/N_0}) \tag{7.77}$$

When $m_0 = 0$ the decision variable is

$$U_0 = \frac{T}{2} \sum_{k=-L}^{L} |X_{0k}|^2 \tag{7.78}$$

which has an ordinary chi-square distribution with with $2(2L+1)$ degrees of freedom. Its moment-generating function is (7.74) with $m = 0$

$$\Psi_0(s) = \frac{1}{(1 - N_0 s)^{2L+1}} \tag{7.79}$$

corresponding to a probability density, see e.g. Proakis (1989)

$$p_0(u) = \frac{1}{(2L)!N_0} \left(\frac{u}{N_0}\right)^{2L} \exp\left(-\frac{u}{N_0}\right) \tag{7.80}$$

The integral of this $p_0(u)$ can be evaluated, see Proakis (1989), with the result

$$\int_\alpha^\infty p_0(u)\,du = \exp\left(-\frac{\alpha}{N_0}\right) \sum_{k=0}^{2L} \frac{1}{k!} \left(\frac{\alpha}{N_0}\right)^k \tag{7.81}$$

The distributions (7.75) and (7.38) are not identical. In the additive noise model the decision variable is continuous and in the BDI model it is a discrete stochastic variable.

Figure 7.11 compares the cumulative probabilities $P_0(\alpha)$ and $P_1(\alpha)$ for the BDI model and the additive noise model for a system with gain $G = 180$. The difference is small and both models result in practically the same numerical error probability.

As will be shown below the discrepancy between the two models is caused by modeling the photodetector as a simple square law device in the additive noise receiver. The shot noise produced by a physical photodiode is missing.

The mean and variance of U are obtained from the derivatives of $\Psi(s)$ through (5.134)

$$E\{U\} = \Psi'_u(0) = mG + (2L+1)\mathcal{N}_0 \tag{7.82}$$

and

$$\mathrm{Var}\{U\} = \Psi''_u(0) - \Psi'_u(0) = 2mG\mathcal{N}_0 + (2L+1)\mathcal{N}_0^2 \tag{7.83}$$

The mean (7.82) and variance (7.83) differ from the expressions (7.43) and (7.44) for the BDI model. For large G, however, the Gaussian approximations are asymptotically the same for both models. For $n_{sp} = 1$ and $L = 0$ substitution of (7.82) and (7.83) into (7.45) gives, after replacing $\mathcal{N}_0$ by $G - 1$ using (7.59), the signal-to-noise ratio

$$\rho = \frac{m_1 G/\sqrt{G-1}}{\sqrt{2m_1 G + G - 1} + \sqrt{G-1}} \tag{7.84}$$

For large G

$$\rho \approx \frac{m_1}{\sqrt{2m_1 + 1} + 1} \tag{7.85}$$

which is identical to the expression (7.51) for the BDI model, and consequently the quantum limit estimated by the Gaussian approximation is exactly the same 42.0 photons/bit for both models.

To operate at the quantum limit the filter H_1 should be chosen to reduce the noise as much as possible but should also be such that the signal can pass without severe degradation. A fundamental result of detection theory states that, with additive white noise, the best filter in terms of minimizing the bit error probability is a filter matched to the signal $s(t)$, see e.g. Proakis (1989). The impulse response of such an H_1 is in complex notation

$$f_1(t) = A^*(T - t) \tag{7.86}$$

As shown in the block diagram Figure 7.9b, a receiver with a matched optical filter (7.86) does not contain any postdetector filter. The decision variable is the square envelope of the output of the matched filter at $t = T$, the end of the signal interval

$$V = \left| \int_0^T Y(t)A^*(t)\, dt \right|^2 \tag{7.87}$$

Substitution of (7.58) gives using (7.57)

$$\int_0^T Y(t)A^*(t)\, dt = \int_0^T [\sqrt{G}A(t) + X(t)]A^*(t)\, dt$$
$$= 2m\sqrt{G} + \int_0^T X(t)A^*(t)\, dt \tag{7.88}$$

The last integral can be evaluated in terms of Fourier coefficients by Parseval's theorem

$$Z = \int_0^T X(t)A^*(t)\, dt = T \sum_{k=-\infty}^{\infty} X_k a_k \qquad (7.89)$$

The complex stochastic variable Z is Gaussian with zero mean. From (7.69) and (7.65)

$$\mathrm{Var}\{Z_c\} = \mathrm{Var}\{Z_s\} = \mathrm{Var}\{X_{kc}\}T^2 \sum a_k^2 = (\mathcal{N}_0/T)T^2(2m/T) = 2m\mathcal{N}_0 \quad (7.90)$$

The decision variable can be written as

$$V = |2m\sqrt{G} + Z|^2 = 4m|\sqrt{mG} + X|^2 \qquad (7.91)$$

where X is a Gaussian variable with with variance $\mathcal{N}_0/2$.

Comparison with (7.71) reveals that the decision variable V is proportional to the decision variable U of the original receiver with $L = 0$. This means that the two receivers have identical performance. The equivalence between the general prefilter receiver with $L = 0$ and the matched filter receiver can be explained by the fact that letting $L = 0$ in (7.64) makes $A(t)$ a rectangular pulse and the optical filter $A(t)$ is then a rectangular pulse and the optical filter (7.63) a corresponding matched filter.

The quantum limit according to the additive noise model is obtained by letting $n_{sp} = 1$ in which case the additive noise takes its smallest value $\mathcal{N}_0 = G - 1$. Calculation of the m_1 needed to achieve $P_e = 10^{-9}$ using (7.76) gives the same value

$$m_1/2 = 38.4 \; photons/bit \qquad (7.92)$$

as for the BDI model, at least within two decimal places.

For small values of G the two models give different results. The quantum limit $m_1/2$ obtained for a preamplifier with a specified gain is shown in Figure 7.8 for the BDI and the additive noise models. For values of the gain below $G = 10$ the difference is noticeable. In the limiting case $G = 1$ the additive noise is equal to zero and the receiver in Figure 7.9 works with deterministic signals and is error-free. The BDI receiver is a photon (photoelectron) counting receiver and $G = 1$ is identical to direct detection which has a well known quantum limit of 10 photons per bit.

7.5   Modified Additive Noise Model

The differences in output signal statistics from the BDI model and the additive noise model of Figure 7.9 are caused by the absence of shot noise in the latter. In Figure 7.12 the quadratic detector is replaced by a photodetector, establishing a

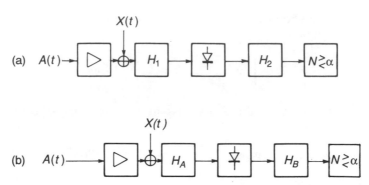

Figure 7.12 Receivers with photodetector in the complete additive noise model. (a) Receiver with an optical bandpass filter H_1 and a postdetector filter H_2 integrating the signal over the data symbol interval [0,T]. (b) A matched filter receiver achieving the quantum limit at high gain. The post-detector filter H_B is a short time integrator.

complete model. We will show that the modified receiver will produce a photoelectron process with counting statistics identical to those of the BDI model. The derivation also includes the effect of the photodetector quantum efficiency.

The decision variables (7.87) and (7.87) are the intensity function of the optical field averaged over the bit time interval $[0, T]$. It is not possible to mesure an optical field directly and a realistic receiver contains a photodetector and the decision is based on its output current, which amounts to photoelectron counting. The generating function for the number of photoelectrons observed in the time interval $[0, T]$ is related to the moment-generating function $\Psi_m(s)$ for the stochastic variable

$$\mathcal{M} = \frac{1}{2} \int_0^T |Y(t)|^2 dt \tag{7.93}$$

The quantity $\mathcal{M}$ is the integral of the optical intensity, represented by the square envelope of the optical field, over the signaling interval.

The number N of photoelectrons generated by a photodetector is a doubly stochastic Poisson variable. Its statistics are determined by a Poisson distribution with a random mean $\eta \mathcal{M}$ where η is the quantum efficiency of the photodetector. See Saleh (1978) for the theoretical background.

The generating function for N is determined in two steps, first conditional on $\mathcal{M}$ and then forming the expectation over $\mathcal{M}$.

$$\Phi_n(z) = E\{z^N\} = E_m\{E\{z^N|\mathcal{M} = m\}\} \tag{7.94}$$

The expectation $E\{z^N\}$ with fixed $\mathcal{M} = m$ is the generating function (7.11) of an ordinary Poisson distribution with mean m.

$$E\{z^N|\mathcal{M} = m\} = \exp[m(z - 1)] \tag{7.95}$$

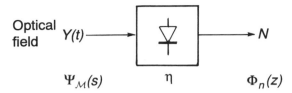

Figure 7.13 The Poisson transform photodetector model. The generating function $\Phi_N(z)$ is determined by the moment-generating function $\Psi_M(s)$ of the integrated optical field intensity M.

Substitution into (7.94) gives

$$\Phi_n(z) = E_m\{\exp[m(z-1)]\} = \Psi_M(z-1) \tag{7.96}$$

with $\Psi_M(\ )$ the moment-generating function of M. The relation (7.96) between the stochastic variables N and M is called the Poisson transform, see Saleh (1978).

The effect of the photodetector quantum efficiency is included by reducing the optical intensity by a factor $\eta \leq 1$, and the Poisson transform including the quantum efficiency is

$$\Phi_n(z) = \Psi_M(\eta[z-1]) \tag{7.97}$$

This is illustrated in Figure 7.13.

The decision variable U in the additive noise model is an integral of the type (7.93), and substitution of (7.74) into (7.97) yields

$$\Phi(z) = \frac{1}{[1 - \eta N_0(z-1)]^M} \exp\left[\frac{\eta m G(z-1)}{1 - \eta N_0(z-1)}\right] \tag{7.98}$$

For $\eta = 1$ the generating function (7.98) is exactly the same as (7.35) for the BDI model. Comparison between the two equations shows that

$$K = \eta N_0$$

and

$$M = 2L + 1$$

The parameter M is equal the ratio of the bandwidth B_o of the optical bandpass filter (7.63) and the data rate $B = 1/T$

$$M = B_o/B \tag{7.99}$$

A Gaussian optical field corresponds to thermal photons, and an equivalent description of the additive noise model is that it represents a mixture of coherent and thermal light.

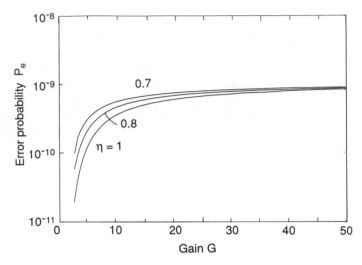

Figure 7.14 The effect of the quantum efficiency η on the performance of an on-off system. The diagram shows the bit error probability P_e as a function of the preamplifier gain G for various values of η. The system operates at the quantum limit with $m_0 = 0$ and $m_1 = 76.8$ photons per bit.

The effect of η is noticeable at low gain only. Figure 7.14 illustrates how the performance deteriorates for decreasing η as a function of G.

We have shown that the BDI models of Section 7.3 and the complete additive noise model of Section 7.4 are equivalent and result in the same decision variable and identical error probability.

The quantum limit for the approximate additive noise model is achieved by the receiver Figure 7.9b with a matched optical filter, corresponding to $M = 1$, but without any postfilter. In the complete receiver Figure 7.12, however, a postfilter is needed, since the shot noise from the photodetector has the character of white noise and cannot be sampled without prior filtering.

For low values of the optical filter bandwidth parameter M the lowpass filter H_2 in Figure 7.12a causes intersymbol interference which increases the error probability. However, if G is very large and the filter H_B in Figure 7.12b is a short time integrator the quantum limit corresponding to $M = 1$ can be achieved. In such a case the output of H_B, with sufficient accuracy, represents the optical intensity which is the decision variable of the approximate receiver in Figure 7.9b and both receivers will have the same performance.

The mean and variance of the number of photoelectrons emitted by the photodetector are obtained from the derivatives of (7.98) in the same way as (7.43)

$$E\{N\} = \eta mG + \eta M \mathcal{N}_0 \tag{7.100}$$

and analogously to (7.44)

$$\text{Var}\{N\} = \eta mG + \eta M \mathcal{N}_0 + 2\eta^2 mG \mathcal{N}_0 + \eta^2 \mathcal{N}_0^2 \tag{7.101}$$

In the literature, see e.g. Olsson (1989), the two first terms in (7.101) are usually referred to as shot noise. The third and fourth term are named signal-spontaneous beat noise and spontaneous-spontaneous beat noise, respectively.

The error probability of an optical preamplifier receiver can be calculated by evaluation of the sums in (7.42) after substitution of the probability (7.38). For a realistic value of G the threshold α is large and so is the number of terms to be calculated. The Laguerre polynomial (7.39) may also cause numerical difficulties.

An alternative with computational advantages is to use a saddlepoint approximation technique to calculate P_e from the generating function (7.36). In principle P_e is obtained by evaluating an inverse z-transform based on $\Phi(z)$. See Appendix D for further details. The terms $P_0(\alpha)$ and $P_1(\alpha)$ in the error probability expression (7.42) can, with good accuracy, be replaced by saddlepoint approximations.

$$P_i(\alpha) \approx [2\pi\phi''(z_i)]^{-1/2} \exp[\phi(z_i)]; \quad i = 0, 1 \tag{7.102}$$

where $\phi(z)$ is the function

$$\phi(z) = \ln[\Phi(z)] - (\alpha + 1)\ln z - \ln|z - 1| \tag{7.103}$$

and $\phi''(\ )$ is its second derivative with respect to z.

The saddlepoint parameters z_0 and z_1 are solutions of

$$\phi'(z_i) = 0; \quad i = 0, 1 \tag{7.104}$$

The saddlepoint approximation does not require any summation or numerical integration. To estimate the error probability for a detector with a specified threshold α, two nonlinear equations (7.104) must be solved, usually by numerical means. A method that works well is the Newton-Raphson algorithm.

Example 7.1   Optical preamplifier receiver

Consider an EDFA operating at an optical wavelength $\lambda = 1.55$ μm with gain $10 \log G = 25$ dB and a noise factor $10 \log F = 5$ dB. It is used as a preamplifier in an optical fiber system with on-off modulation. The data rate is $B = 2.5$ Gb/s and the received optical signal levels at the amplifier input are $p_0 = 50$ nW and $p_1 = 200$ nW, for 'zero' and 'one', respectively. The amplifier is followed by an optical filter with bandwidth $B_o = 75$ GHz. The quantum efficiency of the photodetector is $\eta = 0.8$, and the receiver is an integrate-and-dump filter followed by a sampler and a threshold device.

Determine the optimal decision threshold and calculate the error probability using the saddlepoint technique.

Solution

The average number of photons in the received 'zero' and 'one' signals are, cf. Chapter 5

$$m_0 = \frac{1}{hf} p_0 T = \frac{1.55 \times 10^{-6}}{6.626 \times 10^{-34} \times 2.998 \times 10^8} \times 50 \times 10^{-9} \times \frac{1}{2.5 \times 10^9} = 156.1$$

and

$$m_1 = \frac{1}{hf} p_0 T = \frac{1.55 \times 10^{-6}}{6.626 \times 10^{-34} \times 2.998 \times 10^8} \times 200 \times 10^{-9}$$

$$\times \frac{1}{2.5 \times 10^9} = 624.2$$

The amplifier gain is

$$G = 10^{25/10} = 316.2$$

and the filter parameter M is

$$M = B_o T = \frac{75 \times 10^9}{2.5 \times 10^9} = 30$$

The noise factor of an optical amplifier is defined as $F = 2n_{sp}$ and

$$n_{sp} = \frac{10^{F_{dB}/10}}{2} = \frac{10^{0.5}}{2} = 1.58$$

The parameter $\mathcal{N}_0$ is defined in (7.59)

$$\mathcal{N}_0 = n_{sp}(G - 1) = 1.58 \times 315.2 = 498.0$$

Substitution of (7.36) into (7.103) yields the saddlepoint function (E.17) in Appendix E

$$\phi(z) = -M \ln[1 - \eta \mathcal{N}_0(z - 1)] + \frac{\eta m G(z - 1)}{1 - \eta \mathcal{N}_0(z - 1)} - (\alpha + 1) \ln z - \ln |z - 1| \tag{7.105}$$

The saddlepoint equation (7.104) is obtained by forming the derivative of (7.105)

$$\phi'(z) = \frac{\eta \mathcal{N}_0 M}{1 - \eta \mathcal{N}_0(z - 1)} + \frac{\eta m G}{[1 - \eta \mathcal{N}_0(z - 1)]^2} - \frac{\alpha + 1}{z} - \frac{1}{z - 1} = 0 \tag{7.106}$$

The saddlepoint parameter z_0 is the root $z_0 > 1$ of (7.106) with $m = m_0$ and z_1 is the root $0 < z_1 < 1$ with $m = m_1$. The decision threshold must fall in the interval $m_0 G < \alpha < m_1 G$ and as an example let $\alpha = 300G = 9.5 \times 10^4$. Numerical solution of (7.106) then produces

$$z_0 = 1.00056 \quad \text{and} \quad z_1 = 0.99685$$

The error probability estimate (7.102) contains the second order derivative of (7.105)

$$\phi''(z) = \frac{(\eta \mathcal{N}_0)^2 M}{[1 - \eta \mathcal{N}_0(z - 1)]^2} + \frac{2\eta^2 m \mathcal{N}_0 G}{[1 - \eta \mathcal{N}_0(z - 1)]^3} + \frac{\alpha + 1}{z^2} + \frac{1}{(z - 1)^2} \tag{7.107}$$

Substitution of z_0 together with $m = m_0$ into (7.107) and (7.105) and combining them to (7.102) gives the saddlepoint estimate of $P_0(\alpha)$ for $\alpha = 300G$. In the same way z_1 and $m = m_1$ determine $P_1(\alpha)$. The result is the error probability estimate

$$P_e = \frac{1}{2}[P_0(\alpha) + P_1(\alpha)] = 2.853 \times 10^{-10}$$

The best threshold minimizing the error probability has to be determined either by a joint optimization over the three parameters z_0, z_1 and α, or by varying α and repeating the steps above for each value. The procedure is facilitated by the fact that P_e, as a function of α, is a simple convex function with no local minima. The optimal threshold turns out to be

$$\alpha = 319.6G$$

and the minimum error probability

$$P_e = 5.298 \times 10^{-12} \qquad\qquad \square$$

At a gain $G = 25$ dB the photodetector quantum efficiency η has little effect on the error probability. If the calculations in Example 7.1 are repeated with $\eta = 1$ instead of the correct value $\eta = 0.8$, the calculated error probability becomes

$$P_e = 5.264 \times 10^{-12}$$

which is almost identical to the value obtained in the example.

A Gaussian approximation based on the mean and variances obtained from the probability generating function (7.98) can be derived in the same way as (7.46) and (7.47). When the gain is large the expressions for the mean (7.43) and the variance (7.44) of the decision variable can be replaced by the simpler expressions

$$E\{N\} = \eta G(m + n_{sp}M) \tag{7.108}$$

and

$$\mathrm{Var}\{N\} = \eta^2 n_{sp} G^2(2m + n_{sp}M) \tag{7.109}$$

The signal-to-noise ratio (7.45) becomes

$$\rho = \frac{m_1 - m_0}{\sqrt{(2m_1 + n_{sp}M)n_{sp}} + \sqrt{(2m_0 + n_{sp}M)n_{sp}}} \tag{7.110}$$

Note that neither G nor η appear in this expression. Substitution of $m_0 = 156.1$ and $m_1 = 624.2$ gives

$$\rho = 6.78$$

and from (7.48)

$$P_e \approx 6.26 \times 10^{-12}$$

to be compared with the exact $P_e = 5.30 \times 10^{-12}$.

If the calculations of the Gaussian approximation are made with the exact expressions (7.43) and (7.44) for the mean and variance

$$P_e \approx 5.90 \times 10^{-12}$$

is obtained.

7.6 Frequency-shift Keying

An alternative to on-off keying is frequency-shift keying (FSK). The transmitter sends either of two signals representing data symbols 'zero' and 'one'

$$\left. \begin{array}{l} s_0(t) = \text{Re}\{A(t)\exp(j\omega_0 t)\} \\[2mm] s_1(t) = \text{Re}\{A(t)\exp(j\omega_1 t)\} \end{array} \right\} \quad 0 \le t \le T \qquad (7.111)$$

The receiver, shown in Figure 7.15, contains two optical filters H_0 and H_1 centered on optical frequencies f_0 and f_1, respectively. The prefilters are assumed to be Fourier filters with equivalent baseband impulse response (7.63), and the postdetector filters H_2 are integrate-and-dump filters (7.60). It is assumed that the frequencies are sufficiently separated so that s_0 will not give any contribution to the output of H_1, and vice versa. This condition is usually termed wideband FSK.

Each branch of the FSK receiver is equal to the configuration appearing in the OOK receiver Figure 7.9. The results obtained for that receiver can therefore be used to evaluate the FSK receiver. Assume that s_1 is transmitted. The integrated field intensities after the optical filters H_1 and H_0 are, analogously to (7.71)

$$U_1 = \mathcal{M}_1 = \frac{T}{4} \sum_{k=-L}^{L} |a_k\sqrt{G} + X_{1k}|^2 \qquad (7.112)$$

and

$$U_0 = \mathcal{M}_0 = \frac{T}{4} \sum_{k=-L}^{L} |X_{0k}|^2 \qquad (7.113)$$

The additional factor of $1/2$ compared to (7.71), is caused by the power splitting in the FSK receiver.

The moment-generating functions of $\mathcal{M}_1$ and $\mathcal{M}_0$ are equal to (7.74) and (7.79), with m replaced by $m/2$ and $\mathcal{N}_0$ by $\mathcal{N}_0/2$, respectively.

The integer-valued stochastic variables N_1 and N_0 are the the numbers of photoelectrons observed in the two branches of the receiver during the

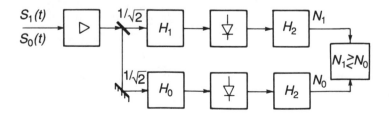

Figure 7.15 Dual filter optical preamplifier receiver for wideband frequency-shift keying (FSK).

signaling interval. Application of the Poisson transform with $\eta = 1$ yields a probability distribution for N_1 analogous to (7.38)

$$P_n^{(1)} = \frac{2^M \mathcal{N}_0^n}{(2 + \mathcal{N}_0)^{n+M}} \exp\left(-\frac{mG}{2 + \mathcal{N}_0}\right) L_n^{(M-1)}\left[-\frac{2mG}{\mathcal{N}_0(2 + \mathcal{N}_0)}\right] \qquad (7.114)$$

Transformation of $\mathcal{M}_0$ gives the distribution of N_0 analogously to (7.41)

$$P_n^{(0)} = \binom{M + n - 1}{n} \frac{2^M \mathcal{N}_0^n}{(2 + \mathcal{N}_0)^{n+M}} \qquad (7.115)$$

The detector decides that a 'one' was transmitted if $N_1 > N_0$ and a 'zero' if $N_0 > N_1$. To obtain a symmetric rule a random decision of 'one' or 'zero' is made when $N_1 = N_0$.

Assume that a 'one' was transmitted and that the number N_1 of photons in branch 1 is equal to k. A decision error occurs with certainty if $N_0 > k$ and with probability $1/2$ when $N_0 = k$. The error probability conditional on $N_1 = k$ is

$$P_0(k) = \sum_{n=k+1}^{\infty} P_n^{(0)} + \frac{1}{2} P_k^{(0)} \qquad (7.116)$$

The error probability is the average of $P_0(k)$ over k

$$P_e = \sum_{k=0}^{\infty} P_k^{(1)} P_0(k) \qquad (7.117)$$

It is possible to obtain a closed-form formula for P_e. We present the special case $M = 1$; an expression for arbitrary M has been derived, with the same technique, by Humblet and Azizoglu (1991).

Substitution of (7.115) with $M = 1$ into (7.116) gives

$$P_0(k) = \sum_{n=k+1}^{\infty} \frac{2\mathcal{N}_0^n}{(2 + \mathcal{N}_0)^{n+1}} + \frac{\mathcal{N}_0^k}{(2 + \mathcal{N}_0)^{k+1}} = \frac{1 + \mathcal{N}_0}{\mathcal{N}_0}\left(\frac{\mathcal{N}_0}{2 + \mathcal{N}_0}\right)^{k+1} \qquad (7.118)$$

and from (7.117)

$$P_e = \frac{1 + \mathcal{N}_0}{\mathcal{N}_0} \sum_{k=0}^{\infty} P_k^{(1)} \left(\frac{\mathcal{N}_0}{2 + \mathcal{N}_0}\right)^{k+1} = \frac{1 + \mathcal{N}_0}{2 + \mathcal{N}_0} \Phi(z) \qquad (7.119)$$

with $z = \mathcal{N}_0/(2 + \mathcal{N}_0)$ and where $\Phi(z)$ is the generating function for N_1, cf. (7.36)

$$\Phi(z) = \frac{2}{2 - \mathcal{N}_0(z - 1)} \exp\left[\frac{mG(z - 1)}{2 - \mathcal{N}_0(z - 1)}\right] \qquad (7.120)$$

Substitution of (7.120) into (7.119) gives the result

$$P_e = \frac{1}{2} \exp\left(-\frac{mG}{2(1 + \mathcal{N}_0)}\right) \qquad (7.121)$$

For an ideal amplifier with $n_{sp} = 1$, in combination with a matched filter receiver with $M = 1$, the spontaneous emission noise density (eq 60.7.3.3.1) is equal to $\mathcal{N}_0 = G - 1$ and for arbitrary gain

$$P_e = \frac{1}{2}\exp\left(-\frac{m}{2}\right) \tag{7.122}$$

The value of m resulting in $P_e = 10^{-9}$ is

$$m = 40.1 \text{ photons/bit} \tag{7.123}$$

which is the quantum limit for a direct detection FSK receiver with an optical preamplifier.

A decision rule $N_1 \geq N_0$ is equivalent to $N_1 - N_0 \geq 0$, corresponding to a decision variable $N = N_1 - N_0$ operating with a zero threshold. The stochastic variables N_1 (7.112) and N_0 (7.113) are independent, and the mean and variance of N are obtained from the derivatives of the generating function (7.120) in the same way as (7.43) and (7.44)

$$E\{N\} = E\{N_1\} - E\{N_0\} = \pm mG/2 \tag{7.124}$$

where the sign $+$ or $-$ depends on whether a 'one' or a 'zero' was transmitted. The variance does not depend on the transmitted data symbol

$$\text{Var}\{N\} = mG(1 + \mathcal{N}_0)/2 + M\mathcal{N}_0(1 + \mathcal{N}_0/2) \tag{7.125}$$

The signal-to-noise ratio is obtained from (5.35). For $n_{sp} = 1$ and large gain $G \gg 1$

$$\rho = \frac{m}{\sqrt{2(m + M)}} \tag{7.126}$$

Letting $M = 1$ and solving for m gives

$$m = \rho^2\left(1 + \sqrt{1 + 2/\rho^2}\right) \tag{7.127}$$

For $\rho = 6.0$

$$m = 73.0 \text{ photons/bit} \tag{7.128}$$

which is the quantum limit for FSK calculated by the Gaussian approximation.

The quantum limit obtained by the Gaussian approximation differs markedly from the exact value (7.123). For $m = 40.1$ the signal-to-noise ratio (7.126) becomes $\rho = 4.4$ and the Gaussian approximation (5.34) produces the estimate $P_e = 5.1 \cdot 10^{-6}$, which is inaccurate to more than three orders of magnitude compared with the correct $P_e = 10^{-9}$.

For FSK the detection is symmetrical and the decision variable changes sign depending on which data symbol was transmitted, but it has the same form in both cases. Figure 7.16 shows the cumulative distributions for N when 'zero' and 'one' was transmitted.

For detectors with a symmetric decision variable the accuracy of the Gaussian approximation depends on how well the Gaussian distribution resembles the true distribution. It is evident from Figure 7.16 that the low accuracy is caused by a large deviation between the distributions.

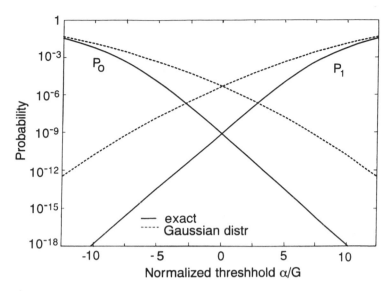

Figure 7.16 Probabilities $P_0(\alpha)$ and $P_1(\alpha)$ for a wideband FSK system with preamplifier gain $G = 100$ at the quantum limit $P_e = 10^{-9}$. The dashed curves represent Gaussian distributions with the same mean and variance as the correct distributions.

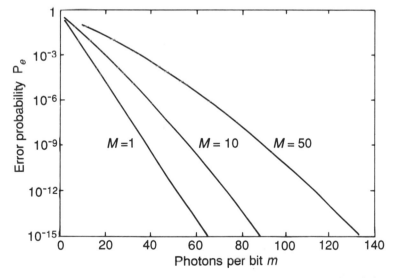

Figure 7.17 Error probability P_e for FSK modulation as a function of the average number of received photons per bit m for various values of the optical filter bandwidth parameter M. The spontaneous noise parameter $n_{sp} = 1$ and the amplifier gain $G = 180$.

For unsymmetric cases, however, the deviation from the true distribution can be large and the Gaussian approximation can produce a reasonable estimate. An on-off detector, as illustrated by Figure 5.19, is not symmetrical. The Gaussian approximation overestimates the probability P_1 and underestimates

P_0, but the result, equal to the average of these, is approximately correct. The performance of the Gaussian approximation is studied by Einarsson and Sundelin (1995).

The closed-form expression (7.121) can be used to verify the quality of the saddlepoint approximation. The exact error probability for an input signal level $m = 40.1$ is

$$P_{e1} = 9.803 \times 10^{-10}$$

Calculation of P_e as $\Pr\{U \geq 0\}$, when a 'zero' was transmitted, by the continuous saddlepoint approximation yields

$$P_{e2} = 9.826 \times 10^{-10}$$

which corresponds to a relative accuracy of

$$(P_{e2} - P_{e1})/P_{e1} = 2.4 \times 10^{-3}$$

which certainly is sufficient for practical system analysis. This agrees with the investigations by Helstrom (1979) and Einarsson (1989).

Note that it is only at a threshold $\alpha = 0$ that a simple expression for the error probability exists. The curves $P_0(\alpha)$ and $P_0(\alpha)$ in Figure 7.16 are determined by the saddlepoint approximation.

7.7 Differential Phase-shift Keying

In differential phase-shift keying (DPSK) the information is coded as phase shifts between successive signal intervals. Let ϕ_{-1} and ϕ_0 denote the phases of the optical field in the symbol intervals $[-T, 0]$ and $[0, T]$, respectively. The data symbol 'zero' is characterized by a phase shift $\phi_0 - \phi_{-1} = \pi$, and 'one' corresponds to $\phi_0 - \phi_{-1} = 0$. The receiver uses the signal in the previous time interval as a phase reference in the detection process.

In DPSK the information is modulated on the phase, which makes it sensitive to phase noise in the received signal. We first study the idealized situation with a coherent light source without phase noise at the transmitter.

The received signal in the time interval [0,T] is

$$s(t) = \text{Re}\{\sqrt{G}A(t)\exp[j(\omega_0 t + \phi_0)] + X(t)\}; \quad 0 < t \leq T \tag{7.129}$$

and the signal in the preceding interval is

$$s(t) = \text{Re}\{\sqrt{G}A(t+T)\exp[j(\omega_0 t + \phi_{-1})] + X(t)\}; \quad -T < t \leq 0 \tag{7.130}$$

The signals passes an optical bandpass filter H_1 assumed to be a Fourier filter specified by (7.63). Two auxiliary signals Y_+ and Y_- are formed by adding or subtracting the output signal and a delayed copy of it. In complex notation, assuming that $\omega_0 T$ is an integer multiple of 2π

$$Y_{\pm}(t) = \frac{1}{2}\left[\sqrt{G}A(t)e^{j\phi_0} + X_1(t) \pm [\sqrt{G}A(t)e^{j\phi_{-1}} + X_1(t-T)]\right] \tag{7.131}$$

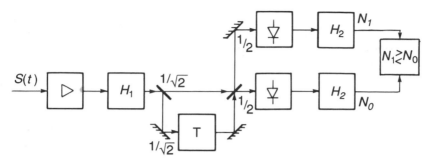

Figure 7.18 An optical preamplifier receiver with a Mach-Zehnder interferometer for differential phase-shift keying (DPSK).

This part of the receiver can be instrumented by a Mach-Zehnder interferometer, as shown in Figure 7.18. The factor $1/2$ is caused by the power splitting in the interferometer. This receiver differs from the one in Figure 5.29 for ideal direct detection DPSK, which has one photodetector. It can be shown that the configuration in Figure 7.18 is the best when $X(t)$ is present.

Each signal is fed to a separate photodetector, producing integrated intensities analogously to (7.71)

$$M_{\pm} = \frac{T}{8} \sum_{k=-L}^{L} \left| a_k \sqrt{G}(e^{j\phi_0} + e^{j\phi_{-1}}) + X_{0k} \pm X_{1k} \right|^2 \tag{7.132}$$

The quantities X_{0k} and X_{1k} are the Fourier coefficients (7.67) of $X(t)$ and $X(t - T)$, respectively. They are equally distributed, complex-valued, Gaussian variables with variances (7.69). They are statistically independent since they originate from white noise in nonoverlapping time intervals.

The factor

$$e^{j\phi_0} \pm e^{j\phi_{-1}} = e^{j\phi_{-1}}(e^{j(\phi_0 - \phi_{-1})} \pm 1)$$

can, without changing the result, be replaced by

$$e^{j(\phi_0 - \phi_{-1})} \pm 1$$

since a phase factor $e^{j\phi_{-1}}$ can be incorporated into the noise where it does not affect the statistics of the absolute value in (7.132).

Assume that the symbol 'one' is transmitted, which means that $\phi_0 - \phi_{-1} = 0$. The upper branch of the receiver corresponds to the $+$ sign and $\exp[j(\phi_0 - \phi_{-1})] + 1 = 2$. The outputs from the postdetector filters H_2, for the upper and lower branches, respectively, are equal to

$$M_1 = \frac{T}{8} \sum_{k=-L}^{L} |2a_k \sqrt{G} + Z_{1k}|^2 \tag{7.133}$$

and

$$M_0 = \frac{T}{8} \sum_{k=-L}^{L} |Z_{0k}|^2 \tag{7.134}$$

with $Z_{0k} = X_{0k} + X_{1k}$ and $Z_{1k} = X_{0k} - X_{1k}$. The prefilter H_1 has a finite impulse response which makes X_{0k} and X_{1k} independent, which in turn makes Z_{0k} and Z_{1k} independent. They are complex Gaussian stochastic variables with quadrature component variance equal to twice the value in (7.69)

$$\text{Var}\{Z_{ck}\} = \text{Var}\{Z_{sk}\} = 2\mathcal{N}_0/T \tag{7.135}$$

Comparison with (7.112) and (7.113) shows that the statistics of the decision variables N_1 and N_0 are of the same form for DPSK as for FSK, but the signal energy in the DPSK detector is twice of that for FSK. The error probability for DPSK is obtained from FSK by changing m to $2m$.

The error probability for DPSK is from (7.121)

$$P_e = \frac{1}{2} \exp\left(-\frac{mG}{1 + \mathcal{N}_0}\right) \tag{7.136}$$

The quantum limit for DPSK is therefore exactly half that for FSK

$$m = 20.0 \text{ photons/bit} \tag{7.137}$$

A comparison between DPSK, FSK and OOK modulation is shown in Figure 7.19 in terms of the error probability as a function of the average number of photons per bit.

The fact that DPSK is sensitive to phase noise in the transmitting laser might make it difficult to achieve the limit of 20.0 photons/bit. In Chapter 8 it is shown that there is an equivalence between heterodyne reception and receivers with optical preamplifiers. The relation is valid for the approximate receiver with envelope detector in Figure 7.9b. This means that the influence of phase noise on a high gain optical preamplifier can be determined from the analysis of

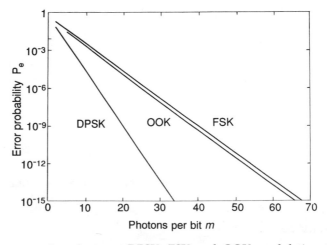

Figure 7.19 Comparison between DPSK, FSK and OOK modulation. The diagram shows the error probability P_e as a function of the average number of received photons per bit. The optical filter has bandwidth $M = 1$. The spontaneous noise parameter $n_{sp} = 1$ and the amplifier gain $G = 1000$.

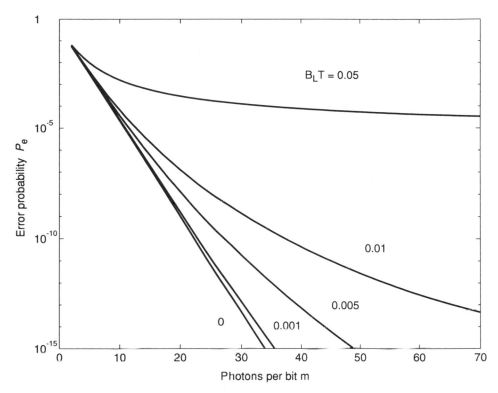

Figure 7.20   The error probability P_e as a function of the average number of received photons per bit for DPSK with phase noise at large preamplifier gain. The parameter $B_L T$ is the laser linewidth B_L multiplied by the bit time T.

a heterodyne receiver. The error probability of an optical preamplifier receiver for DPSK estimated in this way is shown in Figure 7.20. From the diagram it follows that $B_L T$, which is equal to the ratio between the laser linewidth and the system bit rate, must be smaller than 0.005 for DPSK to have a performance that is superior to that of OOK and FSK in Figure 7.19.

The alternative modulation methods for direct detection, OOK and FSK, are much less sensitive to phase noise, since the envelope or the frequency is modulated.

The decision variable for DPSK has a symmetric distribution like FSK, and the Gaussian approximation does not give an accurate error probability estimate.

7.8   Cascaded Amplifiers

An important application of optical amplifiers is to make long-distance optical communication possible by compensating for the fiber attenuation.

The attenuation **a** of an optical fiber is usually expressed in dB per km, see

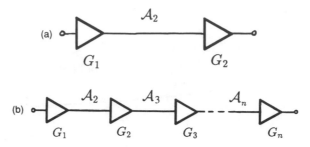

Figure 7.21 Fiber optical system with optical amplifiers. The first amplifier G_1 is a booster and the last amplifier G_n is preamplifier in an optical receiver. The amplifiers G_2, G_3, G_{n-1} are in-line amplifiers compensating for the fiber loss A_k.

Section 4.9. In the calculation below it is convenient to define the transmittance

$$A = 10^{a/10} \tag{7.138}$$

Consider a chain of amplifiers according to Figure 7.21.

The first amplifier acts as a booster. It has gain G_1 and spontaneous emission noise with spectral density $\mathcal{N}_{01}$.

The signal and the noise are attenuated by the fiber link with transmittance A_2 between the first and the second amplifier. At the output of the second amplifier the input signal is amplified by

$$G^{(2)} = A_2 G_1 G_2 \tag{7.139}$$

The noise from the first amplifier is attenuated by A_2, amplified by G_2 and then added to the spontaneous emission noise in the second amplifier:

$$\mathcal{N}_0^{(2)} = A_2 G_2 \mathcal{N}_{01} + \mathcal{N}_{02} \tag{7.140}$$

The combined noise (7.140) is a Gaussian white noise process since it is the sum of two such independent processes. Therefore, the combination in Figure 7.21a of two optical amplifiers connected with an optical fiber is equivalent to a single amplifier with gain $G^{(2)}$ and spontaneous emission noise spectral density $\mathcal{N}_0^{(2)}$. If G_2 is preamplifier in an optical receiver the bit error probability of the complete system follows directly from the theory presented previously in this chapter.

The results are easily extended to a chain of amplifiers, arbitrary in number, Figure 7.21b. Adding another link to Figure 7.21a is equivalent to cascading $G^{(2)}$ with A_3 and G_3. The result for a chain of n amplifiers is

$$\mathcal{N}_0^{(n)} = \sum_{i=1}^{n} \left(\mathcal{N}_{0i} \prod_{k=i+1}^{n} A_k G_k \right) \tag{7.41}$$

and

$$G^{(n)} = G_1 \prod_{k=2}^{n} A_k G_k \tag{7.142}$$

In-line amplifiers are needed to compensate for the fiber attenuation and it is natural to let

$$A_k G_k = 1$$

Then

$$G^{(n)} = G_1$$

and

$$\mathcal{N}_0^{(n)} = \sum_{k=1}^{n} \mathcal{N}_{0k}$$

The analysis above uses the optical field description of the amplifier and the spontaneous emission noise. Identical results were obtained by Li and Teich (1992) with an analysis based on the BDI model in the following way.

The generating function for the number of photons at the output of amplifier G_1 in Figure 7.21a is given by (7.36)

$$\Phi^{(1)}(z) = \frac{1}{[1 - \mathcal{N}_{01}(z - 1)]^M} \exp\left[\frac{mG(z - 1)}{1 - \mathcal{N}_{01}(z - 1)}\right] \tag{7.143}$$

The fiber attenuation A acts as a thinning or random deletion process on the stream of photons. Let X be a binary random variable with probability

$$\left.\begin{array}{l} \Pr\{X = 1\} = A \\[2mm] \Pr\{X = 0\} = 1 - A \end{array}\right\} \tag{7.144}$$

A random variable with a probability distribution (7.144) is called a Bernoulli variable. Its generating function is

$$\Phi_x(z) = (1 - A)z^0 + Az^1 = 1 + A(z - 1) \tag{7.145}$$

For each photon a random experiment is performed. If $X = 1$ the photon is kept intact, and if $X = 0$ it is deleted. Let N_1 denote the random number of input photons. The number of photons at the output of the optical fiber can be expressed as a sum of independent Bernoulli variables, see e.g. Feller (1957).

$$N_2 = \sum_{k=1}^{N_1} X_k \tag{7.146}$$

The generating function of a sum of independent variables is the product of the generating functions of the individual variables, and for a specific value of $N_1 = n_1$

$$\Phi_{n2}(z|N_1 = n_1) = \Phi_x^{n_1}(z) \tag{7.147}$$

The generating function for N_2 is obtained by forming the average of (7.147) over the variable N_1:

$$\Phi_{n2}(z) = \mathrm{E}\{\Phi_x^{N_1}(z)\} = \Phi_{n1}(\Phi_x(z)) = \Phi_{n1}(1 + A(z - 1)) \tag{7.148}$$

The generating function of the number of photons at the input of amplifier G_2 is obtained by applying (7.148) to (7.143):

$$\Phi^{(1)}(1 + A(z-1)) = \frac{1}{[1 - A\mathcal{N}_{01}(z-1)]^M} \exp\left[\frac{mAG(z-1)}{1 - A\mathcal{N}_{01}(z-1)}\right] \qquad (7.149)$$

The relation between the input and output generating functions of an optical amplifier is specified by (7.35) and:

$$\Phi^{(2)}(z) = \Phi^{(1)}(1 + A(z-1))[\Phi_{BD}(z)]\Phi_I(z)$$

$$= \frac{1}{[1 - \mathcal{N}_0^{(2)}(z-1)]^M} \exp\left[\frac{mG^{(2)}(z-1)}{1 - \mathcal{N}_0^{(2)}(z-1)}\right] \qquad (7.150)$$

with $G^{(2)}$ and $\mathcal{N}_0^{(2)}$ equal to (7.139) and (7.140), respectively, Equation (7.150) is identical to the result of the complete additive noise model which proves that both models give the same result for cascaded amplifiers.

An exact analysis is straightforward, since the complete system is equivalent to a single amplifier. Here we make a simplified analysis based on the Gaussian approximation, illustrating the main features of a system with optical amplifiers compensating for fiber attenuation.

Consider a system with a chain of n identical amplifiers. For on-off modulation the mean and variance of the decision variable are equal to (7.100) and (7.101), respectively, with $\mathcal{N}_0$ replaced by $n\mathcal{N}_0$. For simplicity let the quantum efficiency $\eta = 1$

$$E\{N\} = mG + Mn\mathcal{N}_0 \qquad (7.151)$$

and

$$\text{Var}\{N\} = mG(1 + 2n\mathcal{N}_0) + Mn\mathcal{N}_0(1 + n\mathcal{N}_0) \qquad (7.152)$$

Assuming an ideal amplifier with $n_{sp} = 1$ and $M = 1$, together with ideal OOK with $m_0 = 0$, the signal-to-noise ratio (7.45) becomes

$$\rho = \frac{m_1\sqrt{G}}{\sqrt{m_1(1 + 2n(G-1)) + \sigma_0^2 + \sigma_0}} \qquad (7.153)$$

with $\sigma_0^2 = n(G-1)(1 + n(G-1))$.

For large gain $G \gg 1$ the signal-to-noise ratio approaches

$$\rho = \frac{m_1}{\sqrt{2nm_1 + n^2} + n} \qquad (7.154)$$

Solving for m_1 yields

$$m_1 = 2n\rho(\rho + 1) \qquad (7.155)$$

As an example for $\rho = 6.0$ corresponding to $P_e = 10^{-9}$, a system with $n = 10$ requires

$$m_1/2 = 420 \text{ photons/bit} \qquad (7.156)$$

which is a quantum limit, calculated by the Gaussian approximation. Note that m_1 is the output from a light source connected to the input of the booster amplifier G_1.

In the analysis above we have assumed linear amplifiers with, in principle, an infinite dynamic range in combination with an ideal fiber without any limitation on input power. Amplifiers with as large a gain as possible should then be used to optimize the performance by minimizing the number of amplifiers in the system.

In practice saturation effects and nonlinearities in the amplifiers and fiber have to be taken into account in the system design. We refer to Gordon and Mollenauer (1991) and Lichtman (1993) for a further discussion.

8
Heterodyne Systems

8.1 Introduction

In the previous chapters we have dealt with incoherent light. The stochastic signal model (4.60), used in the dispersion analysis, contains random frequency components. Optical signals with this model add in intensity; see (4.62). The analysis of optical receivers in Chapter 5, including the effect of intersymbol interference, was based on that assumption.

Coherent light, in contrast with incoherent, consists ideally of one single optical frequency, and the optical field quantities, in complex notation, are of the form

$$s(t) = A \cos(\omega_0 t + \phi_0) = \text{Re}\{B(t)\} \tag{8.1}$$

with

$$B(t) = A \exp[j(\omega_0 t + \phi_0)] \tag{8.2}$$

A more realistic model of laser light is the Lorentzian spectrum model (4.92)

$$B(t) = A \exp[j(\omega_0 t + \theta_0(t) + \phi_0)] \tag{8.3}$$

The phase noise $\theta(t)$ is a random walk process generated by a frequency noise $\mu(t)$ which is a white Gaussian stochastic process; see (4.94) and Appendix G

$$\theta(t) = 2\pi \int_0^t \mu(s) \, ds \tag{8.4}$$

The frequency noise has a double-sided spectral density R_μ which is related to the 3-dB spectral bandwidth B_L of the signal by (4.96)

$$B_L = 2\pi R_\mu \tag{8.5}$$

See Figure 4.16.

The border between coherent and incoherent optical signals is diffuse. For a laser with small R_μ the phase function $\theta(t)$ becomes negligible, and (8.3) approximates (8.2), while for larger R_μ it gradually approaches an incoherent signal. The intermediate case is sometimes called weakly coherent.

There are several motives for employing coherent signals in optical communication systems. In addition to intensity modulation, which is the only alternative for incoherent signals, the phase and frequency of a coherent signal can be modulated. The narrow band character of coherent signals allows wavelength division multiplexing (WDM), which means that several communication systems centered at different optical frequencies can operate in the same fiber.

An important concept in conjunction with optical systems with coherent or weakly coherent signals is heterodyning and homodyning techniques. The received signal is mixed with a locally generated optical signal and the combined signals are the input to a photodetector. See Figure 8.1. In the homodyne case the received signal and the local oscillator have the same frequency and the output from the photodetector is a baseband (lowpass) signal. Heterodyning corresponds to a situation with the received signal at a frequency f_0 and the local oscillator at a different frequency f_1. The signal after the photodetector is now a bandpass signal located at the heterodyne frequency $f_h = |f_1 - f_0|$.

Homodyning can be considered as a special case of heterodyning with $f_1 = f_0$, and in a broader context we let the term heterodyning denote the general situation, where the received signal is mixed with a locally generated signal before detection. In the analysis, however, it is necessary to distinguish between the two cases since they differ in performance.

The local oscillator frequency f_1 is in general chosen such that the heterodyne frequency falls in the microwave region. This allows sophisticated signal processing, such as estimation of phase and frequency, to be done electrically. The repertoire for optical signal processing is limited and it would in general be much more difficult to perform these tasks optically. The demultiplexing of a WDM optical signal is easily done in a heterodyne receiver through a proper choice of the local oscillator frequency.

The amplitude of a heterodyned signal is proportional to the strength of the local oscillator, which provides amplification of a weak received signal. Perhaps the most important advantage of heterodyning is that it can improve the sensitivity of optical receivers.

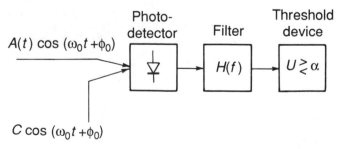

Figure 8.1 Ideal coherent homodyne receiver. The local oscillator provides an exact reference for phase and frequency.

In this Chapter we present the fundamental principles of optical heterodyne systems together with a basic performance analysis. A more comprehensive treatment is available in Jacobsen (1994).

8.1.1   Photoelectron Intensity

We consider optical heterodyne systems where the received optical signal is converted to an electrical signal by a photodetector. The basic principles of photodetection were presented in Chapter 5. The fundamental relation is (5.3) which states that the electron density of the output signal from a photodetector is proportional to the instantaneous intensity (power) of the incoming light.

Consider an optical field signal (8.3), amplitude-modulated by a low-pass real-valued function $A(t)$

$$B(t) = A(t) \exp[j(\omega_0 t + \theta_0(t) + \phi_0)] \tag{8.6}$$

The output from the photodetector is a shot noise process characterized by a photoelectron intensity $\Gamma(t)$. To avoid carrying the factor η/hf along in the analysis it is convenient, in the same way as in Section 7.4, to normalize the optical field such that the photon intensity becomes

$$\Gamma(t) = \frac{1}{2}|B(t)|^2 \tag{8.7}$$

For the signal (8.6) the square envelope $|B(t)|^2 = A^2(t)$ and

$$m = \int_0^T \Gamma(t)\, dt = \frac{1}{2}\int_0^T A^2(t)\, dt \tag{8.8}$$

is the average number of photoelectrons generated by the signal $A(t)$ during the time interval $[0, T]$.

The functional dependence between $\Gamma(t)$ and $B(t)$ is quadratic. If an electrical signal $A \cos \omega t$ is applied to a quadratic device, the output contains the mean square value $A^2/2$ but also a double frequency term $A^2 \cos(2\omega t)$. For optical signals, however, no double frequency component appears in $\Gamma(t)$. For a verification see the quantum-field-theory analysis of photon absorption presented by Mandel (1967). Note that the complex representation of the field gives the correct result and the $\Gamma(t)$ defined by (8.7) contains no double frequency term.

For a single optical signal (8.6) the relation (8.7) shows that the intensity function $\Gamma(t)$ depends on the intensity modulation $A^2(t)$ only. It is not affected by any variations in the optical frequency or the phase function $\theta(t)$. For direct detection of an intensity-modulated signal it does not matter if the received signal is coherent or not.

In heterodyning the received signal $A(t)$ is mixed with a locally generated optical signal $C(t)$ and the sum of these signals is fed into the photodetector. Let the combined optical field in complex notation be

$$B(t) = C(t) \exp[j(\omega_1 t + \theta_1(t) + \phi_1)] + A(t) \exp[j(\omega_0 t + \theta_0(t) + \phi_0)] \tag{8.9}$$

The magnitude squared is

$$|B(t)|^2 = C^2(t) + A^2(t) + 2C(t)A(t) \cos[(\omega_1 - \omega_0)t + \theta_1(t) - \theta_0(t) + \phi_1 - \phi_0] \tag{8.10}$$

which when substituted into (8.7) gives the intensity of the Poisson process produced by the photodetector. The intensity function $\Gamma(t)$ now contains a component at the heterodyne (angular) frequency $\omega_1 - \omega_0$.

The quantity $|B(t)|$ is the envelope of the signal $B(t)$. It can be shown that the relation (8.7) is also valid for more complicated optical signals than those characterized by (low-pass) stochastic phase noise $\theta(t)$. A general representation of an optical field is the complex representation of a bandpass signal (B.22) used in Appendix B

$$B(t) = \tilde{B}(t)e^{j\omega_0 t} \tag{8.11}$$

The optical (angular) frequency ω_0 in (8.11) can be specified arbitrarily, and different ω_0 give different envelopes $|B(t)|$. The correct choice turns out to let ω_0 be equal to the center of gravity of the optical power spectral density, c.f. (4.82). This choice of ω_0 produces an envelope of minimum variation in the mean square sense, as shown by Papoulis (1984). Note that ω_0 for a Lorentzian spectrum satisfies this condition.

8.1.2   Phase Noise Statistics

The operation of a heterodyne receiver is based on the output from the photodetector, which is a shot noise process with intensity proportional to (8.10). To simplify the analysis we assume that the front end filter processing this shot noise signal is of the integrate-and-dump type. More general filtering operation has been studied in the literature, but the effect on system performance is not great; see Jacobsen (1994).

The key quantity determining the transmission error probability in the presence of phase noise is the stochastic variable

$$y = \frac{1}{T} \int_0^T e^{j\theta(t)} \, dt \tag{8.12}$$

representing filtered phase noise.

The phase noise $\theta(t)$ is a simple Markov random walk process, a Wiener process. However, the filtered noise process

$$Z(t) = \frac{1}{T} \int_0^t e^{j\theta(s)} \, ds \tag{8.13}$$

is not Markov since the evolution of the process depends not only on the present position but also on the direction of its progress. An example of $Z(t)$ is shown in Figure 8.2a.

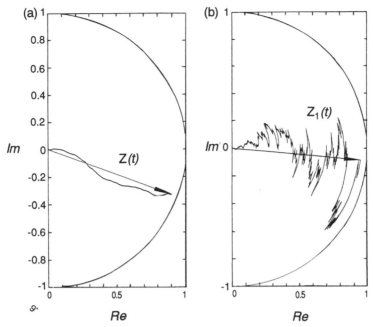

Figure 8.2   Examples of filtered phase noise with $B_L T = 1/\pi = 0.32$:   (a) The evolution of the phase noise integral $Z(t)$ when t increases from 0 to T. (b) The related integral $Z_1(t)$ generated by the same phase noise sequence $\theta(t)$. Note that $\mid Z(T) \mid = \mid Z_1(T) \mid = 0.95$

The related process

$$Z_1(t) = \frac{1}{T}\int_0^t e^{j[\theta(s)-\theta(t)]} \, ds \tag{8.14}$$

illustrated in Figure 8.2b, is Markov and $Z_1(T)$ has the same statistics as $Z(T)$.

The exact statistics of y is difficult to determine. Garrett *et al.* (1990) solved the Fokker-Planck equation associated with $Z_1(t)$ but the method posed significant numerical problems. An analytic expression for the probability density of $|Z(T)|$ is given by Xiaopin (1993). It contains an infinite summation with a slow rate of convergence, making it difficult to use for numerical evaluation.

We use an approach introduced by Foschini and Vannucci (1998b). The phase noise factor $\exp[j\theta(s)]$ is expanded in a Taylor series, and a few terms are retained which make the calculations feasible.

8.2   Ideal Coherent Systems

To show the basic properties of heterodyning in combination with different modulation methods we first analyse the idealized situation with perfect reception where the received signal is completely coherent without any phase noise and an exact phase reference is available in the receiver.

The optical signals are represented by a field (8.2), with ω_0 and ϕ_0 being constants, known both at the transmitter and the receiver.

8.2.1 Homodyne PSK

Phase-shift keying (PSK) means that the binary 'zero' and 'one' are represented by signals identical in shape but with a difference of 180-degrees in carrier phase. Consider signals of time duration T and let the received optical field signals be

$$\left.\begin{aligned}
s_0(t) &= A(t)\cos(\omega_0 t + \phi_0) \\
s_1(t) &= A(t)\cos(\omega_0 t + \phi_0 + \pi) = -A(t)\cos(\omega_0 t + \phi_0)
\end{aligned}\right\} \; 0 \le t \le T \qquad (8.15)$$

The receiver is shown in Figure 8.1. The received signal is mixed (added) optically to a local oscillator signal with constant amplitude and frequency and with a phase identical to the 'zero' signal

$$c(t) = C\cos(\omega_0 t + \phi_0)$$

The combined signal is focused on a photodetector which produces a shot noise process with an intensity function given by (8.7) and (8.10) with $\omega_1 = \omega_0$ and $\phi_1 = \phi_0$. The result can be written as

$$\Gamma(t) = \frac{1}{2}[C \pm A(t)]^2 \qquad (8.16)$$

where the plus sign represents the data symbol 'zero' and the minus sign the symbol 'one'.

The receiver is assumed to be of the photon-counting type analyzed in Chapter 5, which can be implemented as a rectangular (integrate-and-dump) filter followed by a decision threshold.

The decision variable after the receiver filter is the number of photoelectrons produced by the photodetector in the time interval $[0, T]$. It is a Poisson variable with a mean

$$E = \int_0^T \Gamma(t)dt = \frac{1}{2}\int_0^T [C^2 + A^2(t)]dt \pm C\int_0^T A(t)dt \qquad (8.17)$$

In Appendix D an upper bound (D.25) on P_e is derived for the Poisson channel with general binary intensity modulation. It is expressed in terms of the parameters μ and ε defined by

$$E = \mu(1 \pm \varepsilon) \qquad (8.18)$$

The amplitude C of the local oscillator signal is usually much stronger than the received signal amplitude $A(t)$, and (8.17) gives for $C \gg A$

$$\mu = \frac{1}{2} \int_0^T [C^2 + A^2(t)] \, dt \approx \frac{C^2 T}{2} \qquad (8.19)$$

and

$$\varepsilon = 2C \int_0^T A(t) \, dt / \int_0^T [C^2 + A^2(t)] \, dt \approx \frac{2}{CT} \int_0^T A(t) dt \qquad (8.20)$$

For large C the parameter μ is large and ε is small. From (D.29) in the Appendix follows that asymptotically for $\mu \gg 1$ and $\varepsilon \ll 1$

$$P_e < \frac{1}{\sqrt{2\pi x^2}} \exp(-x^2/2) \qquad (8.21)$$

with

$$x^2 = \mu \varepsilon^2 = \frac{4}{T} \left(\int_0^T A(t) dt \right)^2 \qquad (8.22)$$

The average number of photoelectrons produced by a received signal element is from (8.8)

$$m = \frac{1}{2} \int_0^T A^2(t) dt \qquad (8.23)$$

Schwartz inequality, see for instance Papoulis (1984), states that

$$\left(\int_a^b f(t)g(t) dt \right)^2 \leq \int_a^b f^2(t) dt \int_a^b g^2(t) dt \qquad (8.24)$$

with equality if and only if $f(t) = k g(t)$. Letting $f(t) = 1/T$ and $g(t) = A(t)$ yields

$$\left(\frac{1}{T} \int_0^T A(t) dt \right)^2 \leq \frac{1}{T} \int_0^T A^2(t) dt \qquad (8.25)$$

Let $x = 6$ be the value which, substituted into (8.21), makes the right-hand side equal to the system specification $P_e < 10^{-9}$. Combination of (8.22), (8.23) and (8.25) then gives

$$m \geq x^2/4 = 36/4 = 9 \qquad (8.26)$$

Equality is achieved in (8.25) if and only if $A(t)$ is equal to a constant, i.e. the received optical pulses are rectangular.

The ideal situation with $\eta = 1$ constitutes a quantum limit requiring that at least 9 received photons per bit are needed for homodyne PSK to achieve an error probability of $P_e = 10^{-9}$. This is only slightly less than the corresponding value of 10 photons per bit needed for an incoherent on-off modulated system.

Our derivation is based on an upper bound (8.21) on P_e. It can be shown that (8.21) represents the correct asymptotic error probability for large local oscillator signal amplitude C, and $m = 9$ is the coherent quantum limit.

The assumption that the local oscillator signal is much greater than the received signal creates a strong output signal from the photodetector and thermal noise in the electrical circuitry after the detector can usually be neglected. In direct detection systems the limiting factor is often the thermal noise, and the quantum limit has less practical significance. In heterodyning the optical noise determines the performance and there is a real possibility of reaching the quantum limit.

In the incoherent case the quantum limit is achieved by a rectangular filter receiver for arbitrary shape of the received pulses. For a coherent system a photon-counting receiver achieves the quantum limit for rectangular received pulses only. For other shapes a larger number of photons per bit is needed.

Example 8.1   System with triangular pulses

Consider a homodyne PSK optical system with triangular pulses

$$A(t) = A(1 - 2|t|/T); \quad |t| \le T/2$$

The integrals in (8.25) become

$$\frac{1}{T} \int_{-T/2}^{T/2} A(t)\, dt = A/2 \tag{8.27}$$

and

$$\frac{1}{T} \int_{-T/2}^{T/2} A^2(t)\, dt = A^2/3 \tag{8.28}$$

The ratio between (8.28) and the square of (8.27) is

$$\frac{A^2/3}{(A/2)^2} = 4/3$$

After multiplication of the right-hand side of (8.26) with this factor, the average number of photons per bit for $\rho = 6$ becomes

$$m = (4/3)(\rho^2/4) = 12.0 \qquad \square$$

The difference in performance between rectangular and triangular pulses is due to the fact that a rectangular filter is optimal for rectangular pulses only. The optimal receiver filter for a general Poisson channel is (5.143). Substitution of (8.16) after expansion of the logarithm using the fact that $C \gg A(t)$ determines the optimal receiver filter function

$$v(t) = \ln \frac{\Gamma_1}{\Gamma_0} = \ln \left(\frac{C + A(t)}{C - A(t)} \right)^2 \approx \frac{4 A(t)}{C} \tag{8.29}$$

which means that the receiver filter should be matched to the signal $A(t)$.

For a receiver with filter $v(t) = A(t)/C$ the mean and variance of the decision variable at the filter output are, from Campbell's Theorem (5.131)

$$E = \int \Gamma(t)v(t)\, dt = \frac{1}{2C} \int_0^T (C \pm A(t))^2 A(t)\, dt$$

$$\approx \frac{C}{2} \int_0^T A(t)\, dt \pm 2 \int_0^T A^2(t)\, dt \tag{8.30}$$

and

$$\sigma^2 = \int \Gamma(t)v^2(t)\, dt = \frac{1}{2C^2} \int_0^T (C \pm A(t))^2 A^2(t)\, dt$$

$$\approx \frac{1}{2} \int_0^T A^2(t)\, dt \tag{8.31}$$

The signal-to-noise ratio (5.133) at the decision point is, with use of (8.23)

$$\rho = \frac{E_1 - E_0}{\sigma_1 + \sigma_0} = \sqrt{2}\left(\int_0^T A^2(t)\, dt \right)^{1/2} = 2\sqrt{m} \tag{8.32}$$

The error probability bound (8.21) is valid for a decision variable with a Poisson distribution and it applies for a receiver with rectangular filter. For other situations the error probability can be calculated by the saddlepoint approximation as described in Chapter 5.

The error probability can be estimated from the signal-to-noise ratio (8.32) by the Gaussian approximation (5.132). An error probability of $P_e = 10^{-9}$ requires $\rho = 6$ and substitution into (8.32) gives

$$m = \rho^2/4 = 9 \tag{8.33}$$

This is the same result as (8.26) obtained for rectangular pulses by the exact Poisson distribution.

The results of this section indicates that a homodyne optical channel with a strong local oscillator can be modeled as an additive white Gaussian noise channel. The optimal optical receiver contains a matched filter (8.29) and the error probability can be obtained from the Gaussian approximation.

8.2.2   *Homodyne ASK*

In amplitude-shift keying (ASK) binary 'zero' and 'one' are represented by signals with different amplitudes

$$\left.\begin{array}{l} s_0(t) = A_0\, a(t) \cos(\omega_0 t + \phi_0) \\ s_1(t) = A_1\, a(t) \cos(\omega_0 t + \phi_0) \end{array}\right\} \quad 0 \leq t \leq T \tag{8.34}$$

without any loss of generality we let $A_1 > A_0 \geq 0$.

The pulses are assumed to be normalized such that

$$\frac{1}{T}\int_0^T a^2(t)dt = 1 \tag{8.35}$$

The intensity functions for the photodetector output are

$$\Gamma_i(t) = \frac{1}{2}[C + A_i a(t)]^2 \; ; \quad i = 0, 1 \tag{8.36}$$

For a receiver with an integrate-and-dump filter the decision variable has a Poisson distribution with the mean

$$E_i = \int_0^T \Gamma_i(t) \, dt = \frac{1}{2}\int_0^T [C^2 + A_i^2 a^2(t)] \, dt + C\int_0^T A_i a(t) \, dt \tag{8.37}$$

The parameters μ and ε are, for $C \gg A_i$

$$\mu = (E_1 + E_0)/2 \approx \frac{C^2 T}{2} \tag{8.38}$$

and

$$\varepsilon = \frac{E_1 - E_0}{E_1 + E_0} \approx \frac{A_1 - A_0}{CT}\int_0^T a(t)dt \tag{8.39}$$

The error probability is bounded by (8.21) with

$$x^2 = \mu\varepsilon^2 = \frac{(A_1 - A_0)^2}{2T}\left(\int_0^T a(t)dt\right)^2 \tag{8.40}$$

From the inequality (8.25) in combination with (8.35) it follows that

$$\left(\frac{1}{T}\int_0^T a(t)dt\right)^2 \leq 1 \tag{8.41}$$

with equality if and only if the pulses $a(t)$ have rectangular shape. Application of (8.41) to (8.40) yields

$$x^2 \leq (A_1 - A_0)^2 T/2 \tag{8.42}$$

The number of received photoelectrons for binary 'one' and 'zero' are $m_1 = A_1^2 T/2$ and $m_0 = A_0^2 T/2$, respectively. The average $(m_1 + m_0)/2$ is minimized when $A_0 = 0$ which gives

$$m_1/2 = A_1^2 T/4 \tag{8.43}$$

The error probability can be estimated by (8.21) which is the same as letting ρ be equal to x. From (8.43) and (8.42) it follows that to meet the system requirement on P_e specified by ρ

$$m_1/2 \geq \rho^2/2 \qquad (8.44)$$

with equality if the optical pulses are rectangular.

For $P_e = 10^9$, $\rho = 6.0$ and $m_1/2 \geq 18$, which means that ideal ASK requires 18 photons per bit, which is twice as much as for ideal PSK. In terms of signal power this corresponds to a difference of 3 dB.

For pulses $a(t)$ of arbitrary shape the optimal receiver contains a matched filter; see (8.36). The decision variable does not have a Poisson distribution but the Gaussian approximation (8.32) can be used, which yields a result identical to (8.44).

8.2.3   Heterodyne PSK

In heterodyne phase-shift keying (PSK) the local oscillator signal has a frequency different from ω_0:

$$C(t) = C \cos(\omega_1 t + \phi_1)$$

The receiver has the same appearance as Figure 8.1. The intensity functions at the photodetector output are from (8.10)

$$\Gamma(t) = \frac{C^2 + A^2(t)}{2} \pm CA(t) \cos(\omega_h t + \phi_h) \qquad (8.45)$$

where $\omega_h - \omega_1 - \omega_0$ and $\phi_h = \phi_1 - \phi_0$.

A matched filter function is now

$$v(t) = \frac{A(t)}{C} \cos(\omega_h t + \phi_h) \qquad (8.46)$$

The filter can be realized as a bandpass integrator centered on the heterodyne frequency ω_h, which usually is located in the microwave band. This means that it can be implemented by conventional electrical circuit devices.

The mean of the decision variable at the filter output is for $C \gg A(t)$

$$E = \frac{1}{2C} \int_0^T [C^2 + A^2(t) \pm 2CA(t) \cos(\omega_h t + \phi_h)] A(t) \cos(\omega_h t + \phi_h) \, dt$$

$$\approx \pm \frac{1}{2} \int_0^T A^2(t) \, dt \qquad (8.47)$$

under the assumption that the intermediate frequency ω_h is sufficiently large, that the integrals containing $\cos \omega_h t$ and $\cos 2\omega_h t$ can be neglected. The variances are under the same conditions

$$\sigma_0^2 = \sigma_1^2 = \frac{1}{2C^2} \int_0^T [C^2 + A^2(t) \pm 2CA(t) \cos(\omega_h t + \phi_h)] A^2(t) \cos^2(\omega_h t + \phi_h) \, dt$$

$$\approx \frac{1}{4} \int_0^T A^2(t) \, dt \qquad (8.48)$$

The signal-to-noise ratio is

$$\rho = \frac{E_1 - E_0}{\sigma_1 + \sigma_0} = \left(\int_0^T A^2(t) \, dt \right)^{1/2} = \sqrt{2m} \tag{8.49}$$

For $\rho = 6.0$, $m = 18$, which is twice as much as for homodyne PSK and the same as for homodyne ASK.

8.2.4   Heterodyne ASK

The analysis of heterodyne ASK is completely analogous to that of heterodyne PSK.

The intensity function at the photodetector output is, cf. (8.45)

$$\Gamma_i(t) = \frac{C^2 + A_i^2 a^2(t)}{2} + CA_i a(t) \cos(\omega_h t + \phi_h) \tag{8.50}$$

A matched filter function is

$$v(t) = \frac{a(t)}{C} \cos(\omega_h t + \phi_h) \tag{8.51}$$

The difference of the means of the decision variables at the filter output is for $C \gg A_i$, c.f. (8.47)

$$E_1 - E_0 \approx \frac{(A_1 - A_0)}{2} \int_0^T a^2(t) \, dt \tag{8.52}$$

and the variances

$$\sigma_0^2 = \sigma_1^2 \approx \frac{1}{4} \int_0^T a^2(t) \, dt \tag{8.53}$$

With $a(t)$ normalized according to (8.35) the signal-to-noise ratio is

$$\rho = \frac{E_1 - E_0}{\sigma_1 + \sigma_0} = \frac{(A_1 - A_0)\sqrt{T}}{2} \tag{8.54}$$

For $A_0 = 0$ which minimizes the average number $(m_1 + m_0)/2$ of received photoelectrons

$$\rho^2 = A_1^2 T/4 = m_1/2 \tag{8.55}$$

which for $P_e = 10^{-9}$ and $\rho = 6.0$ means $m_1/2 = 36$ photons per bit, twice the value required for homodyne ASK (8.44).

For PSK and ASK, a heterodyne receiver requires 3 dB more signal power than a homodyne receiver.

8.2.5 *Heterodyne FSK*

For frequency-shift keying (FSK) the signals have the same shape but different optical frequencies

$$\left.\begin{array}{l} s_0(t) = A(t)\cos(\omega_0 t + \phi) \\ s_1(t) = A(t)\cos(\omega_1 t + \phi) \end{array}\right\} \quad 0 \le t \le T \tag{8.56}$$

Let the local oscillator have frequency ω_2. To simplify the analysis it is advantageous to let it be shifted $\pi/2$ compared with the signals, i.e.

$$C(t) = C\sin(\omega_2 t + \phi)$$

The intensity functions at the photodetector output are

$$\Gamma_i(t) = \frac{C^2 + A^2(t)}{2} + CA(t)\sin\Omega_i t; \quad i = 0, 1 \tag{8.57}$$

with $\Omega_0 = \omega_2 - \omega_0$ and $\Omega_1 = \omega_2 - \omega_1$.

The optimal receiver filter $v(t) = \ln(\Gamma_1/\Gamma_0)$ becomes for $C \gg A(t)$ a matched filter

$$v(t) = \frac{A(t)}{C}(\sin\Omega_1 t - \sin\Omega_0 t) \tag{8.58}$$

The difference of the means of the decision variables is

$$E_1 - E_0 = \int (\Gamma_1(t) - \Gamma_0(t))v(t)\,dt \approx \int_0^T A^2(t)(\sin\Omega_1 t - \sin\Omega_0 t)^2\,dt \tag{8.59}$$

and for $C \gg A(t)$

$$\sigma_0^2 = \sigma_1^2 = \frac{1}{2}\int_0^T A^2(t)(\sin\Omega_1 t - \sin\Omega_0 t)^2\,dt \tag{8.60}$$

The signal-to-noise ratio becomes

$$\rho = \frac{E_1 - E_0}{\sigma_0 + \sigma_1} = \sqrt{I/2} \tag{8.61}$$

with

$$I = \int_0^T A^2(t)(\sin\Omega_1 t - \sin\Omega_0 t)^2\,dt \tag{8.62}$$

The signal-to-noise ratio is maximized when ω_2 is chosen such that the integral (8.62) assumes its maximal value. This occurs when the local oscillator frequency ω_2 is equal to $(\omega_0 + \omega_1)/2$ and $\Omega_0 = -\Omega_1$. Then

$$I = \int_0^T A^2(t)(2\sin\Omega_1 t)^2\,dt = 2\int_0^T A^2(t)\,dt \tag{8.63}$$

under the assumption that Ω_1 is large compared with the variation of $A^2(t)$.

The signal-to-noise ratio squared becomes

$$\rho^2 = I/2 = \int_0^T A^2(t)\,dt = 2m \tag{8.64}$$

The number of photons per bit to achieve $P_e = 10^{-9}$, corresponding to $\rho = 6.0$, is $m = 18$.

The particular choice $w_2 = (w_0 + w_1)/2$ makes

$$\sin \Omega_0 t = -\sin \Omega_1 t$$

and (8.57) becomes identical to (8.45) for heterodyne PSK. This explains why the same quantum limit $m = 18$ is obtained in both cases.

For FSK systems with well-separated frequencies, so-called wide-band FM, the receiver is often designed with separate filters for the two signal alternatives

$$v_i(t) = \frac{A(t)}{C} \sin \Omega_i t; \quad i = 0, 1 \tag{8.65}$$

The decision is based on which of the filters that produces the largest output. When $|\Omega_0| \neq |\Omega_1|$, and the difference is not too small, the signals are orthogonal and

$$\int_0^T \sin \Omega_0 t\ \sin \Omega_1 t\ dt \approx 0$$

The integral (8.62) is now

$$I = \int_0^T A^2(t)(\sin^2 \Omega_1 t + \sin^2 \Omega_0 t)\,dt = \int_0^T A^2(t)\,dt \tag{8.66}$$

The signal-to-noise ratio squared is

$$\rho^2 = I/2 = \frac{1}{2}\int_0^T A^2(t)\,dt = m \tag{8.67}$$

and the number of photons per bit becomes $m = 36$ for $P_e = 10^{-9}$. This value is commonly referred to as the quantum limit for coherent heterodyne FSK.

8.3 Phase-locked Loop Receiver

The ideal situation studied in the preceding sections assumed coherent optical signals of constant frequency and phase. For coherent detection these parameters must to be known at the receiver. The optical frequency can usually be considered as fixed, but in practice a phase estimator is needed in the receiver.

In this section we consider the more realistic situation where the optical signals are not perfect but contain phase noise, and where the phase is estimated by a phase-locked loop (PLL). A block diagram of an optical PLL is shown in Figure 8.3. The voltage-controlled oscillator (VCO) provides negative feedback.

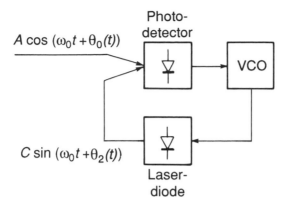

Figure 8.3   Optical phase-locked loop (PLL).

We consider PSK modulation, which generates signals of constant envelope which in turn facilitates the operation of the PLL. The analysis assumes that the PLL is driven by a continuous phase signal accomplished by a 'modulation wipe-out' circuit which eliminates the phase reversals in the demodulated signal. It is controlled by the detected data symbols, which are assumed to be correct.

The phase reference signal is produced by a local laser in the PLL and its phase noise will affect the quality of the estimate.

An exact analysis is complicated and a simplified approach based on a Gaussian approximation is presented here. The operation of an phase-locked loop is described and the influence of phase noise on the quality of the reference signal and on detector performance are studied.

Alternative approaches can be found in Salz (1985), Kazovsky (1985) and Hodgkinson (1987).

8.3.1   Optical Phase-locked Loop

Let the input optical signal to the PLL in Figure 8.3 be the continuous signal

$$r(t) = A\cos(\omega_0 t + \theta_0(t)) \tag{8.68}$$

It is added to the feedback signal

$$f(t) = C\sin(\omega_0 t + \theta_2(t)) \tag{8.69}$$

and the sum is focused on the photodetector surface.

The output $y(t)$ from the photodetector in the PLL is a shot noise process with an intensity function determined by the complex envelope for the combined signals (8.68) and (8.69). In the same way as in (8.10)

$$\Gamma_y(t) = \frac{C^2 + A^2}{2} + CA\sin\psi(t) \tag{8.70}$$

where $\psi(t)$ is the phase error

$$\psi(t) = \theta_2(t) - \theta_0(t) \tag{8.71}$$

The laser diode, driven by the voltage-controlled oscillator (VCO), produces an optical signal with an instantaneous angular frequency proportional to the input $y(t)$

$$\omega = -a\,y(t) + b \tag{8.72}$$

where a and b are constants to be determined later. The negative sign of a indicates negative feedback.

The phase function $\theta_2(t)$ of the feedback signal (8.69) is equal to the time integral of the angular frequency difference $\omega - \omega_0$ plus the phase noise $\theta_1(t)$ of the laser diode

$$\theta_2(t) = \theta_1(t) - a \int_0^t y(u)\, du + (b - \omega_0)t \tag{8.73}$$

It is shown in Appendix C that when $C \gg A$ the shot noise can be approximated by additive white Gaussian noise. The signal $y(t)$ can, analogously to (C.49), be modeled as

$$y(t) = C[C/2 + A \sin \psi(t) + n_1(t)] \tag{8.74'}$$

where $n_1(t)$ is a Gaussian white noise process with a spectral density expressed as an electron intensity

$$\mathcal{N}_1 = 1/2 \tag{8.75}$$

For homodyne reception the frequency from the VCO should be centered on ω_0. This is accomplished by choosing b in (8.72) such that

$$b = \omega_0 + a\,C^2/2 \tag{8.76}$$

which eliminates the bias term $C^2/2$ in (8.74) and yields

$$\omega = \omega_0 - K[A \sin \psi(t) + n_1(t)] \tag{8.77}$$

where $K = aC$ is a constant replacing a.

The derivative of (8.71) is, after substitution of (8.73)

$$\frac{d\psi(t)}{dt} = \omega - \omega_0 + \frac{d\theta_1(t)}{dt} - \frac{d\theta_0(t)}{dt} \tag{8.78}$$

which after substitution of (8.77) can be written as

$$\frac{d\psi(t)}{dt} + A\,K \sin \psi(t) = 2\pi\mu_1(t) - 2\pi\mu_0(t) + K\,n_1(t) = n(t) \tag{8.79}$$

where $\mu_0(t)$ and $\mu_1(t)$ are the frequency noise of the transmitting and the local oscillator laser, respectively. From (8.5) and (8.75) it follows that $n(t)$ is white Gaussian noise with a double-sided power spectral density

$$R_n = 2\pi B_L + K^2/2 \tag{8.80}$$

where $B_L = B_0 + B_1$ is the sum of the 3 dB spectral bandwidths of the transmitting and the local laser.

The phase error in the reference signal produced by the PLL is the solution $\psi(t)$ of the first-order nonlinear differential equation (8.79). When the noise is small compared with the signal amplitude A the phase error will be small and $\sin \psi \approx \psi$, which transforms (8.79) into a linear equation

$$\frac{d\varphi(t)}{dt} + AK\varphi(t) = n(t) \tag{8.81}$$

The phase error obtained from the linearized equation (8.81) is denoted by $\varphi(t)$ to distinguish it from $\psi(t)$, the solution of (8.79). The linearized PLL equation is easy to solve using linear system theory. Equation (8.81) represents a linear filter with input signal $n(t)$. The frequency function of the filter is

$$H(\omega) = \frac{1}{AK + j\omega} \tag{8.82}$$

The output $\varphi(t)$ is filtered white noise, which means that it is a stationary Gaussian stochastic process with zero mean and power spectral density

$$S_\varphi(\omega) = R_n |H(\omega)|^2 = \frac{R_n}{(AK)^2 + \omega^2} \tag{8.83}$$

The variance of φ is

$$\sigma_\varphi^2 = \frac{1}{2\pi} \int_{-\infty}^{\infty} S_\varphi(\omega)\, d\omega = \frac{R_n}{2AK} \tag{8.84}$$

The factor K is next chosen such that σ_φ is minimized. Substitution of N from (8.80) into (8.84) and letting the derivative with respect to K be equal to zero gives

$$\frac{d\sigma_\varphi^2}{dK} = -\frac{\pi B_L}{A K^2} + \frac{1}{4A} = 0 \tag{8.85}$$

The solution $K^2 = 4\pi B_L$ yields a phase error of minimal variance

$$\sigma_\varphi^2 = \left(\frac{\pi B_L}{A^2}\right)^{1/2} = \left(\frac{\pi B_L T}{2m}\right)^{1/2} \tag{8.86}$$

where $m = A^2 T/2$ is the average number of photoelectrons in the input PLL signal (8.68) during the signalling time interval T, see (8.23). The power spectral density (8.83) becomes

$$S_\varphi(\omega) = \frac{2\beta/T}{(\beta/T)^2 + \omega^2} \sigma_\varphi^2 \tag{8.87}$$

where $\beta = \sqrt{8\pi m B_L T} = 4m\sigma_\varphi^2$.

In the derivation above the phase φ is assumed to be a real variable taking values between $-\infty$ and ∞. From a physical point of view $\varphi = \varphi + 2\pi n$ and it is more appropriate to define the phase error in the interval $-\pi < \tilde{\varphi} \leq \pi$. The probability density of $\tilde{\varphi}$ is

$$p(\tilde{\varphi}) = \sum_{n=-\infty}^{\infty} p(\varphi + 2\pi n) \tag{8.88}$$

It is shown in Figure 8.4 for various values of $\delta = 1/\sigma_{\varphi}^2$. For $B_L T \gg m$, when σ_{φ}^2 is large, $p(\tilde{\varphi})$ approaches a uniform distribution and the variance of $\tilde{\varphi}$ takes its largest value $\pi/3$. It is clear from (8.88) that $\sigma_{\tilde{\varphi}}^2$ is always smaller than σ_{φ}^2.

The phase error $\varphi(t)$ of the linearized PLL is a stationary Gaussian process which is completely determined by the spectral density (8.83). The original non-linear equation (8.79) is more difficult to solve and the analysis presented here is limited to a derivation of the phase error variance.

The theory of random differential equations shows that the solution of a first-order differential equation driven by white noise will be a first-order Markov process. This means that it is a random walk process, where the statistics of the process at a certain time are determined by their values at that time, and are not influenced by the past, i.e. of the path the process has taken to reach the present state.

The probability density $p(\psi, t)$ of $\psi(t)$ at time t can be derived from the fundamental properties of Markov processes; see for instance Viterbi (1966). The result is the so-called Fokker-Planck equation, which for a first-order PLL takes the form

$$\frac{\partial p(\psi, t)}{\partial t} = \frac{\partial}{\partial \psi}[AK p(\psi, t) \sin \psi(t)] + \frac{R_n}{2} \frac{\partial^2 p(\psi, t)}{\partial \psi^2} \tag{8.89}$$

The stationary probability distribution for ψ, i.e. of the phase error in steady state operation after the initial synchronization transients have died out, is obtained from (8.89) by letting $\partial p/\partial t$ be equal to zero which yields

$$\frac{\partial}{\partial \psi}\left[AK p(\psi) \sin \psi + \frac{R_n}{2}\frac{\partial p(\psi)}{\partial \psi}\right] = 0 \tag{8.90}$$

Integration with respect to ψ gives

$$2AK p(\psi) \sin \psi + R_n \frac{\partial p(\psi)}{\partial \psi} = c_1 \tag{8.91}$$

The probability density $p(\psi)$ is defined in the region $-\pi \leq \psi \leq \pi$. The requirement that $p(-\pi) = p(\pi)$ makes it an even function which determines $c_1 = 0$. With $c_1 = 0$ the equation (8.91) is separable and can be written as

$$\delta \sin \psi \, d\psi = -\frac{dp}{p} \tag{8.92}$$

which after integration gives

$$\delta \cos \psi = \ln p + c_2$$

or equivalently

$$p(\psi) = c_3 \exp(\delta \cos \psi)$$

where $\delta = 2AK/N$.

The integration constant c_3 is determined from

$$\int_{-\pi}^{\pi} p(\psi)\, d\psi = 1$$

which gives

$$p(\psi) = \frac{\exp(\delta \cos \psi)}{2\pi I_0(\delta)} \tag{8.93}$$

where $I_0(\delta)$ is the modified Bessel function of zero order.

The probability density function (8.93) is shown in Figure 8.4 for various values of the parameter δ. In the literature it is referred to as the Tikhonov distribution after the Russian mathematician V. I. Tikhonov (1959).

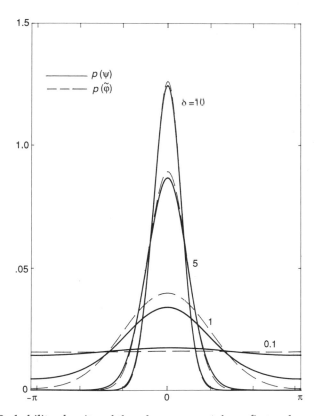

Figure 8.4 Probability density of the phase error ψ for a first-order optical PLL. The dashed curves are the probability densities of the phase error $\tilde{\varphi}$ from the linear approximation.

A large value of δ gives a distribution that is peaked around zero phase error. A comparison with (8.84) shows that $\delta = 1/\sigma_\varphi^2$ and that the value of K obtained from (8.85) will maximize δ. In the sequel we therefore use; see (8.86)

$$\delta = \left(\frac{2m}{\pi B_L T} \right)^{1/2} \tag{8.94}$$

8.3.2 Homodyne PSK with PLL

A block diagram of a homodyne PSK-PLL receiver is shown in Figure 8.5. The incoming signal is divided by a beam splitter into two parts, one of which is used in a conventional coherent PSK receiver to demodulate the signal. The other part is the input to the PLL which produces a phase estimate for the receiver. Note that the feedback signal of the PLL and the reference signal for the receiver differ by $\pi/2$ in phase.

We consider PSK modulation (8.15) with rectangular received pulses. The intensity function after the photodetector is

$$\Gamma(t) = \frac{C^2 + kA^2}{2} \pm C\sqrt{k}A \cos \psi(t) \tag{8.95}$$

where $\psi(t)$ is the phase error in the PLL reference signal. The factor k determines how the power of the received optical signal is divided between the receiver and the PLL.

The receiver contains a rectangular (integrate-and-dump) filter (5.49) followed by a decision threshold. The phase function $\psi(t)$ is a stochastic

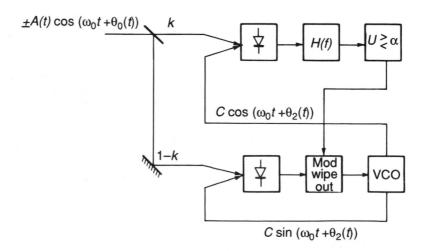

Figure 8.5 Homodyne PSK receiver with an optical PLL. The mod-wipe-out device eliminates the modulation and delivers a continuous signal to the VCO.

process and $\Gamma(t)$ is a doubly stochastic Poisson process with a random mean; see Section C.3 in Appendix C.

$$M = \int_0^T \Gamma(t)\, dt = (C^2 + kA^2)T/2 \pm C\sqrt{kATX} \qquad (8.96)$$

with

$$X = \frac{1}{T}\int_0^T \cos\psi(t)\, dt \qquad (8.97)$$

To calculate the signal-to-noise ratio at the decision point the mean and variance of the stochastic variable U are needed.

From (C.41)

$$E\{U\} = E\{M\} = (C^2 + kA^2)T/2 \pm C\sqrt{kATE\{X\}} \qquad (8.98)$$

and from (C.42)

$$\begin{aligned} \text{Var}\{U\} &= \text{Var}\{M\} + E\{M\} \\ &= C^2kA^2T^2\text{Var}\{X\} + (C^2 + kA^2)T/2 \pm C\sqrt{kATE\{X\}} \end{aligned} \qquad (8.99)$$

For $C \gg A$

$$\sigma_0^2 \approx \sigma_1^2 \approx C^2kA^2T^2\text{Var}\{X\} + C^2T/2 \qquad (8.100)$$

and the detection signal-to-noise ratio becomes after substitution of $m = A^2T/2$

$$\rho = \frac{E_1 - E_0}{\sigma_1 + \sigma_0} = \frac{2\sqrt{kmE\{X\}}}{(1 + 4km\text{Var}\{X\})^{1/2}} \qquad (8.101)$$

For a situation where the phase error $\psi(t)$ varies slowly within the symbol interval $[0, T]$ the integral (8.97) is approximately equal to

$$X = \frac{1}{T}\int_0^T \cos\psi(t)\, dt \approx \cos\psi \qquad (8.102)$$

The mean and the variance of the approximation (8.102) are easily obtained from the moment-generating function for $\cos\psi$. From the Tikhonov distribution (8.93)

$$\begin{aligned} \Psi(s) &= \int_{-\pi}^{\pi} \exp(s\cos\psi)p(\psi)\, d\psi \\ &= \frac{1}{2\pi I_0(\delta)}\int_{-\pi}^{\pi} \exp[(\delta + s)\cos\psi]\, d\psi = \frac{I_0(\delta + s)}{I_0(\delta)} \end{aligned} \qquad (8.103)$$

The first and second order derivatives of $\Psi(s)$ are

$$\Psi'(s) = \frac{I_1(\delta + s)}{I_0(\delta)}$$

and

$$\Psi''(s) = \frac{1}{I_0(\delta)} \left[I_0(\delta + s) - \frac{I_1(\delta + s)}{\delta + s} \right]$$

respectively.

The mean of $\cos \psi$ is

$$E\{\cos \psi\} = \Psi'(0) = I_1(\delta)/I_0(\delta) \tag{8.104}$$

and the variance

$$\text{Var}\{\cos \psi\} = \Psi''(0) - [\Psi'(0)]^2 = 1 - I_1(\delta)/\delta I_0(\delta) - [I_1(\delta)/I_0(\delta)]^2 \tag{8.105}$$

In the linear model of the PLL the phase error φ is a Gaussian stochastic variable and the mean of $\cos \varphi$ is

$$E\{\cos \varphi\} = \text{Re } E\{e^{j\varphi}\} = \text{Re } \frac{1}{\sqrt{2\pi} \, \sigma_\varphi} \int_{-\infty}^{\infty} \exp(j\varphi - \varphi^2/2\sigma_\varphi^2) \, d\varphi = e^{-\sigma_\varphi^2/2} \tag{8.106}$$

where from (8.86) and (8.94)

$$\sigma_\varphi^2 = \left(\frac{\pi B_L T}{2m(1 - k)} \right)^{1/2} = \frac{1}{\delta} \tag{8.107}$$

In the same way

$$E\{(\cos \varphi)^2\} = E\{(1 + \cos 2\varphi)/2\} = (1 + e^{-2\sigma_\varphi^2})/2$$

which yields

$$\text{Var}\{\cos \varphi\} = (1 - e^{-\sigma_\varphi^2})^2/2 \tag{8.108}$$

The exact mean $E\{\cos \psi\}$ and the approximate $E\{\cos \varphi\}$ are shown in Figure 8.6a as a function of the parameter σ_φ^2, and the variances (8.105) and (8.108) are shown in Figure 8.6b. The linear model overestimates the mean and underestimates the variance, but it constitutes a reasonable approximation at low values of σ_φ^2 which corresponds to a low phase noise parameter $B_L T$. At large values of $B_L T$ the phase error is uniformly distributed $[-\pi, \pi]$ and for both models the mean approaches zero and the variance the value 0.5.

The key statistic in the signal-to-noise ratio (8.101) is the stochastic variable $\mathcal{X}$ defined in (8.97). To calculate its variance a statistical model of the stochastic process $\cos \psi(t)$ is needed, and not only the first-order probability distribution (8.3.1).

For the linear approximation

$$\mathcal{X}_\ell = \frac{1}{T} \int_0^T \cos \varphi(t) \, dt \tag{8.109}$$

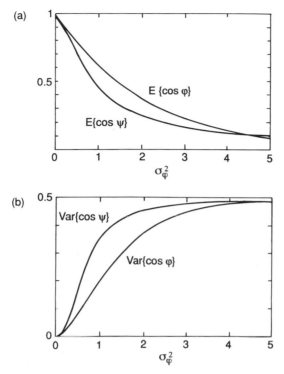

Figure 8.6 (a) The average of cos ψ for a first-order PLL and of cos φ from the linear approximation (b) The variance of cos ψ for a first-order PLL and of cos ψ from the linear approximation.

the analysis is easy due to the Gaussian statistics involved. The mean is, from (8.105)

$$E\{\mathcal{X}_\ell\} = \frac{1}{T} \int_0^T E\{\cos \varphi(t)\}\, dt = E\{\cos \varphi\} = e^{-\sigma_\varphi^2/2} \tag{8.110}$$

The mean square value becomes

$$E\{\mathcal{X}_\ell^2\} = \frac{1}{T^2} \int_0^T \int_0^T E\{\cos \varphi(s) \cos \varphi(t)\}\, ds\, dt$$

$$= \frac{2}{T^2} \int_0^T (T - \tau) R(\tau)\, d\tau \tag{8.111}$$

where

$$R(\tau) = E\{\cos \varphi(\tau + t) \cos \varphi(t)\}$$

is the correlation function of cos $\varphi(t)$. It is related to the correlation function of $\varphi(t)$ in the following way.
Consider

$$2 \cos \varphi(\tau + t) \cos \varphi(t) = \cos[\varphi(\tau + t) + \varphi(t)] + \cos[\varphi(\tau + t) - \varphi(t)] \tag{8.112}$$

The arguments of the cosine functions are Gaussian since they are the sum and the difference of Gaussian variables. Their variances are

$$E\{[\varphi(\tau + t) + \varphi(t)]^2\} = 2[R_\varphi(0) + R_\varphi(\tau)]$$

and

$$E\{[\varphi(\tau + t) - \varphi(t)]^2\} = 2[R_\varphi(0) - R_\varphi(\tau)]$$

respectively. The correlation function $R_\varphi(\tau)$ is the Fourier transform of the spectral density (8.87)

$$E\{\varphi(\tau + t)\varphi(t)\} = R_\varphi(\tau) = \sigma_\varphi^2 e^{-\beta|\tau|/T} \qquad (8.113)$$

where

$$\beta = 4m(1 - k)\sigma_\varphi^2 \qquad (8.114)$$

Forming the average of the cosine terms on the right-hand side of (8.112), see (8.106), gives the correlation function

$$R(\tau) = \frac{1}{2}\exp(-[R_\varphi(0) + R_\varphi(\tau)]) + \frac{1}{2}\exp(-[R_\varphi(0) - R_\varphi(\tau)])$$
$$= \exp(-\sigma_\varphi^2)\cosh[R_\varphi(\tau)] \qquad (8.115)$$

The mean square value of $\mathcal{X}_\ell$ is obtained by substitution of (8.115) into the last expression in (8.111) and carrying out the integration, which has to be done numerically. A useful and rapidly converging series expansion results when cosh[ ] is expanded in a power series

$$E\{\mathcal{X}_\ell^2\} = \frac{2e^{-\sigma_\varphi^2}}{T^2}\int_0^T (T - \tau)\left[1 + \frac{R_\varphi^2(\tau)}{2!} + \frac{R_\varphi^4(\tau)}{4!}\cdots\right] d\tau \qquad (8.116)$$

Substitution of $R_\varphi(\tau)$ from (8.113) and performing the integration yields

$$E\{\mathcal{X}_\ell^2\} = 2e^{-\sigma_\varphi^2}\left[\frac{1}{2} + \frac{\sigma_\varphi^4 r(2\beta)}{2!} + \frac{\sigma_\varphi^8 r(4\beta)}{4!}\cdots\right] \qquad (8.117)$$

where

$$r(x) = \frac{e^{-x} + x - 1}{x^2}$$

The series converges rapidly when $\sigma_\varphi < 1$, or equivalently $\delta > 1$, and the first two terms are often sufficient, which results in

$$\text{Var}\{\mathcal{X}_\ell\} = \frac{\sigma_\varphi^4 \exp(-\sigma_\varphi^2)[\exp(-2\beta) + 2\beta - 1]}{4\beta^2} \qquad (8.118)$$

with β defined in (8.114).

Substitution of (8.110) and (8.118) into (8.101) gives the signal-to-noise ratio for the linear model. Error probability curves obtained from the Gaussian

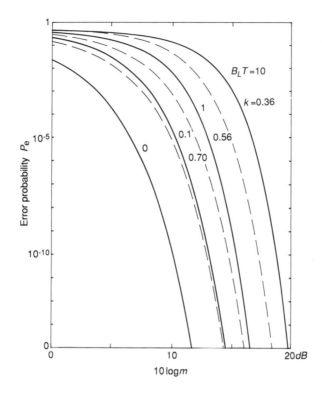

Figure 8.7 Approximate bit error probability for homodyne PSK with PLL-receiver for various values of the phase noise parameter $B_L T$ as a function of the average number of received photoelectrons per bit. The parameter k denotes the power division in the beam splitter. The dashed curves are results from the linear approximation of the PLL.

approximation $P_e = Q(\rho)$ are presented in Figure 8.7 for various values of the phase noise parameter $B_L T$.

The linear approximate model overestimates the mean and underestimates the variance of the decision variable, which means that it produces an error probability which is too low. An approximate estimate of the error probability for the nonlinear PLL is obtained by using the correct mean (8.103) together with an approximate variance obtained by increasing the linear variance (8.118) by a factor $\mathrm{Var}\{\cos\psi\}/\mathrm{Var}\{\cos\varphi\}$. This is equivalent to using the correct mean and variance for $\cos\psi$ in combination with spectral properties derived from the linear model. See Figure 8.7 for results.

In Figure 8.7 the parameter k is optimized for best performance. For very low phase noise it is easy to obtain a good estimate of the phase error and only a small part of the received optical power is needed for the PLL. For large phase errors more power is directed to the PLL since the signal-to-noise ratio benefits more from a better phase estimate than from increased signal power in the detector.

8.3.3 *Heterodyne PSK with PLL*

A block diagram for a heterodyne system for reception of PSK modulated signals using a PLL phase reference is shown in Figure 8.8. The phase-locked loop is operating on the heterodyne frequency which is located in the microwave band. Conventional electrical circuitry techniques can be used and there is no need to split the received signal power into two parts as in the homodyne receiver. As before the analysis assumes that the PLL is driven by a continuous signal.

The received signal $x(t)$ after the photodetector is a shot-noise process centered on the heterodyne frequency ω_h with intensity function

$$\Gamma_x(t) = \frac{C^2 + A^2}{2} \pm CA \cos[\omega_h t + \theta(t)] \tag{8.119}$$

where $\theta(t) = \theta_1(t) - \theta_0(t)$ is the combined phase noise of the received signal and the local oscillator, and $\omega_h = \omega_1 - \omega_0$.

For $C \gg A$ the stochastic process $x(t)$ can be modeled as the sum of a signal component and white Gaussian noise, see (8.74)

$$x(t) \approx C/2 \pm A \cos[\omega_h t + \theta(t)] + n_1(t) \tag{8.120}$$

We denote the feedback signal by

$$B \sin[\omega_h t + \theta(t) + \psi(t)]$$

with $\psi(t)$ equal to the phase error.

The signal $x(t)$ is multiplied by the feedback signal and the product is passed through a lowpass filter. The result, after the lowpass filter, is a baseband signal

$$y(t) = BC\left[\pm\frac{A}{2}\sin \psi(t) + n_2(t)\right] \tag{8.121}$$

where $n_2(t)$ has a spectral density $\mathcal{N}_2 = 1/4$ equal to half the value of (8.75) due

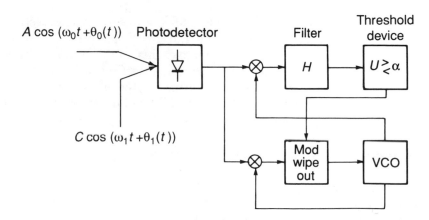

Figure 8.8   Heterodyne PSK receiver with electronic PLL.

to the effect of the demodulation operation. The bias C in (8.74) and modulation products at $2\omega_h$ are eliminated by the filter.

The analysis of the PLL is completely analogous to that of the homodyne case. The differential equation (8.79) becomes

$$\frac{\psi(t)}{dt} + \frac{AK}{2}\sin\psi(t) = n(t) \tag{8.122}$$

where $n(t)$ is white Gaussian noise with double-sided power spectral density

$$R_n = 2\pi B_L + K^2/4 \tag{8.123}$$

The optimal value of K is obtained in the same way as in Section 8.3.1 and the result is

$$\sigma_\varphi^2 = 1/\delta = \left(\frac{\pi B_L T}{m}\right)^{1/2} \tag{8.124}$$

A comparison with (8.107) shows that a heterodyne PLL produces a synchronization signal with a phase error identical to that of a homodyne PLL with $1 - k = 0.5$.

The detector uses the estimated phase reference signal

$$\cos[\omega_h t + \theta(t) + \psi(t)]$$

to demodulate the heterodyned received signal. The signal after the lowpass filter is analogous to (8.121)

$$z(t) = BC\left[\pm\frac{A}{2}\cos\psi(t) + n_3(t)\right] \tag{8.125}$$

To obtain the decision variable U the signal $z(t)$ is integrated over the symbol interval T, which is equivalent to passing it through a filter with rectangular impulse response (integrate-and-dump).

$$U = \int_0^T z(t)\,dt = BC\left[\pm\frac{A}{2}\int_0^T \cos\psi(t)\,dt + \int_0^T n_3(t)\,dt\right]$$
$$= BC[AT\mathcal{X}/2 + \xi] \tag{8.126}$$

where $\mathcal{X}$ is the phase error integral (8.97) and ξ is a Gaussian stochastic variable with variance $\mathcal{N}_2 T = T/4$.

The mean and variance of U are easily determined and the detection signal-to-noise ratio becomes

$$\rho = \frac{E_1 - E_0}{\sigma_1 + \sigma_0} = \frac{\sqrt{2m}E\{\mathcal{X}\}}{(1 + 2m\text{Var}\{\mathcal{X}\})^{1/2}} \tag{8.127}$$

A comparison with (8.102) reveals that the error probability of a heterodyne receiver is equal to that of a homodyne receiver with $k = 0.5$ since the heterodyne technique gives a 3 dB loss in both the detection and the PLL part of the receiver.

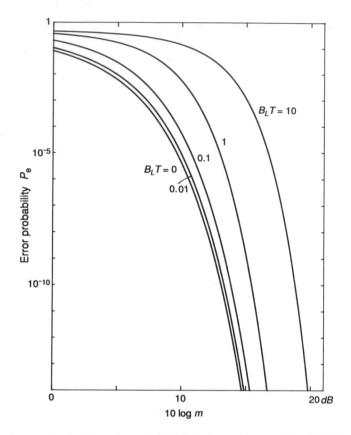

Figure 8.9 Approximate bit error probability for heterodyne PSK with PLL-receiver for various values of the phase noise parameter $B_L T$ as a function of the average number of received photoelectrons per bit.

The P_e as a function of the channel signal-to-noise ratio in dB is shown in Figure 8.9. The heterodyne is inferior to the homodyne by a factor of 3 dB at low values of the phase noise parameter $B_L T$. This is in accordance with the analysis of coherent systems in Section 8.2 which shows a 3 dB difference between homodyne and heterodyne systems. At higher $B_L T$, where the homodyne receiver requires a k-value around $k = 0.5$, the difference practically vanishes.

The analysis of phase-locked loop receivers show that phase noise can severely degrade system performance. The result is higher system costs both in terms of receiver sensitivity and equipment complexity. We have studied the PLL under the simplifying assumption that an unmodulated continuous-phase signal can be established for its operation. A practical system under less idealized conditions will produce a higher error probability than the one calculated here. An alternative would be to send a separate phase reference signal for the PLL. Transmitted-reference schemes are analyzed by Azizoglu and Humblet (1995).

8.4   Incoherent Receiver

The phase-locked loop receiver is a natural approach to coherent detection. However, the phase noise prevents the PLL from producing a perfect phase estimate, and it is only when $B_L T$ is small that the performance is close to the theoretical limits set by ideal coherent transmission.

The necessity to establish a phase reference complicates the receiver. An alternative approach studied by Salz (1985), Foschini *et al.* (1988), Garrett *et al.* (1990), and others, is to accept the fact that the phase is not known at the receiver, and to base the detection on the envelope of the received signal. Amplitude modulation (ASK) and frequency-shift keying (FSK) are possible alternatives. Ordinary phase-shift modulation (PSK) cannot be used since the envelope is the same for both signals. An alternative, however, is differential phase-shift keying (DPSK) where the signal in the preceding time interval is used as a phase reference.

The notion of an envelope presumes a carrier frequency, and envelope detection is possible for heterodyne reception only.

8.4.1   Heterodyne ASK

A block diagram of a heterodyne ASK receiver is shown in Figure 8.10. It contains an envelope detector together with a pre- and a postfilter which will be specified later.

The received optical signals for ASK are specified in (8.34). We consider a system with on-off intensity modulation with rectangular pulses of duration T and amplitudes $A_1 = A$ and $A_0 = 0$. The analysis is easily modified for $A_0 > 0$.

The output from the photodetector is a shot noise process with intensity, cf. (8.50)

$$\Gamma(t) = \frac{C^2 + A^2}{2} + CA \cos(\omega_h t + \theta(t)) \tag{8.128}$$

where $\theta(t) = \theta_2(t) - \theta_0(t)$ is the difference between the phase noise of the local oscillator and of the transmitting laser. The phase noise $\theta(t)$ is assumed to vary slowly compared with the heterodyne angular frequency ω_h.

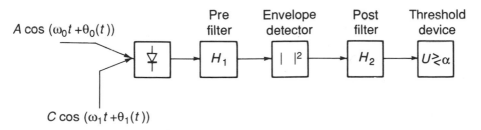

Figure 8.10   Heterodyne ASK receiver with envelope detector. The predetector filter H_1 is a bandpass filter at the heterodyne frequency and the postdetector filter H_2 is a lowpass filter.

The shot noise current $I(t)$ is modeled as the sum of the intensity and additive Gaussian noise, assuming that the local oscillator amplitude C is much greater than A, see Appendix C.

$$I(t)/C \approx C/2 + A\cos(\omega_h t + \theta(t)) + X_1(t) \tag{8.129}$$

where $X_1(t)$ is white Gaussian noise with intensity $\mathcal{N}_1 = 1/2$.

The prefilter H_1 is a bandpass integrator centered on the heterodyne frequency ω_h. Its impulse response is

$$h_1(t) = \begin{cases} 2\cos\omega_h t; & 0 < t < T' \\ 0; & \text{otherwise} \end{cases} \tag{8.130}$$

The time duration $T' = T/L$ with L a positive integer.

It is assumed that $\omega_h T' \gg 1$ which makes it convenient to use the complex representation of narrowband signals. The direct current term $C/2$ in (8.129) will be eliminated by the bandpass filter (8.130) and the baseband representation of the remaining part is

$$Y(t) = Ae^{j\theta(t)} + X(t) \tag{8.131}$$

where $X(t)$ is complex-valued white Gaussian noise with intensity $2\mathcal{N}_1 = 1$. The equivalent baseband impulse response for (8.130) is

$$f_1(t) = \begin{cases} 1; & 0 < t < T' \\ 0; & \text{otherwise} \end{cases} \tag{8.132}$$

During the data symbol interval the prefilter output is sampled L times at $t = kT'$, $k = 1, 2, \ldots, L$, generating a sequence of complex-valued stochastic variables. Let

$$V_k = \frac{1}{T'}\int_{(k-1)T'}^{kT'} Y(t)\,dt = \frac{1}{T'}\int_{(k-1)T'}^{kT'} [Ae^{j\theta(t)} + X(t)]\,dt \tag{8.133}$$

This can be expressed as

$$V_k = A\mathcal{Y}_k + X_k \tag{8.134}$$

where

$$\mathcal{Y}_k = \frac{1}{T'}\int_{(k-1)T'}^{kT'} e^{j\theta(t)}\,dt \tag{8.135}$$

and

$$X_k = \frac{1}{T'}\int_{(k-1)T'}^{kT'} X(t)\,dt \tag{8.136}$$

is a complex-valued, zero-mean Gaussian variable. Its quadrature components have equal variance

$$\sigma^2 = 2\mathcal{N}_1/T' = 1/T' \tag{8.137}$$

The envelope detector forms the square of the magnitude (absolute value) of V_k and the lowpass postfilter H_2 is modeled as a discrete time integrator. The decision variable is

$$U = \frac{T'}{2} \sum_{k=1}^{L} |V_k|^2 \tag{8.138}$$

where the factor $T'/2$ is included to simplify the results.

The fact that $X(t)$ is white Gaussian noise and the phase noise $\theta(t)$ is a random walk process with independent increments makes $|V_k|^2$ a sequence of independent and equally distributed random variables.

The probability distribution of U is related in a simple way to the moment-generating function (MGF) of

$$|V|^2 = |\mathcal{Y} + X|^2 = (\mathcal{Y}_c + X_c)^2 + (\mathcal{Y}_s + X_s)^2 \tag{8.139}$$

where the index k is dropped. The MGF of $|V|^2$ is

$$\Psi_V(s) = \mathrm{E}\{\exp(|V|^2 s)\} = \mathrm{E}\{\exp([\mathcal{Y}_c^2 + 2\mathcal{Y}_c X_c + X_c^2 + \mathcal{Y}_s^2 + 2\mathcal{Y}_s X_s + X_s^2]s) \tag{8.140}$$

The variables X_c, X_s and $\mathcal{Y}$ are independent, and the averaging in (8.140) can be carried out for each variable separately.

$$\Psi_V(s) = \mathrm{E}_y\Big\{[\exp(|\mathcal{Y}|^2 s)\mathrm{E}\{\exp(g_c(X_c)s)\}\mathrm{E}\{\exp(g_s(X_s)s)\}\Big\} \tag{8.141}$$

with

$$\left.\begin{array}{l} g_c(X_c) = 2\mathcal{Y}_c X_c + X_c^2 \\ g_s(X_s) = 2\mathcal{Y}_s X_s + X_s^2 \end{array}\right\} \tag{8.142}$$

The average over X_c is

$$\mathrm{E}\{\exp(g_c(X_c)s)\} = \frac{1}{\sqrt{2\pi\sigma^2}} \int_{-\infty}^{\infty} \exp([2\mathcal{Y}_c x_c + x_c^2]s) \exp\left(-\frac{x_c^2}{2\sigma^2}\right) dx_c \tag{8.143}$$

The integral can be evaluated by completing the square in the exponent. The result is

$$\mathrm{E}\{\exp(g_c(X_c)s)\} = \frac{1}{\sqrt{1 - 2\sigma^2 s}} \exp\left(\frac{2\mathcal{Y}_c^2 \sigma^2 s^2}{1 - 2\sigma^2 s}\right) \tag{8.144}$$

The average over X_s results in an identical expression as (8.144) with $\mathcal{Y}_c^2$ replaced by $\mathcal{Y}_s^2$. Substitution into (8.141) yields, after some algebra

$$\Psi_V(s) = \frac{1}{1 - 2\sigma^2 s} \Psi_y\left(\frac{s}{1 - 2\sigma^2 s}\right) \tag{8.145}$$

where

$$\Psi_{\mathcal{Y}}(s) = E\{\exp(|\mathcal{Y}|^2 s)\} \tag{8.146}$$

is the MGF of the squared envelope $|\mathcal{Y}|^2$.

The probability distribution of $|\mathcal{Y}|^2$ is difficult to determine. In an approximate approach, introduced by Forchini and Vannucci (1988), the integrand $\exp(j\theta(t))$ in (8.135) is expanded in a Taylor series.

$$\mathcal{Y} \approx \tilde{\mathcal{Y}} = 1 - \frac{1}{2T'} \int_0^{T'} \theta^2(t)\, dt + \frac{j}{T'} \int_0^{T'} \theta(t)\, dt \tag{8.147}$$

The statistical distribution for the square envelope of the variable $\tilde{\mathcal{Y}}$ is studied in Appendix G. The MGF of $|\tilde{\mathcal{Y}}|^2$ is (G.26)

$$\Psi_{\tilde{y}}(s) = \exp(s)\left[\text{sinch}\sqrt{2\beta s}\right]^{-1/2} \tag{8.148}$$

where 'sinch' denotes the hyperbolic sinc-function

$$\text{sinch}\, x = \frac{\sinh x}{x} = \frac{e^x - e^{-x}}{2x} \tag{8.149}$$

The parameter $\beta' = 2\pi B_L T'$ is equal to 2π times the product of B_L and the integration interval T' with $B_L = B_0 + B_1$ the sum of the 3 dB linewidths of the lasers at the transmitter and the local oscillator.

Substitution of (8.148) into (8.145) gives a useful approximate expression for the MGF of $|V_k|^2$. Including the factor A the result is

$$\Psi_V(s) = \frac{1}{1 - 2\sigma^2 s}\exp\left(\frac{A^2 s}{1 - 2\sigma^2 s}\right)\left[\text{sinch}\left(\sqrt{\frac{2\beta' A^2 s}{1 - 2\sigma^2 s}}\right)\right]^{-1/2} \tag{8.150}$$

The decision variable (8.138) is the sum of L equally distributed independent variables $|V_k|^2$, and the MGF of U is (8.151) with s replaced by $sT'/2$ and raised to the power L. From (8.137) it follows that $\sigma^2 T' = 1$ and

$$\Psi_U(s) = [\Psi_V(sT'/2)]^L = \frac{1}{(1-s)^L}\exp\left(\frac{m_1 s}{1-s}\right)\left[\text{sinch}\sqrt{\frac{2\beta m_1 s}{(1-s)L^2}}\right]^{-L/2} \tag{8.151}$$

where

$$\beta = 2\pi B_L T \tag{8.152}$$

and the parameter $m_1 = A^2 T/2 = A^2 L T'/2$ is the expected number of photoelectrons in the received optical pulse. The relation (8.145) and an expression equivalent to (8.151) have been presented by Ribeiro and Ferreira de Rocha (1994).

The error probability can be calculated from $\Psi_U(s)$ using the saddlepoint approximation in the usual way. The result is shown in Figure 8.11 for various values of the combined phase noise parameter $B_L T$. The value of L resulting in the lowest P_e is used in the calculation of P_e.

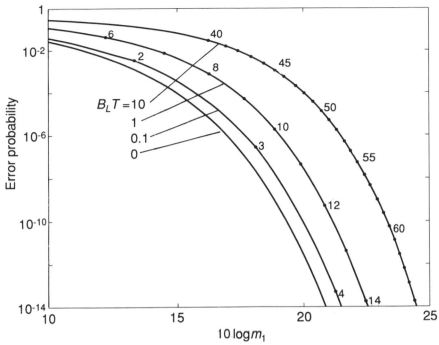

Figure 8.11 Bit error probability for heterodyne ASK (on-off modulation) with envelope detector receiver calculated from the Taylor expansion of the filtered phase noise for various values of $B_L T$. The optimal prefilted bandwidth parameter L is indicated along the curves. The parameter m_1 is the expected number of photoelectrons in the received pulse for symbol 'one'.

The prefilter bandwidth is proportional to $1/T' = B'$ and $L = T/T' = B'/B$ is a measure of the magnitude of B' relative to the rate or bandwidth B of the data signal. The optimal L as a function of $B_L T$ for systems operating at $P_e = 10^{-12}$ is shown in Figure 8.12.

The detector has a nonzero threshold and the Gaussian approximation can be used to obtain an estimate of the bit error probability.

To obtain the mean and variance of the decision variable (8.138), under the hypothesis that binary 'zero' or 'one' was transmitted, we first calculate the mean and variance of $|V|^2$. Evaluating the first derivative of the MGF (8.145) at $s = 0$ gives the mean

$$E\{|V|^2\} = E\{|\mathcal{Y}|^2\} + 2\sigma^2 \tag{8.153}$$

In the same way the second derivative yields

$$E\{|V|^4\} = E\{|\mathcal{Y}|^4\} + 8\sigma^2 E\{|\mathcal{Y}|^2\} + 8\sigma^4 \tag{8.154}$$

Combination of (8.153) and (8.154) gives

$$\{|V|^2\} = \{|\mathcal{Y}|^2\} + 4\sigma^2 E\{|\mathcal{Y}|^2\} + 4\sigma^4 \tag{8.155}$$

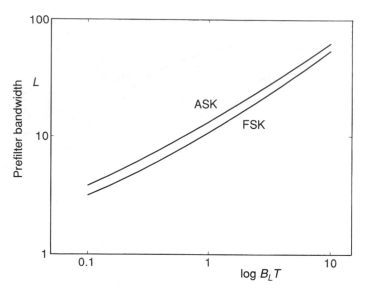

Figure 8.12 The optimal prefiler bandwidth parameter L as a function of the phase noise parameter $B_L T$ for ASK and FSK at $P_e = 10^{-12}$.

For on-off modulation with $A_0 = 0$ and $A_1 = A$ the decision variable for 'one' is

$$\left.\begin{aligned} \mathrm{E}\{U_1\} &= L(A^2\mathrm{E}\{|\mathcal{Y}|^2\} + 2\sigma^2) \\ \mathrm{Var}\{U_1\} &= L(A^4\mathrm{Var}\{|\mathcal{Y}|^2\} + 4A^2\sigma^2\mathrm{E}\{|\mathcal{Y}|^2\} + 4\sigma^4) \end{aligned}\right\} \tag{8.156}$$

for binary 'zero' transmitted $A = 0$ and

$$\left.\begin{aligned} \mathrm{E}\{U_0\} &= L2\sigma^2 \\ \mathrm{Var}\{U_0\} &= L4\sigma^4 \end{aligned}\right\} \tag{8.157}$$

The signal-to-noise ratio becomes

$$\begin{aligned} \rho &= \frac{\mathrm{E}\{U_1\} - \mathrm{E}\{U_0\}}{\sqrt{\mathrm{Var}\{U_1\}} + \sqrt{\mathrm{Var}\{U_0\}}} \\[2mm] &= \frac{LA^2\mathrm{E}\{|\mathcal{Y}|^2\}}{\sqrt{L\left(A^4\mathrm{Var}\{|\mathcal{Y}|^2\} + 4A^2\sigma^2\mathrm{E}\{|\mathcal{Y}|^2\} + 4\sigma^4\right)} + \sqrt{4L\sigma^4}} \end{aligned} \tag{8.158}$$

The integer $L = T/T'$ is a free parameter that should be optimized for the best system performance. It determines the bandwidth of the prefilter relative to the transmission rate.

The average number of received photoelectrons for binary 'one' is

$$m = \frac{A^2 T}{2} \tag{8.159}$$

Substitution of A^2 from (8.159) together with $\sigma^2 = 1/2T'$ and $T = L/T'$ into (8.158) yields

$$\rho = \frac{mE\{|\mathcal{Y}|^2\}\sqrt{L}}{\sqrt{m^2\text{Var}\{|\mathcal{Y}|^2\} + 2mLE\{|\mathcal{Y}|^2\} + L^2} + L} \tag{8.160}$$

As shown in Appendix G the mean and variance of $|\mathcal{Y}|^2$ are

$$E\{|\mathcal{Y}|^2\} = \frac{4(\beta' - 2 + 2e^{-\beta'/2})}{\beta'2} \tag{8.161}$$

and

$$\text{Var}\{|\mathcal{Y}|^2\} = \frac{8(711 - 198\beta' + 18\beta'^2 - 640e^{-\beta'/2} - 192\beta'e^{-\beta'/2} - 72e^{-\beta'} + e^{-2\beta'})}{9\beta'^4} \tag{8.162}$$

with $\beta' = 2\pi B_L T'$.

The error probability calculated by the Gaussian approximation is shown in Figure 8.13 with, for comparison, the result from the saddlepoint approximation. The parameter L is optimized at each point of the curves by choosing the value that maximizes the signal-to-noise ratio ρ.

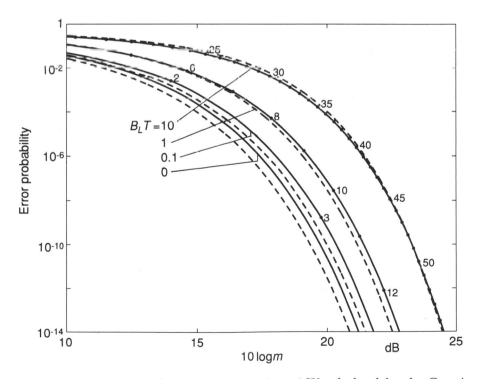

Figure 8.13 Bit error probability for heterodyne ASK calculated by the Gaussian approximation. For comparison the results from Figure 8.11 are shown as dashed lines. The optimal prefilter bandwidth parameter L is indicated along the curves.

Note that the expressions for the mean and variance in the Gaussian approximation are derived from the true phase noise variable $\mathcal{Y}$ and not from the series expansion (8.147) used in the saddlepoint approximation.

8.4.2　Heterodyne FSK

Consider a wideband optical FSK heterodyne system with rectangular pulses. The received optical signals are, c.f. (8.56)

$$\left.\begin{array}{l} s_0(t) = A\cos[\omega_0 t + \theta_{01}(t)] \\ s_1(t) = A\cos[\omega_1 t + \theta_{11}(t)] \end{array}\right\} \quad 0 \le t \le T \tag{8.163}$$

The received signal is mixed with a local oscillator signal of angular frequency ω_2 and it is assumed that $\omega_1 > \omega_2 > \omega_0$.

The intensity function of the shot noise after the photodetector is, analogously to (8.57)

$$\Gamma_i(t) = \frac{C^2 + A^2}{2} + CA\sin[\Omega_i t + \theta_i(t)]; \quad i = 0, 1 \tag{8.164}$$

with $\Omega_0 = \omega_2 - \omega_0$ equal to the heterodyne angular frequency for 'zero' and $\Omega_1 = \omega_2 - \omega_1$ for 'one'. The phase noises $\theta_0(t) = \theta_2(t) - \theta_{01}(t)$ and $\theta_1(t) = \theta_{11}(t) - \theta_2(t)$ represent the combined noise from the lasers at the transmitter and the local oscillator.

A block diagram of the incoherent receiver is shown in Figure 8.14. It contains two prefilters H_0 and H_1 in the form of bandpass integrators centered on the frequencies Ω_0 and Ω_1, respectively. Their impulse responses are

$$h_i(t) = \begin{cases} 2\cos(\Omega_i t); & 0 < t < T'; \\ & \qquad\qquad\qquad i = 0, 1 \\ 0; & \text{otherwise;} \end{cases} \tag{8.165}$$

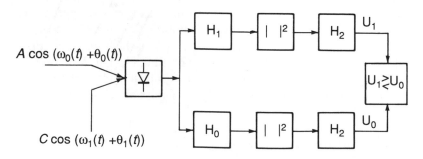

Figure 8.14　Heterodyne FSK receiver with envelope detectors. The predetector filters H_1 and H_0 are bandpass filters at the heterodyne frequencies representing data symbols 'one' and 'zero', respectively. The postdetector filters H_2 are lowpass filters.

Each bandpass filter is followed by an envelope detector and a lowpass postdetector filter H_2 realized as discrete-time summation. The time duration $T' = T/L$, with L a positive integer and it is assumed that $\Omega_i T' \gg 1$.

Each branch of the FSK receiver in Figure 8.14 is equal to the ASK receiver of Figure 8.10. Assume that $S_1(t)$ is transmitted. The stochastic variable obtained by sampling the postfilter in the upper branch of Figure 8.14 is then equal to (8.138)

$$U_1 = T' \sum_{k=1}^{L} |A\mathcal{Y}_k + X_{1k}|^2 \tag{8.166}$$

The moment-generating function of U_1, using the approximation (8.147), is equal to (8.151) raised to the power L.

The signal $S_1(t)$ will not pass the filter H_0 and the output of the lower branch is

$$U_0 = T' \sum_{k=1}^{L} |X_{0k}|^2 \tag{8.167}$$

Its approximate MGF is obtained from (8.151) with $A = 0$

$$\Psi_{U_0}(s) = \left(\frac{1}{1 - 2\sigma^2 s} \right)^L \tag{8.168}$$

The receiver makes its decision by comparing U_1 with U_0, or equivalently

$$U = U_1 - U_0 \tag{8.169}$$

with a threshold $\alpha = 0$. The stochastic variable U is the difference between two independent variables and its moment-generating function is the product

$$\Psi_U(s) = [\Psi_V(s)]^L \Psi_{U_0}(-s) \tag{8.170}$$

When $s_1(t)$ is transmitted the error probability is equal to $P_e = \Pr\{U < 0\}$. Due to the symmetric form of the receiver, the same value is obtained when $s_0(t)$ is transmitted.

Figure 8.15 shows the error probability calculated using the saddlepoint approximation for the linewidth bit time product $B_L T = 0$, 0.1, 1, 10. The abscissa shows the channel signal-to-noise ratio $10 \log m$. The parameter L is chosen in such a way that the bit error probability P_e is minimized.

Since the decision variable U is symmetric the Gaussian approximation does not give a useful estimate for FSK. It is shown by Einarsson and Sundelin (1995) that it can be incorrect by several orders of magnitude. As an example, when the true error probability is 1.03×10^{-9} the Gaussian approximation gives 5.23×10^{-6}.

8.4.3 *Differential Phase-shift Keying*

Differential phase-shift keying (DPSK) has been studied by Garrett *et al.* (1990) and Kaiser *et al.* (1993), among others. The approach presented here is proposed by Einarsson *et al.* (1995).

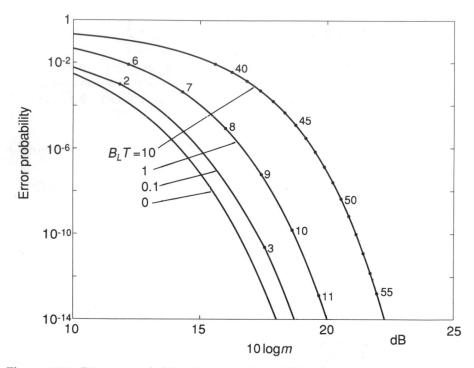

Figure 8.15 Bit error probability for heterodyne FSK with envelope detector receiver calculated from the Taylor expansion of the filtered phase noise for various values of $B_L T$. The optimal prefilter bandwidth parameter L is indicated along the curves. The parameter m is the average number of received photoelectrons per bit.

In DPSK the information is coded as phase shifts between successive signal intervals. Let ϕ_0 and ϕ_1 denote the phase of the optical field in the symbol intervals $[-T, 0]$ and $[0, T]$, respectively. The data symbol 'zero' is characterized by a phase-shift $\phi_1 - \phi_0 = \pi$ and 'one' corresponds to $\phi_1 - \phi_0 = 0$. The receiver uses the signal in the preceding time interval as a phase reference in the detection process. The receiver can be realized by any of the two equivalent configurations shown in Figure 8.16.

The envelope detector receiver Figure 8.16b generates two signals

$$\left.\begin{array}{l} V_+ = (\xi_1 + j\eta_1) + (\xi_0 + j\eta_0) \\ V_- = (\xi_1 + j\eta_1) - (\xi_0 + j\eta_0) \end{array}\right\} \tag{8.171}$$

The envelopes of these signals are compared and the decision is based on

$$\begin{aligned} |V_+|^2 - |V_-|^2 &= (\xi_1 + \xi_0)^2 + (\eta_1 + \eta_0)^2 - (\xi_1 - \xi_0)^2 - (\eta_1 - \eta_0)^2 \\ &= 4(\xi_1\xi_0 + \eta_1\eta_0) = 4\mathrm{Re}\{(\xi_1 + j\eta_1)(\xi_0 + j\eta_0)^*\} \end{aligned} \tag{8.172}$$

where $\mathrm{Re}\{\ \}$ is the real part and $*$ denotes the complex conjugate.

The relation (8.172) shows that the two receivers have identical performance. The filter H_2 eliminates the double frequency component resulting from the multiplication of two bandpass signals.

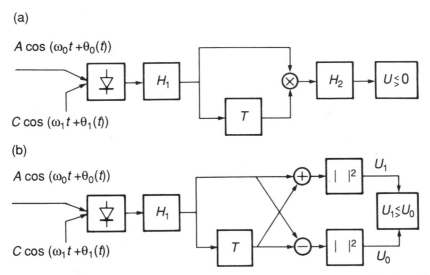

Figure 8.16   (a) Heterodyne DPSK receiver with signal multiplication. The filter H_1 is a bandpass filter at the heterodyne frequency and the filter H_2 is a lowpass filter. (b) Heterodyne DPSK receiver with envelope detectors. The predetector filter H_1 is a bandpass filter at the heterodyne frequency.

In the analysis of DPSK it is convenient to express the phase noise $\theta(t)$ as the deviation from the value θ_0 at time zero. The received signal in the time interval $[0,T]$ is, in equivalent baseband representation

$$s(t) = \text{Re}\{A \exp[j(\omega_0 t + \theta(t) + \theta_0 + \phi_1)] + X(t)\}; \quad 0 < t \le T \tag{8.173}$$

and the signal in the preceding interval

$$s(t) = \text{Re}\{A \exp[j(\omega_0 t + \theta(t) + \theta_0 + \phi_0)] + X(t)\}; \quad -T < t \le 0 \tag{8.174}$$

where $X(t)$ represents the photodetector shot noise.

The front end part of the receiver is the same as in the ASK receiver Figure 8.10. In the analysis presented here the prefilter has a fixed bandwidth and consists of a bandpass integrator over the whole signal interval $[0,T]$. One sample only is generated and no postfilter is needed.

From the output from the bandpass filter H_1 two signal branches V_+ and V_- are formed by adding or subtracting the signal and a delayed copy from the previous time interval. The resulting set of signals at time $t = T$ is, analogously to (8.131)

$$V_\pm = \frac{1}{T}\int_0^T [Ae^{j[\theta(t)+\theta_0+\phi_1]} + X(t)]\, dt \pm \frac{1}{T}\int_{-T}^0 [Ae^{j[\theta(t)+\theta_0+\phi_0]} + X(t)]\, dt \tag{8.175}$$

The complex-valued signals V_+ and V_- are passing through envelope detectors which makes their phases irrelevant. Dividing by a phase factor $\exp[j(\theta_0 + \phi_0)]$ will not change the result. In addition, an arbitrary phase can, without changing the statistics, be included in the noise $X(t)$, and (8.175) can be expressed as

$$V_\pm = b_0 A \mathcal{Y}_1 + X_1 \pm [A \mathcal{Y}_0 + X_0] \tag{8.176}$$

with

$$\left. \begin{aligned} \mathcal{Y}_1 &= \frac{1}{T} \int_0^T e^{j\theta(t)} \, dt \\ \mathcal{Y}_0 &= \frac{1}{T} \int_{-T}^0 e^{j\theta(t)} \, dt \end{aligned} \right\} \tag{8.177}$$

and where $b_0 = e^{j(\phi_1 - \phi_0)} = \pm 1$ represents the transmitted data symbol.

The fact that $\theta(0) = 0$ makes the phase noise integrals (8.177) independent and equally distributed stochastic variables. The quantities X_1 and X_0 are filtered shot noise from the photodetector

$$\left. \begin{aligned} X_1 &= \frac{1}{T} \int_0^T X(t) \, dt \\ X_0 &= \frac{1}{T} \int_{-T}^0 X(t) \, dt \end{aligned} \right\} \tag{8.178}$$

They are independent, complex-valued, zero-mean Gaussian variables with quadrature components of equal variance (8.137):

$$\sigma^2 = 2\mathcal{N}_1/T = 1/T \tag{8.179}$$

Expanding the integrand of (8.177) in a Taylor series and keeping the first three terms gives the following approximations valid for weak phase noise

$$\left. \begin{aligned} \tilde{\mathcal{Y}}_1 &= \frac{1}{T} \int_0^T 1 + j\theta(t) \, dt = 1 + j\tilde{\mathcal{Y}}_{1s} \\ \tilde{\mathcal{Y}}_0 &= \frac{1}{T} \int_{-T}^0 1 + j\theta(t) \, dt = 1 + j\tilde{\mathcal{Y}}_{0s} \end{aligned} \right\} \tag{8.180}$$

The phase noise $\theta(t)$ is related to the laser frequency noise by (4.94)

$$\theta(t) = 2\pi \int_0^t \mu(s) \, ds \tag{8.181}$$

where $\mu(t)$ is white Gaussian noise with two-sided spectral density (4.96)

$$R_\mu = B_L/2\pi \tag{8.182}$$

Integration by parts yields

$$\tilde{\mathcal{Y}}_{1s} = \frac{1}{T} \int_0^T \theta(t) \, dt = \frac{1}{T} \left. t\theta(t) \right|_0^T - \frac{1}{T} \int_0^T t\theta'(t) \, dt = \frac{2\pi}{T} \int_0^T (T-t)\mu(t) \, dt \tag{8.183}$$

Application of the theory for linear operations on stochastic processes gives

$$\mathrm{Var}\{\tilde{\mathcal{Y}}_{1s}\} = \left(\frac{2\pi}{T}\right)^2 \int_0^T R_\mu(T-t)^2 \, dt = \frac{\beta}{3} \tag{8.184}$$

with $\beta = 2\pi B_L T$ defined in (8.152).

The receiver compares the output of the upper and lower branches in Figure

8.16. It is convenient to let the decision variable be

$$U = \frac{T|V|^2}{4} = \frac{T(|V_+|^2 - |V_-|^2)}{4} \tag{8.185}$$

Assume that the symbol 'one' is transmitted, which means that $b_0 = 1$. Expressing the variable V in terms of real and imaginary parts of its components gives

$$|V|^2 = [2A + X_{1c} + X_{0c}]^2 + [A(\tilde{Y}_{1s} + \tilde{Y}_{0s}) + X_{1s} + X_{0s}]^2$$
$$- [X_{1c} - X_{0c}]^2 - [A(\tilde{Y}_{1s} - \tilde{Y}_{0s}) + X_{1s} - X_{0s}]^2 \tag{8.186}$$

This is equal to

$$|V|^2 = (2A + Z_{1c})^2 + (AW_{1s} + Z_{1s})^2 - Z_{0c}^2 - (AW_{0s} + Z_{0s})^2 \tag{8.187}$$

with $Z_{1c} = X_{1c} + X_{0c}$ and $Z_{0c} = X_{1c} - X_{0c}$, etc. All Z-variables in (8.187) are uncorrelated and Gaussian, which means that they are independent. They have equal variance $2/T$ which is twice (8.179).

The variables $W_{1s} = \tilde{Y}_{1s} + \tilde{Y}_{0s}$ and $W_{0s} = \tilde{Y}_{1s} - \tilde{Y}_{0s}$ have variances $2\beta/3$ equal to twice (8.184).

The first term $(2A + Z_{1c})^2$ in (8.187) has a noncentral chi-square distribution with one degree of freedom with an MGF, see (7.74)

$$\Psi_1(s) = \frac{1}{\sqrt{1 - 4s/T}} \exp\left(\frac{4A^2s}{1 - 4s/T}\right) \tag{8.188}$$

The next term $(AW_1 + Z_{1s})^2$ is the square of a zero mean Gaussian variable with variance $2A^2\beta/3 + 2/T$. It has an (ordinary) chi-square distribution with one degree of freedom, and its MGF is, cf. (7.79)

$$\Psi_2(s) = \frac{1}{\sqrt{1 - 4(A^2\beta/3 + 1/T)s}} \tag{8.189}$$

The term Z_{0c}^2 has zero mean and its MGF $\Psi_3(s)$ is equal to (8.189) with $A = 0$. The last term in (8.187) has same distribution as the second term and MGF is given by (8.189).

The MGF of $|V|^2$ is the product of the MGF's of its components, observing the negative sign of the two last terms.

$$\Psi_V(s) = \Psi_1(s)\Psi_2(s)\Psi_3(-s)\Psi_2(-s)$$
$$= \frac{1}{\sqrt{1 - [4(A^2\beta/3 + 1/T)s]^2}\sqrt{1 - (4s/T)^2}} \exp\left(\frac{4A^2s}{1 - 4s/T}\right) \tag{8.190}$$

The MGF for the decision variable (8.185) is obtained from (8.190) by replacing s by $sT/4$:

$$\Psi_U(s) = \Psi_V(sT/4) = \frac{1}{\sqrt{1 - (2m\beta/3 + 1)^2 s^2}\sqrt{1 - s^2}} \exp\left(\frac{2ms}{1 - s}\right) \tag{8.191}$$

where $m = A^2T/2$ is the average number of received photoelectrons per bit.

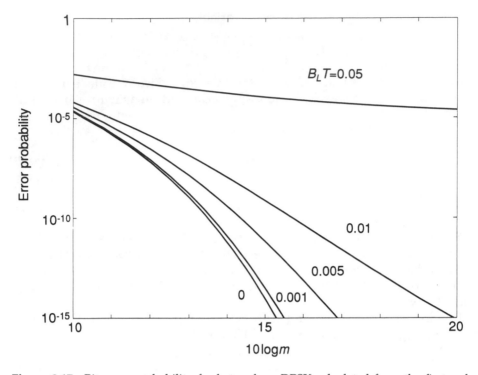

Figure 8.17 Bit error probability for heterodyne DPSK calculated from the first order Taylor expansion of the phase noise for various values of $B_L T$. The parameter m is the average number of received photoelectrons per bit.

The error probability for various values of the phase noise parameter $B_L T$ is shown in Figure 8.17. Comparison with Figure 8.11 and Figure 8.15 shows that DPSK is much more sensitive to phase noise than ASK or FSK.

8.4.4 *Relation to Optical Amplifier Receivers*

There is an interesting equivalence between the error probability analysis of optical systems with heterodyne receivers and direct detection with an optical preamplifier, as pointed out by Tonguz and Wagner (1991).

The signal in an optical preamplifier receiver is equal to (7.58). Dividing by $\sqrt{G}$ and supplementing with phase noise gives

$$Y(t)/\sqrt{G} = Y'(t) = A(t)e^{j\theta(t)} + X'(t) \qquad (8.192)$$

where $X'(t)$ is white Gaussian noise with spectral density from (7.59)

$$\mathcal{N}'_0 = n_{sp}(1 - 1/G) \qquad (8.193)$$

For $n_{sp} = 1$ and $G \gg 1$ the spontaneous emission noise $X'(t)$ and the shot noise in heterodyne detection at large local oscillator amplitude $X(t)$ in (8.131) have the same density $2\mathcal{N}_1 = \mathcal{N}_0 = 1$. The signals on which the demodulation is based are identical. The heterodyne receivers in this section contain envelope detectors. The optical amplifier receiver in Figure 7.9b for the approximate additive noise model has the same configuration. The system performance results obtained for heterodyne reception in the presence of phase noise also apply to optical amplifier receivers. The equivalence is valid for all modulation formats: OOK, FSK and DPSK.

Note that the equivalence is between the approximate additive noise model of the preamplifier receiver and the heterodyne receiver. It should be used with care if applied to the complete receiver model with a photodetector producing shot noise.

Appendix A
Modes in Cylindrical Waveguides

A.1 Relations for Bessel Functions

The following recursive relations can be found in, for example, Handbook of Mathematical Functions edited by Abramowitz and Stegun (1968).

Bessel Functions

$$J_{-\nu}(z) = (-1)^{\nu} J_{\nu}(z)$$

$$J_{\nu-1}(z) + J_{\nu+1}(z) - \frac{2\nu}{z} J_{\nu}(z)$$

$$J_{\nu-1}(z) - J_{\nu+1}(z) = 2J'_{\nu}(z)$$

$$J'_{\nu}(z) = J_{\nu-1}(z) - \frac{\nu}{z} J_{\nu}(z)$$

$$J'_{\nu}(z) = -J_{\nu+1}(z) + \frac{\nu}{z} J_{\nu}(z)$$

Modified Bessel Functions

$$K_{-\nu}(z) = K_{\nu}(z)$$

$$-K_{\nu-1}(z) + K_{\nu+1}(z) = \frac{2\nu}{z} K_{\nu}(z)$$

$$K_{\nu-1}(z) + K_{\nu+1}(z) = -2K'_{\nu}(z)$$

$$-K'_{\nu}(z) = K_{\nu-1}(z) + \frac{\nu}{z} K_{\nu}(z)$$

$$K'_{\nu}(z) = -K_{\nu+1}(z) + \frac{\nu}{z} K_{\nu}(z)$$

A.2   Weakly Guided (LP) Modes

We start from the characteristic equation (4.63) in Okoshi (1982), changing his notations u and w to ua and wa, respectively

$$\left[\frac{J_\nu'(ua)}{ua\,J_\nu(ua)} + \frac{K_\nu'(wa)}{wa\,K_\nu(wa)}\right]\left[\frac{n_1^2}{n_2^2}\frac{J_\nu'(ua)}{ua\,J_\nu(ua)} + \frac{K_\nu'(wa)}{wa\,K_\nu(wa)}\right]$$

$$= \nu^2\left(\frac{1}{(ua)^2} + \frac{1}{(wa)^2}\right)\left(\frac{n_1^2}{n_2^2}\frac{1}{(ua)^2} + \frac{1}{(wa)^2}\right) \tag{A.1}$$

For $\Delta \ll 1$ is $n_1/n_2 \approx 1$ which gives

$$\frac{J_\nu'(ua)}{ua\,J_\nu(ua)} + \frac{K_\nu'(wa)}{wa\,K_\nu(wa)} = \pm\nu\left(\frac{1}{(ua)^2} + \frac{1}{(wa)^2}\right) \tag{A.2}$$

From the relations for Bessel functions in Section A.1 we obtain

$$\frac{J_\nu'(z)}{zJ_\nu(z)} = \frac{J_{\nu-1}(z)}{zJ_\nu(z)} - \frac{\nu}{z^2} = -\frac{J_{\nu+1}(z)}{zJ_\nu(z)} + \frac{\nu}{z^2} \tag{A.3}$$

$$\frac{K_\nu'(z)}{zK_\nu(z)} = -\frac{K_{\nu-1}(z)}{zK_\nu(z)} - \frac{\nu}{z^2} = -\frac{K_{\nu+1}(z)}{zK_\nu(z)} + \frac{\nu}{z^2} \tag{A.4}$$

When $\nu = 0$, solutions are obtained which are circularly symmetrical, i.e. independent of the angle φ in the cylindrical coordinate system. The corresponding modes are of the types TE and TM. Substitution of (A3) and (A4) into (A2) gives for $\nu = 0$

$$\frac{J_1(ua)}{ua\,J_0(ua)} = -\frac{K_1(wa)}{wa\,K_0(wa)} \tag{A.5}$$

For $\nu \geq 1$ and a positive sign on the right-hand side of (A.2)

$$\frac{J_{\nu+1}(ua)}{ua\,J_\nu(ua)} = -\frac{K_{\nu+1}(wa)}{wa\,K_\nu(wa)} \tag{A.6}$$

This gives hybrid modes of the type EH. For $\nu \geq 1$ and a negative sign on the right-hand side of (A.2), hybrid modes are obtained which are of the type HE characterized by the relation

$$\frac{J_{\nu-1}(ua)}{ua\,J_\nu(ua)} = \frac{K_{\nu-1}(wa)}{wa\,K_\nu(wa)} \tag{A.7}$$

Using the relation (A.1), we obtain

$$J_\nu(z) = \frac{2(\nu-1)}{z}\,J_{\nu-1}(z) - J_{\nu-2}(z)$$

$$K_\nu(z) = \frac{2(\nu-1)}{z}\,K_{\nu-1}(z) + K_{\nu-2}(z)$$

Substitution into (A.7) gives

$$\frac{J_{\nu-1}(ua)}{ua\, J_{\nu-2}(ua)} = -\frac{K_{\nu-1}(wa)}{wa\, K_{\nu-2}(wa)} \tag{A.8}$$

All three equations (A.5), (A.6) and (A.8) can be written in the common form

$$\frac{ua\, J_{m-1}(ua)}{J_m(ua)} = -\frac{wa\, K_{m-1}(wa)}{K_m(wa)} \tag{A.9}$$

in which

$$
\begin{array}{lll}
m = 1 & \text{for TE and TM modes} & (\nu = 0) \\
m = \nu + 1 & \text{for EH modes} & (\nu \geq 1) \\
m = \nu - 1 & \text{for HE modes} & (\nu \geq 1)
\end{array}
$$

A particular index m results in a characteristic equation (A.9) which generates a sequence of solutions which are called LP modes. Such a mode, in its turn, is composed of a number of degenerated modes of the types TE, TM, HE or EH. As an example, an LP mode with $m = 1$ contains TE and TM ($\nu = 0$) and HE ($\nu = 2$) modes. The only combination for which $m = 0$ is when $\nu = 1$ for HE modes. From this it follows that $m = 0$ is the only case in which the LP mode contains only one HE mode which still can have two polarization directions.

In order to determine the cutoff frequencies, we let $w \to 0$ in (A.9). For the function $K_m(z)$, the following approximate expressions are valid for $z \ll 1$; see Abramonitz and Stegun (1968).

$$K_0(z) \approx -\ln z$$

$$K_m(z) \approx \frac{1}{2}\, \Gamma(m) \left(\frac{z}{2}\right)^{-m}, \quad m \geq 1$$

with $\Gamma(m)$ the gamma function.

From this it follows that the right-hand side of (A.9) approaches zero when $w \to 0$ for all values of m. The cutoff frequencies are thus given by the ua values which satisfy

$$\frac{ua\, J_{m-1}(ua)}{J_m(ua)} = 0 \tag{A.10}$$

Since $J_{-1}(z) = -J_1(z)$, the cutoff frequencies are the roots to

$$J_{|m-1|}(ua) = 0 \tag{A.11}$$

The numerator in (A.10) is zero for $ua = 0$, but in order for this to be a solution, it is necessary that $J_m(0) \neq 0$. This is the case only when $m = 0$. See Figure 3.7. Therefore, it is only the LP_{01} mode which has the cutoff frequency equal to zero.

Appendix B

Delay and Dispersion in Linear Systems

B.1 Linear Systems

In this section some basic properties of linear systems are presented. For further details, see for instance, Papoulis (1962).

A linear system is characterized by its impulse response $h(t)$ or frequency function $H(\omega)$ which constitute a Fourier transform pair; see Figure 4.1

$$h(t) = \frac{1}{2\pi} \int_{-\infty}^{\infty} H(\omega) e^{j\omega t} d\omega \tag{B.1}$$

$$H(\omega) = A(\omega) e^{j\phi(\omega)} = \int_{-\infty}^{\infty} h(t) e^{-j\omega t} dt \ ; \ \omega = 2\pi f \tag{B.2}$$

where $A(\omega)$ denotes the system's amplitude function and $\phi(\omega)$ its phase function.

From (B.2) it follows that the integral of the impulse response $h(t)$ is

$$A_h = \int_{-\infty}^{\infty} h(t) dt = H(0) = A(0) \tag{B.3}$$

The output $y(t)$ is related to the input $x(t)$ by the relation, cf. Figure 4.1

$$y(t) = \int_{-\infty}^{\infty} x(s) h(t-s) \, ds \tag{B.4}$$

$$Y(\omega) = X(\omega) H(\omega) \tag{B.5}$$

in which $X(\omega)$ and $Y(\omega)$ denote the Fourier transforms of the input and output signals, respectively.

The derivatives of the frequency function $H(\omega)$ are related to the moments of the impulse response $h(t)$. Derivation of both sides of the relation (B.2) with respect to ω gives

$$H'(\omega) = \frac{dH(\omega)}{d\omega} = -j \int_{-\infty}^{\infty} t \, h(t) e^{-j\omega t} dt \tag{B.6}$$

The second derivative becomes

$$H''(\omega) = -\int_{-\infty}^{\infty} t^2 h(t)e^{-j\omega t} dt \tag{B.7}$$

B.2   Baseband Signals

The relations for delay and dispersion presented below are based on (B.6) and (B.7) which are valid for an arbitrary linear system for which the first and second order moments of the impulse response are finite. However, the direct interpretation of these integrals as measures of delay and pulse broadening is meaningful for lowpass signals only. Bandpass signals will be treated in the next section.

A natural measure of the location of a signal $h(t)$ is the center of gravity or mean

$$m_t = \frac{1}{A_h} \int_{-\infty}^{\infty} t\, h(t) dt \tag{B.8}$$

where A_h is the area (B.3) of $h(t)$.

From (B.6) it follows that m_t is related to the derivative

$$H'(\omega) = A'(\omega)e^{j\phi(\omega)} + jA(\omega)\phi'(\omega)e^{j\phi(\omega)} \tag{B.9}$$

For a system with real-valued input and output signals, $h(t)$ is a real function which implies that $A(\omega)$ must be an even function and $\phi(\omega)$ an odd function. Owing to this, $A'(0) = 0$ and $H(0) = A(0)$ and for $\omega = 0$

$$H'(0) = jA(0)\phi'(0) = -j\int_{-\infty}^{\infty} t\, h(t) dt \tag{B.10}$$

which means that

$$m_t = \tau(0) \tag{B.11}$$

with τ the group delay

$$\tau(\omega) = -\frac{d\phi(\omega)}{d\omega} \tag{B.12}$$

Because $y(t) = h(t)$ for $x(t) = \delta(t)$ (Dirac function), the group delay is a measure of how much a narrow pulse has been delayed by passing through the system; see Figure 4.3.

Consider next the mean square pulse width

$$\sigma_t^2 = \frac{1}{A_h} \int_{-\infty}^{\infty} (t - m_t)^2 h(t)\, dt = \frac{1}{A_h} \int_{-\infty}^{\infty} t^2 h(t)\, dt - m_t^2 \tag{B.13}$$

The second derivative of $H(\omega)$ is

$$H''(\omega) = A''(\omega)e^{j\phi(\omega)} + j2A'(\omega)\phi'(\omega)e^{j\phi(\omega)} \tag{B.14}$$
$$+ jA(\omega)\phi''(\omega)e^{j\phi(\omega)} - A(\omega)[\phi'(\omega)]^2 e^{j\phi(\omega)}$$

Substitution of $\omega = 0$ using the fact that $\phi(0) = 0$, $\phi''(0) = 0$ and $A'(0) = 0$, in combination with (B.11), gives

$$H''(0) = A''(0) - A(0)m_t^2 = -\int_{-\infty}^{\infty} t^2 h(t)\, dt \tag{B.15}$$

and

$$\sigma_t^2 = -\frac{1}{A_h} A''(0) \tag{B.16}$$

From (B.5) it follows that

$$\phi_y(\omega) = \phi_x(\omega) + \phi(\omega) \tag{B.17}$$

Derivation gives

$$\tau_y = \tau_x + \tau \tag{B.18}$$

The delay of the output signal is therefore the sum of the delay of the input signal and the system delay.

The amplitude function of the output signal is

$$A_y(\omega) = A_x(\omega)A(\omega) \tag{B.19}$$

Derivating twice gives

$$A_y''(\omega) = A_x''(\omega)A(\omega) + 2A_x'(\omega)A'(\omega) + A_x(\omega)A''(\omega) \tag{B.20}$$

For $\omega = 0$ is $A'(\omega) = 0$ which leads to

$$\sigma_y^2 = \sigma_x^2 + \sigma^2 \tag{B.21}$$

Dispersion measured in terms of σ_t^2 is an additive quantity for linear systems. This means that the resulting delays and pulse broadening for a linear system, which is composed of a cascade of subsystems, are easy to calculate from the characteristics of the subsystems.

B.3   Bandpass Signals

Optical signals do not exist in the baseband, they constitute bandpass signals. The carrier frequency is usually very high compared with the modulation rate which makes them narrow-band.

It is convenient to represent a bandpass signal as the real part of a complex signal, see e.g. Dugundji (1958) or the appendix in part III of van Trees (1971).

$$x(t) = \text{Re}\{\tilde{x}(t)e^{j\omega_0 t}\} \tag{B.22}$$

where ω_0 is the carrier frequency.

It is convenient to define the signal delay and dispersion with respect to the square of the envelope $|x(t)|^2 = |\tilde{x}(t)|^2$ of the signal, which for an optical signal corresponds to signal intensity. The relations are more complicated than for baseband signals and depend in general on the specific form of the signal and system functions. We illustrate with an example.

Example B.1   Gaussian pulses

Consider a bandpass signal

$$x(t) = e^{-t^2/2a^2} \, \cos \omega_0 t \tag{B.23}$$

The corresponding complex signal is

$$\tilde{x}(t) = e^{-t^2/2a^2} \tag{B.24}$$

which has the Fourier transform

$$\tilde{X}(\omega) = a\sqrt{2\pi}e^{-a^2\omega^2/2} \tag{B.25}$$

The area of the envelope square is

$$A_x = \int_{-\infty}^{\infty} |\tilde{x}|^2 dt = \int_{-\infty}^{\infty} e^{-t^2/a^2} dt = a\sqrt{\pi} \tag{B.26}$$

The delay of $|\tilde{x}(t)|^2$ is

$$\tau_x = \frac{1}{a\sqrt{\pi}} \int_{-\infty}^{\infty} t|\tilde{x}(t)|^2 dt = 0 \tag{B.27}$$

and its mean square envelope width is

$$\sigma_x^2 = \frac{1}{a\sqrt{\pi}} \int_{-\infty}^{\infty} t^2|\tilde{x}(t)|^2 dt = a^2/2 \tag{B.28}$$

The signal $x(t)$ is transmitted through an optical fiber which is modeled as a linear system with the complex transfer function, c f. (4.5)

$$\tilde{H}(\omega) = A \, e^{j\phi(\omega)} \tag{B.29}$$

The complex-valued output signal $\tilde{y}(t)$ has the Fourier transform

$$\tilde{Y}(\omega) = aA\sqrt{\pi} \, e^{-a^2\omega^2/2}e^{j\phi(\omega)} \tag{B.30}$$

With the use of Parseval's relation

$$A_y = \int_{-\infty}^{\infty} |\tilde{y}(t)|^2 dt = \int_{-\infty}^{\infty} \frac{1}{2\pi} |\tilde{Y}(\omega)|^2 d\omega = aA^2\sqrt{\pi}/2 \tag{B.31}$$

The delay of the output signal is

$$\tau_y = \frac{2}{aA^2\sqrt{\pi}} \int_{-\infty}^{\infty} t|\tilde{y}(t)|^2 dt = \frac{j}{aA^2\sqrt{\pi^3}} \int_{-\infty}^{\infty} \tilde{Y}(\omega)\tilde{Y}^*(\omega)d\omega \tag{B.32}$$

where $*$ denotes complex conjugate and $\tilde{Y}'$ is the derivative of (B.30)

$$\tilde{Y}'(\omega) = aA\sqrt{\pi}\, e^{-a^2\omega^2/2}[-a^2\omega + j\phi'(\omega)]e^{j\phi(\omega)} \tag{B.33}$$

For a narrow-band signal, the phase function can be expanded in a Taylor series around the carrier frequency ω_0, which is equivalent to expanding $\phi(\omega)$ in a Maclaurin series around $\omega = 0$

$$\phi(\omega) = \phi_0 + \phi_1\omega + \frac{\phi_2}{2}\omega^2 + \frac{\phi_3}{3!}\omega^3 + \cdots \tag{B.34}$$

where ϕ_i is the ith derivative of $\phi(\omega)$ at $\omega = \omega_0$. Substitution of the derivative of (B.34) $\phi(\omega)$ into (B.33) and (B.32) gives

$$\tau_y = j\frac{a}{\sqrt{\pi}}\int_{-\infty}^{\infty}[-a^2\omega + j(\phi_1 + \phi_2\omega + \frac{\phi_3}{2!}\omega^2 + \cdots)e^{j\phi(\omega)}]e^{-a^2\omega^2}e^{-j\phi(\omega)}\,d\omega$$

$$= -\frac{a}{\sqrt{\pi}}\int_{-\infty}^{\infty}(\phi_1 + \frac{\phi_3}{2!}\omega^2 + \cdots)e^{-a^2\omega^2}\,d\omega = -(\phi_1 + \frac{\phi_3}{4a^2} + \cdots) \tag{B.35}$$

To obtain the dispersion, we determine

$$\frac{1}{A_y}\int_{-\infty}^{\infty}t^2|\tilde{y}(t)|^2\,dt = \frac{1}{aA^2\sqrt{\pi^3}}\int_{-\infty}^{\infty}|\tilde{Y}'(\omega)|^2\,d\omega$$

$$= \frac{a}{\sqrt{\pi}}\int_{-\infty}^{\infty}[(a^2\omega)^2 + (\phi_1 + \phi_2\omega + \frac{\phi_3}{2!}\omega^2 + \cdots)^2]e^{-\omega^2 a^2}\,d\omega$$

$$= \phi_1^2 + (a^4 + \phi_2^2 + \phi_1\phi_3)/2a^2 + \cdots \tag{B.36}$$

The mean square width of the received pulse is

$$\sigma_y^2 = \frac{1}{A_y}\int_{-\infty}^{\infty}t^2|\tilde{y}(t)|^2\,dt - \tau_y^2$$

$$= \phi_1^2 + (a^4 + \phi_2^2 + \phi_1\phi_3)/2a^2 + \cdots - (\phi_1 + \frac{\phi_3}{4a^2} + \cdots)^2$$

$$= \frac{a^2}{2} + \frac{\phi_2^2}{2a^2} + \cdots = \sigma_x^2 + \frac{\phi_2^2}{4\sigma_x^2} + \cdots \tag{B.37}$$

$\square$

In the example, it was assumed that the amplitude function $A(\omega)$ was a constant which is true for optical systems. The dispersion is the term $\phi_2^2/4\sigma_x^2$ in (B.37). Note that $\phi_2 = 0$ for a lowpass system, but it is in general nonzero in the bandpass case. For narrow-band signals, when $1/\sigma_x^2$ is small and σ_x^2 is much greater than ϕ_2 and ϕ_3, the delay of the envelope becomes approximately equal to the system group delay and pulse broadening can be neglected.

This indicates that narrow-band signal envelopes exhibit approximately the same properties with regard to delay and pulse broadening as baseband signals. See Figure 4.4.

Appendix C
Shot Noise

C.1 Light with Constant Intensity

The electric current from a photodetector is

$$I(t) = \sum_{k=-\infty}^{\infty} a_k g(t - t_k) \tag{C.1}$$

where a_k represents the avalanche gain and $g(t)$ is the impulse response of of the photodetector.

To determine the statistical properties of $I(t)$ for light of constant intensity $\Gamma(t) = \Gamma$ we consider a long time interval extending from $-T$ to T. Denote by N the number of Poisson events occurring in the interval. The expectations below are calculated in two steps: first the conditional expectation for fixed N, denoted by E_N and then the average with respect to N is formed. The Poisson events in an interval have the same statistics as points generated by N independent trials of an uniformly distributed stochastic variable. A proof can be found in Papoulis (1984). Later we let T approach infinity and obtain results valid at all times t.

Assume that N events t_k have occurred in the time interval $[-T, T]$. The average of $I(t)$ is, assuming a_k and t_k are independent,

$$E_N\{I(t)\} = E\{a_k\} \sum_{k=1}^{N} E\{g(t - t_k)\} \tag{C.2}$$

The times t_k are independent and uniformly distributed in the interval, yielding

$$E\{g(t - t_k)\} = \int_{-T}^{T} g(t - t_k) \frac{1}{2T} \, dt = \frac{1}{2T} \int g(x) \, dx = \frac{q}{2T} \tag{C.3}$$

This is valid for pulses inside the interval. The end effects will be eliminated by letting T approach infinity.

Substitution of (C.3) into (C.2) gives

$$E_N\{I(t)\} = E\{a_k\} N q / 2T \tag{C.4}$$

The average of N is from (5.4)

$$E\{N\} = m = 2T\Gamma \tag{C.5}$$

and the average of the right-hand side of (C.4) becomes

$$E\{I(t)\} = I_0 = E\{a_k\}q\Gamma \tag{C.6}$$

which is valid for arbitrary values of T.

The autocorrelation function of $I(t)$ is

$$r_N(\tau) = E\{I(t+\tau)I(t)\} = \tag{C.7}$$

$$= \sum_{k=1}^{N} \sum_{j=1}^{N} E\{a_k a_j\} E\{g(t - \tau - t_k)g(t - t_j)\}$$

The events t_k and t_j are independent and for $k \neq j$

$$E\{g(t+\tau-t_k)g(t-t_j)\} = E\{g(t+\tau-t_k)\}E\{g(t-t_j)\} \tag{C.8}$$

$$= (q/2T)^2$$

For $k = j$

$$E\{g(t+\tau-t_k)g(t-t_k)\} = \frac{1}{2T}\int_{-T}^{T} g(x)g(x+\tau)\,dx \tag{C.9}$$

The correlation integral in (C.9)

$$\varphi(\tau) = \int_{-T}^{T} g(x)g(x+\tau)\,dx \tag{C.10}$$

depends on the shape of $g(t)$.

The avalanche mechanism in photodiodes has normally no memory and a_k and a_j can be assumed to be uncorrelated and

$$E\{a_k a_j\} = \begin{cases} (E\{a_k\})^2 & k \neq j \\ E\{a_k^2\} & k = j \end{cases} \tag{C.11}$$

Substitution of (C.9) and (C.11) into (C.7) gives

$$r_N(\tau) = E\{a_k^2\}N\varphi(\tau)/2T + (E\{a_k\})^2 N(N-1)(q/2T)^2 \tag{C.12}$$

The average of N is m and from (5.6) follows that

$$E\{N^2\} = \text{Var}\{N\} + m^2 = m + m^2 \tag{C.13}$$

which gives

$$E\{N(N-1)\} = m^2 \tag{C.14}$$

The autocorrelation function for the stochastic process (C.1) is obtained by forming the average over N of (C.12) and letting T approach infinity

$$r(\tau) = E\{a_k^2\}\varphi(\tau)\Gamma + (E\{a_k\}q\Gamma)^2 \tag{C.15}$$

The power spectral density of the shot noise process is the Fourier transform of $r(\tau)$

$$R(f) = \mathrm{E}\{a_k^2\}\Gamma \int_{-\infty}^{\infty} \varphi(\tau)e^{-j2\pi f\tau}d\tau + (\mathrm{E}\{a_k\}q\Gamma)^2\delta(f) \qquad \text{(C.16)}$$

where $\delta(f)$ is a Dirac function.
Applying the rules of Fourier transforms to (C.10) gives

$$R(f) = \mathrm{E}\{a_k^2\}|G(f)|^2\Gamma + I_0^2\delta(f) \qquad \text{(C.17)}$$

where $G(f)$ is the Fourier transform of $g(t)$

$$G(f) = \int_{-\infty}^{\infty} g(t)e^{-j2\pi ft}dt \qquad \text{(C.18)}$$

and I_0 is equal to (C.6), the direct current component of $I(t)$.
For a photodetector with short impulse response, $g(t)$ can be approximated by a Dirac function $q\delta(t)$ and

$$I(t) = \sum_{k=-\infty}^{\infty} a_k\delta(t - t_k) \qquad \text{(C.19)}$$

The frequency function of $\delta(t)$ is $G(f) = 1$ and the shot noise (C.19) is white noise.

C.2 Filtered Shot Noise

The shot noise process from light of constant intensity is an ordinary stationary stochastic process. The effect of filtering or other linear operations follows directly from the theory of random processes.

For communication the light intensity is modulated by the transmitted information and the received shot noise signal is time-varying. This makes it is necessary to analyse the effect of filtering in the time domain.

Consider a shot noise signal

$$I(t) = q\sum_{k}\delta(t - t_k) \qquad \text{(C.20)}$$

with a time-varying intensity $\Gamma(t)$ representing a finite energy optical pulse.
The output from a linear filter with impulse response $h(t)$ is

$$Z(t) = \int_{-\infty}^{\infty} h(t - \tau)I(\tau)\,d\tau \qquad \text{(C.21)}$$

Substitution of (C.20) gives

$$Z(t) = q\sum_{k}h(t - \tau_k) \qquad \text{(C.22)}$$

The statistical distribution of $Z(t)$ at a certain time t is specified by the moment-generating function (MGF)

$$\Psi_z(s) = E\{\exp(Z(t)s)\} \tag{C.23}$$

The Poisson events t_k, considered individually without regard to their internal order, are statistically independent with a probability density

$$p(\tau_k) = \frac{\Gamma(\tau_k)}{m} \tag{C.24}$$

where

$$m = \int_{-\infty}^{\infty} \Gamma(\tau) \, d\tau$$

The expectation in (C.23) is performed first over the τ_k:s for a fixed number N of Poisson events and then over the random variable N. Substitution of (C.22) into (C.23) and using the fact that N has a Poisson distribution and remembering that the Poisson events are independent gives

$$\Psi_z(s) = E_N\left\{ \prod_{k=1}^{N} \left[\int_{-\infty}^{\infty} p(\tau_k) \exp[qh(t-\tau_k)s] \, d\tau_k \right] \right\}$$

$$= \sum_{n=0}^{\infty} e^{-m} \frac{m^n}{n!} \prod_{k=1}^{n} \int_{-\infty}^{\infty} \frac{\Gamma(\tau_k)}{m} \exp[qh(t-\tau_k)s]d\tau_k \tag{C.25}$$

$$= \sum_{n=0}^{\infty} e^{-m} \frac{m^n}{n!} \left[\frac{1}{m} \int_{-\infty}^{\infty} \Gamma(\tau) \, \exp[qh(t-\tau)s] \, d\tau \right]^n$$

The last expression is of the form

$$e^{-m} \sum_{n=0}^{\infty} \frac{m^n}{n!} a^n = e^{-m} e^{am} \tag{C.26}$$

and (C.25) can be written as

$$\Psi_z(s) = \exp\left(\int_{-\infty}^{\infty} \Gamma(\tau)(\exp[qh(t-\tau)s] - 1) \, d\tau \right) \tag{C.27}$$

The moment-generating function for a detector with avalanche gain is obtained by replacing $h(t-\tau_k)$ by $a_k h(t-\tau_k)$ in (C.25):

$$\Psi_{za}(s) = E\left\{ \sum_{n=0}^{\infty} e^{-m} \frac{m^n}{n!} \prod_{k=1}^{n} \int_{-\infty}^{\infty} \frac{\Gamma(\tau_k)}{m} \exp[a_k qh(t-\tau_k)s] \, d\tau_k \right\}$$

$$= \sum_{n=0}^{\infty} e^{-m} \frac{m^n}{n!} \left[\frac{1}{m} \int_{-\infty}^{\infty} \Gamma(t) \, E\{\exp[a_k qh(t-\tau)s]\} \, d\tau \right]^n \tag{C.28}$$

which is equal to

$$\Psi_{za}(s) = \exp\left(\int_{-\infty}^{\infty} \Gamma(t)(\Psi_a[qh(t-\tau)s] - 1)d\tau \right) \tag{C.29}$$

with $\Psi_a(s)$ the moment-generating function for the avalanche gain.

The mean and variance of the filter output (C.21) are easily determined from the moment-generating function (C.27) using the relations

$$\left.\begin{aligned} E\{Z(t)\} &= \Psi'_Z(s=0) \\ E\{Z^2(t)\} &= \Psi''_Z(s=0) \end{aligned}\right\} \tag{C.30}$$

The result is

$$\left.\begin{aligned} E\{Z(t)\} &= q \int_{-\infty}^{\infty} \Gamma(\tau)h(t-\tau)\,dt \\ \mathrm{Var}\{Z(t)\} &= q^2 \int_{-\infty}^{\infty} \Gamma(\tau)h^2(t-\tau)\,dt \end{aligned}\right\} \tag{C.31}$$

The relations (C.31) are known as Campbell's Theorem.

For an avalanche photodetector it follows from (C.29)

$$\left.\begin{aligned} E\{Z(t)\} &= E\{a_k\}q \int_{-\infty}^{\infty} \Gamma(\tau)h(t-\tau)\,dt \\ \mathrm{Var}\{Z(t)\} &= E\{a_k^2\}q^2 \int_{-\infty}^{\infty} \Gamma(\tau)h^2(t-\tau)\,dt \end{aligned}\right\} \tag{C.32}$$

C.3 Doubly Stochastic Poisson Process

In optical amplifiers and other applications is the light intensity $\Gamma(t)$ a random function of time. The point process representing the shot noise current produced by the photodetector is a doubly stochastic Poisson process, see for instance Snyder and Miller (1991).

The moment-generating function for the output $Z(t)$ (C.21) of a filter with impulse response $h(t)$ is obtained from (C.27) by forming the average over $\Gamma(t)$:

$$\Psi_z(s) = E\left\{\exp\left(\int_{-\infty}^{\infty} \Gamma(\tau)(\exp[qh(t-\tau)s]-1)\,d\tau\right)\right\} \tag{C.33}$$

The moments of $Z(t)$ are equal to the derivatives of $\Psi_z(s)$ at $s=0$. The first two moments are

$$\Psi'(0) = E\{Z(t)\} = E\left\{q \int_{-\infty}^{\infty} \Gamma(\tau)h(t-\tau)\,d\tau\right\} \tag{C.34}$$

and

$$\Psi''(0) = E\{Z^2(t)\} = E\left\{\left(q \int_{-\infty}^{\infty} \Gamma(\tau)h(t-\tau)\,d\tau\right)^2 + q^2 \int_{-\infty}^{\infty} \Gamma(\tau)h^2(t-\tau)\,d\tau\right\} \tag{C.35}$$

The variance of $Z(t)$ is $E\{Z^2(t)\} - E^2\{Z(t)\}$

$$\text{Var}\{Z(t)\} = E\left\{q^2 \int_{-\infty}^{\infty} \Gamma(\tau)h^2(t-\tau)\,d\tau\right\} + \text{Var}\left\{q\int_{-\infty}^{\infty} \Gamma(\tau)h(t-\tau)\,d\tau\right\}$$

(C.36)

The decision variable is often the output of a linear filter sampled at $t = T$, the end of the signaling interval. It is then convenient to introduce the filter function

$$v(t) = qh(T-t) \tag{C.37}$$

For the special case when $v(t)$ is a rectangular (integrate-and-dump) filter

$$Z(T) = \int_0^T \Gamma(t)v(t)\,dt = \int_0^T \Gamma(t)\,dt = \mathcal{M} \tag{C.38}$$

The output $Z(T)$ is now a Poisson variable with random mean $\mathcal{M}$ and the filter output is a integer valued stochastic variable N.

Denote the probability density function of $\mathcal{M}$ by $p_m(x)$. The probability-generating function for N is

$$\Phi_n(z) = E\{z^N\} = \int_0^\infty p_m(x)E\{z^x | \mathcal{M} = x\}\,dx = \int_0^\infty p_m(x)\exp[x(z-1)]\,dx \tag{C.39}$$

since N for a fixed $\mathcal{M} = x$ has a Poisson distribution.

This can be expressed as

$$\Phi_n(z) = \Psi_m(z-1) \tag{C.40}$$

where $\Psi_m(\cdot)$ is the moment-generating function of the random mean $\mathcal{M}$. The relation (C.40) is the Poisson Transform; see Saleh (1978).

From the properties of probability-generating functions presented in Chapter 7 it follows that

$$E\{N\} = \Phi_n'(1) = \Psi_m'(0) = E\{\mathcal{M}\} \tag{C.41}$$

and

$$\text{Var}\{N\} = \Phi_n''(1) + \Phi_n'(1) - [\Phi_n'(1)]^2 \tag{C.42}$$
$$= \Psi_m''(0) - [\Psi_m'(0)]^2 + \Psi_m'(0) = \text{Var}\{\mathcal{M}\} + E\{\mathcal{M}\}$$

C.4   Additive Noise Approximation

The variance of a Poisson variable is equal to its mean which makes optical noise signal dependent. However, when a light signal is mixed with light from a strong local oscillator the shot noise can be approximated by an additive Gaussian noise component.

With heterodyne reception the intensity function for the shot noise signal in the receiver has the form, see (8.128)

$$\Gamma(t) = \frac{C^2 + A^2}{2} + CA \cos(\omega_h t + \theta(t)) \tag{C.43}$$

where A^2 and C^2 represent the optical intensity of the signal and the local oscillator, respectively, and $\theta(t)$ is phase noise. Consider the shot noise generated by (C.43) filtered by an integrating filter

$$Z(t) = \frac{1}{qC} \int_0^t I(\tau) \, d\tau \tag{C.44}$$

For a specific realization of the phase noise $\theta(t)$ the stochastic variable $Z(t)$ has the moment-generating function from (C.27)

$$\Psi_z(s) = \exp\left(\int_0^t \Gamma(\tau)[\exp(s/C) - 1] \, d\tau \right) \tag{C.45}$$

Substitution of (C.43) and expansion of $\exp(s/C)$ into a Taylor series gives for $C \gg A$

$$\Psi_z(s) = \exp\left\{ \left(\int_0^t [C/2 + A\cos(\omega_h\tau + \theta(\tau))] \, d\tau \right)(s + s^2/2C + \cdots) \right\}$$
$$\approx \exp\left\{ \int_0^t [C/2 + A\cos(\omega_h\tau + \theta(\tau))]s \, d\tau + \int_0^t (s^2/4) \, d\tau \right\} \tag{C.46}$$

Compare $Z(t)$ with the integral

$$Z_1(t) = \int_0^t [f(\tau) + X(\tau)] \, d\tau \tag{C.47}$$

where $f(\tau)$ is an arbitrary function and $X(\tau)$ is Gaussian white noise with (photon or photoelectron) intensity $\mathcal{N}$. The stochastic variable $Z_1(t)$ has a Gaussian distribution with mean $\int_0^t f(\tau) \, d\tau$ and variance $\sigma^2 = \mathcal{N}t$. Its moment-generating function is, see (5.61),

$$\Psi_{z1}(s) = \exp\left(s \int_0^t f(\tau) \, d\tau + \frac{s^2}{2} \int_0^t \mathcal{N} \, d\tau \right) \tag{C.48}$$

Comparison with (C.46) shows that $I(t)$ can be modeled as

$$I(t) = C[C/2 + A\cos(\omega_h t + \theta(t)) + X_1(t)] \tag{C.49}$$

where $X_1(t)$ is a Gaussian white noise process with intensity

$$\mathcal{N}_1 = 1/2 \tag{C.50}$$

Appendix D

Bounds and Approximations

D.1   Poisson Distribution

D.1.1   Chernoff Bounds

The Chernoff bound is a simple and quite general bound on the tail of a probability distribution

$$P(X \geq \alpha) = \int_{\alpha}^{\infty} p(x)dx \tag{D.1}$$

The integral can be written as

$$\int_{-\infty}^{\infty} u(x - \alpha)p(x)dx \tag{D.2}$$

where $u(x)$ is the unit step function.

Since $u(x) \leq \exp(sx)$ for $s \geq 0$

$$P(X \geq \alpha) \leq \int_{-\infty}^{\infty} \exp[s(x - \alpha)]p(x)dx \tag{D.3}$$

$$= e^{-s\alpha} \int_{-\infty}^{\infty} e^{sx}p(x)dx$$

The last integral in (D.3) is equal to the moment-generating function

$$\Psi(s) = E\{\exp(sX)\} \tag{D.4}$$

and the bound can be written as

$$P(X \geq \alpha) \leq e^{-s\alpha}\Psi(s); \quad s \geq 0 \tag{D.5}$$

The value of s which minimizes the right-hand side of (D.5), and results in the tightest bound, is obtained by setting the derivative of (D.5), with respect to s, equal to zero. This gives the equation

$$\alpha\Psi(s) = \Psi'(s) \tag{D.6}$$

The lower tail is bounded in the same way since for $s \leq 0$

$$P(X \leq \alpha) = \int_{-\infty}^{\alpha} p(x)dx$$

$$= \int_{-\infty}^{\infty} u(\alpha - x)p(x)dx$$

$$\leq \int_{-\infty}^{\infty} \exp[s(x - \alpha)]p(x)dx = e^{-s\alpha}\Psi(s); \quad s \leq 0 \qquad (D.7)$$

For the Poisson distribution

$$\Psi(s) = E\{e^{sN}\} = \sum_{n=0}^{\infty} \frac{m^n e^{-m}}{n!} e^{sn}$$

$$= e^{-m} \sum_{n=0}^{\infty} \frac{(me^s)^n}{n!} = \exp[m(e^s - 1)] \qquad (D.8)$$

Substitution into (D.6) gives

$$\alpha = me^s \qquad (D.9)$$

Substitution of (D.8) and (D.9) into (D.5) and (D.7) gives the Chernoff bounds for the Poisson distribution

$$P(N \geq \alpha) \leq (m/\alpha)^{\alpha} \exp(\alpha - m); \quad \alpha \geq m \qquad (D.10)$$

$$P(N \leq \alpha) \leq (m/\alpha)^{\alpha} \exp(\alpha - m); \quad \alpha \leq m \qquad (D.11)$$

D.1.2   Improved Chernoff Bounds

The Chernoff bounds (D.10) and (D.11) are not tight. An improved bound for the Poisson distribution is obtained as follows:

$$\sum_{n=\alpha}^{\infty} e^{-m} \frac{m^n}{n!} = e^{-m} \frac{m^{\alpha}}{\alpha!} \left(1 + \frac{m}{\alpha+1} + \frac{m^2}{(\alpha+1)(\alpha+2)} + \cdots \right)$$

$$< e^{-m} \frac{m^{\alpha}}{\alpha!} \left(1 + \frac{m}{\alpha+1} + (\frac{m}{\alpha+1})^2 + \cdots \right)$$

$$= e^{-m} \frac{m^{\alpha}}{\alpha!} \frac{\alpha+1}{\alpha-m+1}; \quad \alpha > m \qquad (D.12)$$

The Stirling approximation

$$\alpha! > \sqrt{2\pi\alpha}\, \alpha^{\alpha} e^{-\alpha} \qquad (D.13)$$

gives

$$P(N \geq \alpha) < \frac{1}{\sqrt{2\pi\alpha}} \frac{\alpha+1}{\alpha-m+1} (m/\alpha)^\alpha \exp(\alpha - m); \quad \alpha > m \tag{D.14}$$

In the same way

$$\sum_{n=0}^{\alpha} e^{-m} \frac{m^n}{n!} = e^{-m} \frac{m^\alpha}{\alpha!} \left(1 + \frac{\alpha}{m} + \frac{\alpha(\alpha-1)}{m^2} + \cdots + \frac{\alpha!}{m^\alpha}\right)$$

$$< e^{-m} \frac{m^\alpha}{\alpha!} \left(1 + \frac{\alpha}{m} + (\frac{\alpha}{m})^2 + \cdots + (\frac{\alpha}{m})^\alpha\right)$$

$$= e^{-m} \frac{m^\alpha}{\alpha!} \frac{1 - (\alpha/m)^{\alpha+1}}{1 - \alpha/m} < e^{-m} \frac{m^\alpha}{\alpha!} \frac{m}{m-\alpha}; \quad \alpha < m \tag{D.15}$$

The Stirling approximation (D.13) gives

$$P(N \leq \alpha) < \frac{1}{\sqrt{2\pi\alpha}} \frac{m}{m-\alpha} (m/\alpha)^\alpha \exp(\alpha - m); \quad \alpha < m \tag{D.16}$$

The improved bounds (D.14) and (D.16) have the same exponent as the original bounds (D.10) and (D.11) but they are much closer to the exact values. In the region which is interesting for detection probabilities the relative error is only a few percent. Almost identical bounds are presented by Mazo and Salz (1976).

For numerical calculations it is convenient to express the bounds in the form

$$P(N \leq \alpha) = \sum_{n=0}^{\alpha} \frac{m}{n!} e^{-m} < \frac{1}{\sqrt{2\pi\alpha}} \frac{m}{m-\alpha} \exp[-\Theta(m, \alpha)]; \quad m > \alpha \tag{D.17}$$

$$P(N \geq \alpha) = \sum_{n=\alpha}^{\infty} \frac{m}{n!} e^{-m} < \frac{1}{\sqrt{2\pi\alpha}} \frac{\alpha+1}{\alpha-m+1} \exp[-\Theta(m, \alpha)]; \quad m < \alpha \tag{D.18}$$

where

$$\Theta(m, \alpha) = m - \alpha[1 + \ln(m/\alpha)] \tag{D.19}$$

The error probability of an optical receiver is the sum of two probabilities, see (5.24)

$$P_e = \frac{1}{2}[P(N \geq [\alpha] + 1|m = m_0) + P(N \leq [\alpha]|m = m_1)] \tag{D.20}$$

with $[\alpha]$ the integer-valued receiver threshold.

To obtain an upper bound on P_e we express m_0 as $\mu(1 - \varepsilon)$ and m_1 as $\mu(1 + \varepsilon)$ with

$$\mu = (m_1 + m_0)/2 \tag{D.21}$$

and

$$\varepsilon = \frac{m_1 - m_0}{m_1 + m_0} \tag{D.22}$$

The optimal threshold is the integer part of α (5.26)

$$\alpha = \frac{m_1 - m_0}{\ln m_1 - \ln m_0} = \frac{2\mu\varepsilon}{\ln(1+\varepsilon) - \ln(1-\varepsilon)} \tag{D.23}$$

Since $[\alpha] \leq \alpha$ it follows that

$$P(N \geq [\alpha] + 1 | m = m_0) < P(N \geq \alpha | m = m_0)$$

and

$$P(N \leq [\alpha] | m = m_1) \leq P(N \leq \alpha | m = m_1)$$

The exponents $\Theta(m_1, \alpha)$ and $\Theta(m_0, \alpha)$ become identical for α determined by (D.23). It is convenient to write the common exponent as

$$\Theta(m_1, \alpha) = \Theta(m_0, \alpha) = \frac{\mu\varepsilon^2}{2}\vartheta(\varepsilon) \tag{D.24}$$

The function $\vartheta(\varepsilon)$ is shown in Figure D.1. It is close to one for small ε and approaches four when ε goes to one.

The bound on the error probability obtained from (D.17) and (D.18) can be written as

$$P_e < \frac{c(\varepsilon)}{\sqrt{2\pi\mu\varepsilon^2}}\exp\left[-\frac{\mu\varepsilon^2}{2}\vartheta(\varepsilon)\right] \tag{D.25}$$

The function $c(\varepsilon)$ is shown in Figure D.1. It is always less than 1.23 and approaches 1.0 for small ε.

For $\varepsilon \ll 1$ the functions involved can be expanded in rapidly converging

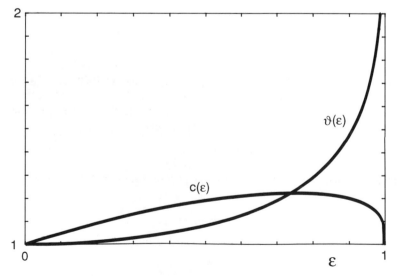

Figure D.1   The functions $\vartheta(\varepsilon)$ and $c(\varepsilon)$ appearing in the bound (D.25).

power series. The threshold α is

$$\alpha = \frac{\mu}{1 + \varepsilon^2/3 + \varepsilon^4/5 + \cdots} = \mu(1 - \varepsilon^2/3 - 4\varepsilon^4/45 + \cdots) \qquad \text{(D.26)}$$

and the exponent $\vartheta(\varepsilon)$

$$\vartheta(\varepsilon) = 1 + 5\varepsilon^2/18 - \cdots \qquad \text{(D.27)}$$

The function $c(\varepsilon)$ is bounded by

$$c(\varepsilon) < \frac{1 + \varepsilon/2}{1 - \varepsilon^2/9} \qquad \text{(D.28)}$$

Combination of these expressions gives the bound

$$P_e < \frac{(1 + \varepsilon/2)}{(1 - \varepsilon^2/9)\sqrt{2\pi\mu\varepsilon^2}} \exp\left[-\frac{\mu\varepsilon^2}{2}(1 + 5\varepsilon^2/18)\right] \qquad \text{(D.29)}$$

The bound (D.29) is valid for all $\mu \geq 0$ and $\varepsilon \leq 1$ and is tight when ε is small.

D.2   Poisson-plus-Gauss Distribution

We consider the sum of two independent stochastic variables

$$U = N + X \qquad \text{(D.30)}$$

The component N has a Poisson distribution (5.5) with mean m, whereas X is Gaussian with zero mean and variance σ^2. The probability density of U is the convolution of the densities of N and X

$$p(u) = \sum_{n=0}^{\infty} \frac{e^{-m}}{\sqrt{2\pi\sigma^2}} \frac{m^n}{n!} \exp[-(u-n)^2/2\sigma^2)] \qquad \text{(D.31)}$$

The infinite summation makes (D.31) difficult to handle analytically and numerically. A more tractable way is to utilize the moment-generating function which, since U is the sum of two independent variables, is the product of the moment-generating functions of the components

$$\Psi_u(s) = \Psi_n(s)\Psi_x(s) \qquad \text{(D.32)}$$

The moment-generating function for a Poisson variable with mean m is from (D.8)

$$\Psi_n(s) = \exp[m(e^s - 1)] \qquad \text{(D.33)}$$

For a Gaussian variable with zero mean and variance σ^2

$$\Psi_x(s) = \frac{1}{\sqrt{2\pi\sigma^2}} \int_{-\infty}^{\infty} e^{sx} e^{-x^2/s\sigma^2} dx = \exp(s^2\sigma^2/2) \qquad \text{(D.34)}$$

Substitution of (D.33) and (D.34) into (D.32) gives

$$\Psi_u(s) = \exp[m(e^s - 1) + s^2\sigma^2/2] \tag{D.35}$$

Chernoff bounds on the tails of the probability distribution (D.31) are easily obtained from the moment-generating function (D.35). Substitution into (D.5) and (D.7) gives

$$P(U \geq \alpha) \leq \exp[\psi(s)]; \quad s \geq 0 \tag{D.36}$$

and

$$P(U \leq \alpha) \leq \exp[\psi(s)]; \quad s \leq 0 \tag{D.37}$$

where

$$\psi(s) = m(e^s - 1) + s^2\sigma^2/2 - s\alpha \tag{D.38}$$

The optimum value $s = s_0$, resulting in the tightest bounds, is obtained by letting the derivative of $\psi(s)$ be equal to zero

$$\psi'(s_0) = me^{s_0} + \sigma^2 s_0 - \alpha = 0 \tag{D.39}$$

The solution of (D.39) has to be obtained numerically. A convenient method is the Newton-Raphson algorithm.

The solution s_0 of (D.39) is positive when $\alpha > m$, which corresponds to the upper tail bound (D.36), and is negative for $\alpha < m$, corresponding to the lower tail bound (D.37).

Appendix E
Saddlepoint Approximation

E.1 Continuous Distribution

Following Helstrom (1978) we derive a numerically simple approximation to the cumulative probability distribution of a continuous stochastic variable with density function $p(x)$.

Let $q_+(\alpha)$ denote the upper tail

$$q_+(\alpha) = \int_\alpha^\infty p(x)\, dx \tag{E.1}$$

and

$$q_-(\alpha) = \int_{-\infty}^\alpha p(x)\, dx \tag{E.2}$$

the lower tail of the probability distribution.

The bilateral Laplace transform of $p(x)$ can be expressed in term of the moment-generating function as

$$\int_{-\infty}^\infty p(x)e^{-sx} dx = \Psi(-s) \tag{E.3}$$

The probability density $p(x)$ is equal to the inverse integral

$$p(x) = \frac{1}{2\pi j} \int_{c-j\infty}^{c+j\infty} \Psi(-s)e^{sx} ds \tag{E.4}$$

where c is in the convergence region of the transform.

Replace $p(x)$ in (F.1) by (E.4) and chose the contour of integration such that $c < 0$ to guarantee convergence of the integral. The result is

$$q_+(\alpha) = \frac{-1}{2\pi j} \int_{c-j\infty}^{c+j\infty} \frac{e^{s\alpha}}{s} \Psi(-s)\, ds$$

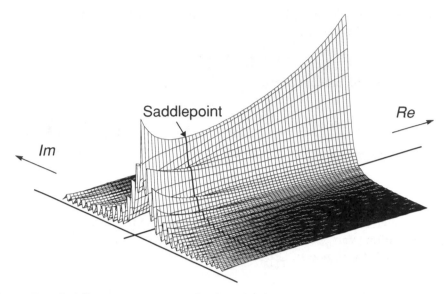

Figure E.1   Saddlepoint integration. The figure shows the integrand for a typical MGF. The integration path and the saddlepoint are indicated.

Changing the integration variable from $-s$ to s gives

$$q_+(\alpha) = \frac{1}{2\pi j} \int_{c-j\infty}^{c+j\infty} \frac{e^{-s\alpha}}{s} \Psi(s)\, ds; \quad c > 0 \tag{E.5}$$

The parameter c is next chosen to be the value of s for which the integrand is minimal. It turns out that this point $s = s_0$ corresponds to a saddlepoint in the complex plane, whence the name of the method. The integrand of (E.5) for a typical MGF is illustrated in Figure E.1. The integration path through the saddlepoint is indicated.

We express the integrand in terms of a 'phase' function $\psi(s)$ defined by the relation

$$\exp[\psi(s)] = |s|^{-1} \exp(-s\alpha) \Psi(s) \tag{E.6}$$

The function $\psi(s)$ is expanded in a Taylor series around the point $s = s_0$

$$\psi(s) = \psi(s_0) + \frac{1}{2}\psi''(s_0)(s - s_0)^2 + \cdots \tag{E.7}$$

The first derivative does not appear since $s = s_0$ is an extremum of $\psi(s)$.

Substitution of (E.7) into (E.5) and neglecting higher-order terms yields the saddlepoint approximation

$$q_+(\alpha) \approx \frac{1}{2\pi} \exp[\psi(s_0)] \int_{-\infty}^{\infty} \exp\left[-\frac{1}{2}\psi''(s_0)y^2\right] dy \tag{E.8}$$

$$= [2\pi\psi''(s_0)]^{-1/2} \exp[\psi(s_0)]$$

The parameter s_0 is the value of s for which $\psi(s)$ has a minimum. It is equal to the positive root of the equation

$$\psi'(s) = 0 \tag{E.9}$$

For the lower tail, analogously to (E.5),

$$q_-(\alpha) = \frac{-1}{2\pi j} \int_{c-j\infty}^{c+j\infty} \frac{e^{-s\alpha}}{s} \Psi(s) ds \tag{E.10}$$

with $c < 0$. Expansion of $\psi(s)$ in a Taylor series and integration gives

$$q_-(\alpha) \approx [2\pi\psi''(s_1)]^{-1/2} \exp[\psi(s_1)] \tag{E.11}$$

with s_1 equal to the negative root of (E.9).

Only two terms in the Taylor expansion are used in the derivation above. The approximation can be improved by including higher-order terms, as demonstrated by Helstrom (1978).

E.2 Discrete Distribution

Let P_n denote the probabilities of a discrete stochastic variable. Its generating function is

$$\Phi(z) = \sum_{n=-\infty}^{\infty} z^n P_n \tag{E.12}$$

We define the upper tail of the distribution as

$$Q_+(\alpha) = \sum_{n=\alpha}^{\infty} P_n \tag{E.13}$$

and the lower tail as

$$Q_-(\alpha) = \sum_{n=-\infty}^{\alpha-1} P_n \tag{E.14}$$

The function $\Phi(z)$ can be interpreted as a double sided z-transform with the inverse

$$P_n = \frac{1}{2\pi j} \oint z^{-n-1} \Phi(z) dz \tag{E.15}$$

Summation of (eq C.17) gives

$$Q_+(\alpha) = \frac{1}{2\pi j} \oint \sum_{\alpha}^{\infty} z^{-n-1} \Phi(z) \, dz = \frac{1}{2\pi j} \oint_{C_1} \frac{z^{-\alpha}}{z-1} \Phi(z) \, dz \tag{E.16}$$

where C_1, the path of integration, is chosen such that $|z^{-1}| < 1$ to ensure that the sum converges.

The path of integration is then changed to become a vertical line in the complex plane passing through the saddlepoint $z = z_0$ with $z_0 > 1$.

The phase function $\phi(z)$ is defined as the logarithm of the integrand

$$\phi(z) = \ln \Phi(z) - \alpha \ln z - \ln |z - 1| \tag{E.17}$$

Expanding $\phi(z)$ in a Taylor series around the point $z = z_0$ and integrating yields the approximation

$$Q_+(\alpha) \approx [2\pi\phi''(z_0)]^{-1/2} \exp[\phi(z_0)] \tag{E.18}$$

with $\phi(z)$ defined by (E.17) and $z_0 > 1$ a root of the equation

$$\phi'(z) = \frac{\Phi'(z)}{\Phi(z)} - \frac{\alpha}{z} - \frac{1}{|z - 1|} = 0 \tag{E.19}$$

In the same way for $Q_-(\alpha)$

$$Q_-(\alpha) = \frac{1}{2\pi j} \oint \sum_{-\infty}^{\alpha-1} z^{-n-1} \Phi(z) \, dz = \frac{1}{2\pi j} \oint_{C_2} \frac{z^{-\alpha}}{1 - z} \Phi(z) \, dz \tag{E.20}$$

where the path C_2 is such that $|z| < 1$.

The approximation obtained after a Taylor series expansion and integration is

$$Q_-(\alpha) \approx [2\pi\phi''(z_1)]^{-1/2} \exp[\phi(z_1)] \tag{E.21}$$

with $\phi(z)$ defined by (E.17) and $0 < z_1 < 1$ a root of (E.19).

Appendix F

Optimal Linear Receivers

F.1 Max SNR Receiver

Consider an optical system without intersymbol interference and with a receiver containing a filter followed by a threshold. We want to determine a filter function $v(t)$ such that the signal-to-noise ratio (5.133) used in the Gaussian approximation (5.132) is maximized.

Instead of maximizing ρ it is convenient to minimize $1/\rho$. From (5.133) and (5.134) it follows that

$$\delta(v) = \frac{1}{\rho} = \frac{\left(\int_0^T \Gamma_1(t)v^2(t)\,dt\right)^{1/2} + \left(\int_0^T \Gamma_0(t)v^2(t)\,dt\right)^{1/2}}{\int_0^T [\Gamma_1(t) - \Gamma_0(t)]v(t)\,dt} \tag{F.1}$$

The receiver filter minimizing $\delta(v)$ is determined by using the calculus of variations. For $v(t)$ to be optimum it must constitute a stationary point and any small variation of $v(t)$ must result in an increased $\delta(v)$. Replace $v(t)$ in (F.1) by $v(t) + \varepsilon f(t)$ with $f(t)$ an arbitrary function and ε equal to a small positive or negative constant.

The first-order variation in $\delta(v)$ is obtained by expanding of $\delta(v + \varepsilon f) - \delta(v)$ in terms of powers of ε and keeping terms of first and lower order. The first term in the numerator of $\delta(v + \varepsilon f)$ is

$$\left(\int_0^T \Gamma_1(t)[v(t) + \varepsilon f(t)]^2\,dt\right)^{1/2}$$

$$= \left(\int_0^T \Gamma_1(t)v^2(t)\,dt + 2\varepsilon \int_0^T \Gamma_1(t)v(t)f(t)\,dt + \varepsilon^2 \int_0^T \Gamma_1(t)f^2(t)\,dt\right)^{1/2}$$

$$\approx \left(\int_0^T \Gamma_1(t)v^2(t)\,dt\right)^{1/2} \left(1 + \frac{\varepsilon \int_0^T \Gamma_1(t)v(t)f(t)\,dt}{\int_0^T \Gamma_1(t)v^2(t)\,dt}\right)$$

$$= \sigma_1 \left(1 + \frac{\varepsilon}{\sigma_1^2} \int_0^T \Gamma_1(t)v(t)f(t)\,dt\right) \tag{F.2}$$

An analogous expression is obtained for the term containing $\Gamma_0(t)$.

In the same way for the denominator

$$\left(\int_0^T [\Gamma_1(t) - \Gamma_0(t)][v(t) + \varepsilon f(t)] \, dt \right)^{-1}$$

$$= (E_1 - E_0)^{-1} \left(1 - \frac{\varepsilon}{E_1 - E_0} \int_0^T [\Gamma_1(t) - \Gamma_0(t)] f(t) \, dt \right) \tag{F.3}$$

Combination of these expressions yields

$$\delta(v + \varepsilon f) - \delta(v) \tag{F.4}$$

$$= \frac{\varepsilon}{E_1 - E_0} \int_0^T f(t) \left\{ \frac{1}{\sigma_1} \Gamma_1(t) v(t) + \frac{1}{\sigma_0} \Gamma_0(t) v(t) - \frac{\sigma_1 + \sigma_0}{E_1 - E_0} (\Gamma_1(t) - \Gamma_0(t)) \right\} dt$$

The expression within braces in (F.4) must be equal to zero since otherwise it would be possible to specify $f(t)$ and ε such that $\delta(v + \varepsilon f)$ is less than $\delta(v)$. This gives the following equation for $v(t)$

$$v(t) = \text{const.} \, \frac{\Gamma_1(t) - \Gamma_0(t)}{\sigma_0 \Gamma_1(t) + \sigma_1 \Gamma_0(t)} \tag{F.5}$$

Since σ_0 and σ_1 depend on $v(t)$, (F.5) provides an implicit specification of the optimum receiver filter maximizing the signal-to-noise ratio ρ.

The result is easily modified to include thermal noise in the receiver. White noise with intensity $\mathcal{N}$ increases the variances σ_1^2 and σ_0^2 by (5.110)

$$\sigma^2 = \int_0^T \mathcal{N} v^2(t) \, dt$$

Comparison with (F.1) shows that the optimum filter is then obtained by replacing $\Gamma_1(t)$ by $\mathcal{N} + \Gamma_1(t)$ and $\Gamma_0(t)$ by $\mathcal{N} + \Gamma_0(t)$ resulting in

$$v(t) = \text{const.} \frac{\Gamma_1(t) - \Gamma_0(t)}{\sigma_0[\mathcal{N} + \Gamma_1(t)] + \sigma_1[\mathcal{N} + \Gamma_0(t)]} \tag{F.6}$$

For

$$\left. \begin{array}{l} \Gamma_0(t) = \gamma_0 \\ \Gamma_1(t) = \gamma_0 + \gamma(t) \end{array} \right\} \tag{F.7}$$

the filter function $v(t)$ becomes

$$v(t) = \frac{\gamma(t)}{\mathcal{N} + \gamma_0 + a\gamma(t)} \tag{F.8}$$

with

$$a = \frac{\sigma_0}{\sigma_0 + \sigma_1}$$

F.2   MSE Receiver

Consider an optical system using binary on-off modulation with pulses of shape $p(t)$. The received optical signal power is

$$\mathcal{P}(t) = p_0 + \sum_{k=-\infty}^{\infty} b_k p(t - kT) \tag{F.9}$$

where b_k represents the data symbols which are assumed to be uncorrelated stochastic variables taking the value 0 or 1 with equal probability.

The intensity function for the photo electrons produced by the detector is, cf. (5.3) and (5.43)

$$\Gamma(t) = i_0/q + \gamma_0 + \sum_{k=-\infty}^{\infty} b_k \gamma(t - kT) \tag{F.10}$$

The receiver is assumed to consist of a linear filter with impulse response $h(t)$ followed by a sampler operating periodically at times $t = kT$ and a decision threshold. In what follows we will study the receiver operation at an arbitrary time $t = kT$ and to simplify the notation we introduce the filter function (5.105)

$$v(t) = q^{-1} h(kT - t)$$

The output signal is the sum of two uncorrelated terms

$$u(t) = u_d(t) + u_n(t) \tag{F.11}$$

where $u_d(t)$ is a filtered shot noise signal produced by the input optical signal and $u_n(t)$ is stationary noise with variance (5.110)

$$E\{u_n^2\} = \mathcal{N} \int v^2(t)\, dt \tag{F.12}$$

The receiver can be viewed as a linear discrete-time system with inputs $\{b_k\}$ and outputs $u(kT)$. One way of designing the receiver is to minimize the variance of the error $e_k = u(kT) - b_k$ between output and input

$$\sigma_e^2 = E\{[e_k - B_e]^2\} = E\{[u(kT) - b_k - B_e]^2\} \tag{F.13}$$

where $B_e = E\{e_k\}$.

The average has to be performed over all possible input sequences as well as with respect to the optical and thermal noise.
The mean-square error (F.13) can be expressed as

$$\sigma_e^2 = E\{e_k^2\} - B_e^2 \tag{F.14}$$

We start with evaluation of the term

$$\begin{aligned} E\{e_k^2\} &= E\{[u(kT) - b_k]^2\} \\ &= E\{u_d^2(kT)\} + E\{u_n^2(kT)\} - 2E\{b_k\, u_d(kT)\} + E\{b_k^2\} \end{aligned} \tag{F.15}$$

For the further derivation it is convenient to express the intensity function (F.10) as

$$\Gamma(t) = \psi(t) + \zeta(t) = \psi(t) + \sum_{k=-\infty}^{\infty} c_k \gamma(t - kT) \tag{F.16}$$

where

$$\psi(t) = i_0/q + \gamma_0 + \frac{1}{2} \sum_{k=-\infty}^{\infty} \gamma(t - kT) \tag{F.17}$$

is the average of $\Gamma(t)$ and c_k, which replaces b_k, assumes the values $-1/2$ or $1/2$ with equal probability.

The mean-square value $E\{u_d^2(kT)\}$ for a fixed input data sequence can be determined with the aid of Campbell's Theorem (5.131) together with the relation

$$E\{u_d^2(kT)\} = \text{Var}\{u_d(kT)\} + (E\{u_d(kT)\})^2$$

The result is

$$E\{u_d^2(kT)\} = \int E\{\Gamma(t)\}v^2(t)\,dt + \int\int E\{\Gamma(t)\Gamma(s)\}v(t)v(s)\,dt\,ds$$

$$= \int \psi(t)v^2(t)\,dt + \left[\int\int \psi(t)v(t)\,dt\right]^2$$

$$+ \int\int E\{\zeta(t)\zeta(s)\}v(t)v(s)\,dt\,ds \tag{F.18}$$

where the averaging over the input data remains to be performed. The last term in (F.18) is equal to

$$E\{\zeta(t)\zeta(s)\} = \sum_k \sum_m E\{c_k c_m\}\gamma(t - kT)\gamma(s - mT)$$

$$= \frac{1}{4}\sum_k \gamma(t - kT)\gamma(s - kT) \tag{F.19}$$

In the same way

$$E\{b_k u_d(kT)\} = \int E\{b_k \gamma(t)\}v(t)\,dt$$

$$= \frac{1}{2}\int \psi(t)v(t)\,dt + \frac{1}{4}\int \gamma(t)v(t)\,dt \tag{F.20}$$

From the results above it follows that the bias B_e is equal to

$$B_e = E\{u(kT) - b_k\} = \int \psi(t)v(t)\,dt - 1/2 \tag{F.21}$$

Combination of the results (F.15) to (F.21) and substitution into (F.14) yields the mean-square error

$$\sigma_e^2 = \int \psi(t)v^2(t)\,dt + \int N v^2(t)\,dt + \frac{1}{4}\left[\sum_{k\neq 0} r_k^2 + (1 - r_0)^2\right] \tag{F.22}$$

where

$$r_k = \int \gamma(t + kT)v(t) \, dt \tag{F.23}$$

The optimum receiver filter minimizing σ_e^2 is determined by the calculus of variations. For $v(t)$ to be optimal it must constitute a stationary point and any small variation of $v(t)$ must result in an increased σ_e^2. If $v(t)$ in (F.22) is replaced by $v(t) + \varepsilon f(t)$ with $f(t)$ an arbitrary function and ε equal to a small positive or negative constant, the first-order variation in σ_e^2 becomes

$$
\begin{aligned}
&\sigma_e^2(v + \varepsilon f) - \sigma_e^2(v) \\
&= \int (\psi(t) + \mathcal{N}) 2\varepsilon f(t) v(t) \, dt \\
&\quad + \frac{1}{4}\left[\sum_k r_m \int \gamma(t + kT) 2\varepsilon f(t) \, dt - 2\int \gamma(t) 2\varepsilon f(t) \, dt \right] \\
&= 2\varepsilon \int f(t) \left\{ (\psi(t) + \mathcal{N})v(t) + \frac{1}{4}\left(\sum_k r_k \gamma(t + kT) - \gamma(t) \right) \right\} dt
\end{aligned}
\tag{F.24}
$$

The expression within braces in (F.24) must be equal to zero since otherwise it would be possible to specify $f(t)$ and ε such that $\sigma_e^2(v + \varepsilon f)$ is less than $\sigma_e^2(v)$. This gives the following equation for $v(t)$

$$(\psi(t) + \mathcal{N})v(t) + \frac{1}{4}\left(\sum_k r_k \gamma(t + kT) - \gamma(t) \right) - 0 \tag{F.25}$$

A $v(t)$ satisfying (F.25) is

$$v(t) = \sum_{m=-\infty}^{\infty} C_m g(t + mT) \tag{F.26}$$

with

$$g(t) = \frac{1}{4} \frac{\gamma(t)}{\psi(t) + \mathcal{N}} \tag{F.27}$$

This is proved by direct substitution into (F.25), remembering that $\psi(t)$ is a periodic function. The following relations for the coefficients $\{C_m\}$ are obtained

$$C_m = \begin{cases} 1 - r_0 ; & m = 0 \\ -r_m ; & m \neq 0 \end{cases} \tag{F.28}$$

The conclusion of (F.26) is that the optimum MSE receiver contains a filter $g(t)$ specified by (F.27) followed by a transversal filter with tap coefficients $\{C_m\}$.

To determine the coefficients $\{C_m\}$, multiply both sides of (F.26) by $\gamma(t + kT)$ and integrate to obtain the relation

$$r_k = \sum_m C_m S_{k-m} \tag{F.29}$$

where

$$S_k = \int_{-\infty}^{\infty} \gamma(t + kT)g(t) \, dt \tag{F.30}$$

The relation (F.29) is a discrete-time convolution and the double-sided z-transform of r_k is

$$R(z) = \sum_{k=-\infty}^{\infty} r_k z^{-k} = C(z)S(z) \tag{F.31}$$

The tap gains $\{C_m\}$ are obtained by solving (F.31) for $C(z)$ utilizing (F.28). The result is

$$C(z) = \frac{1}{1 + S(z)} \tag{F.32}$$

The minimum mean-square error is related to $C(z)$ in a simple manner. Multiplication of both sides of (F.25) by $v(t)$ and integration gives, after substitution into (F.22),

$$\sigma_e^2 = \frac{1}{4} \left[\sum_m C_m r_m + \sum_{m \neq 0} r_m^2 + (1 - r_0)^2 \right] \tag{F.33}$$

which by the use of (F.28) is simplified to

$$\sigma_e^2 = C_0/4 \tag{F.34}$$

Consider the unbiased output signal

$$u^*(kT) = u(kT) - B_e$$

The mean values of $u^*(kT)$ for input data symbols $b_k = 0$ and $b_k = 1$, respectively are

$$\left. \begin{aligned} E\{u^*(kT)/b_k = 0\} &= 1/2 - r_0/2 \\ E\{u^*(kT)/b_k = 1\} &= 1/2 + r_0/2 \end{aligned} \right\} \tag{F.35}$$

The corresponding variances are

$$\left. \begin{aligned} \sigma_0^2 &= \sigma_{mse}^2 - \frac{1}{2} \int \gamma(t) v^2(t) \, dt \\ \sigma_1^2 &= \sigma_{mse}^2 + \frac{1}{2} \int \gamma(t) v^2(t) \, dt \end{aligned} \right\} \tag{F.36}$$

where

$$\sigma_{mse}^2 = \int \psi(t) v^2(t) \, dt + \int \mathcal{N} v^2(t) \, dt + \frac{1}{4} \sum_{k \neq 0} r_k^2 = \sigma_e^2 - C_0^2/4 \tag{F.37}$$

is the average variance caused by optical and thermal noise and intersymbol interference.

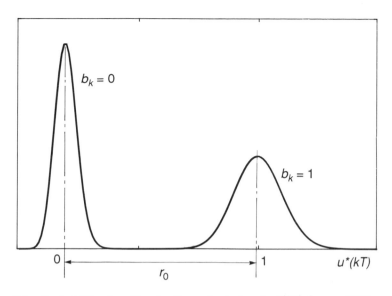

Figure F.1 Probability densities for the output signal $u^*(kT)$ from a MSE-receiver.

The probability density for $u^*(kT)$ is illustrated in Figure F.1.
The average signal-to-noise ratio at the receiver output is

$$\rho_{mse} = \frac{r_0/2}{\sigma_{mse}} \tag{F.38}$$

It is related to the mean-square error (F.34) through

$$\rho_{mse}^2 = \frac{1 - 4\sigma_e^2}{4\sigma_e^2} = \frac{1 - C_0}{C_0} \tag{F.39}$$

The mean-square error σ_e^2 and the signal-to-noise ratio ρ_{mse} are easy to calculate but they do not constitute sufficient information to determine the detection error probability.

The receiver derived above achieves the smallest mean-square error by allowing a small amount of intersymbol interference in the output signal. An alternative solution is to specify that $u(kT)$ should be free from intersymbol interference, i. e. $r_m = 0, m \neq 0$. Equation (F.31) is now modified to

$$R(z) = r_0 = C(z)S(z) \tag{F.40}$$

which gives the tap gains for the zero forcing equalizer

$$C(z) = \frac{r_0}{S(z)} \tag{F.41}$$

It can be shown that its mean square error, which is greater than σ_e^2, is still determined by (F.34).

At time $t = mT$ the receiver has made decisions on the previous data symbols b_k, $k = m - 1, m - 2, \ldots$. When the error probability is low the decisions are likely to be correct most of the time, and they can be used to improve the equalization. The method is called decision feedback and the principle is to reduce the mean-square error by subtracting the influence of the detected data symbols. The mean-square error for decision feedback equalization is

$$\sigma_e^2 = E\{[u(kT) - \sum_{m=1}^{\infty} d_m b_{k-m} - b_k - B_e]^2\} \qquad (F.42)$$

The equalizing filter that minimizes this expression can be determined by a simple extension of the analysis of the linear MSE equalizer. The optimal structure turns out to be the same as for the linear MSE equalizer: the filter $g(t)$ specified by (F.27) followed by a transversal filter. A detailed derivation can be found in Messerschmitt (1978) or Salz (1973).

Appendix **G**

Phase Noise

G.1   Lorentzian Laser Light

The optical field for light of constant intensity is in the complex notation

$$B(t) = A \exp[j(2\pi f_0 t + \theta(t) + \phi)] \tag{G.1}$$

where A is the amplitude and $\theta(t)$ is the phase noise. The phase angle ϕ is assumed to be independent of $\theta(t)$ and equally distributed in the interval $0 \le \phi \le 2\pi$.

The phase noise is modeled as a Wiener process

$$\theta(t) = 2\pi \int_0^t \mu(s)\, ds \tag{G.2}$$

with $\mu(s)$ a white Gaussian stochastic process with power spectral density R_μ (double-sided).

The autocorrelation function for $B(t)$ is

$$r(\tau) = \frac{1}{2} E\{B(t+\tau)B^*(t)\}$$

$$= \frac{A^2}{2} \exp(j2\pi f_0 \tau) E\{\exp[j\psi(\tau)]\} \tag{G.3}$$

where

$$\psi(\tau) = \theta(t+\tau) - \theta(t) = 2\pi \int_t^{t+\tau} \mu(s)\, ds \tag{G.4}$$

is a zero mean Gaussian variable with variance

$$\sigma_\psi^2 = E\{[\psi^2(\tau)]\} = E\left\{ 2\pi \int_t^{t+\tau} \mu(s)\, ds \right\}^2$$

$$= (2\pi)^2 \int_t^{t+\tau} \int_t^{t+\tau} E\{\mu(s)\mu(r)\}\, ds\, dr$$

$$= (2\pi)^2 \int_t^{t+\tau} \int_t^{t+\tau} R_\mu\, \delta(s-r)\, ds\, dr$$

$$= (2\pi)^2 \int_t^{t+\tau} R_\mu\, ds = (2\pi)^2 R_\mu |\tau| \tag{G.5}$$

The average in (G.3) is

$$E\{e^{j\psi}\} = \frac{1}{\sqrt{2\pi\sigma_\psi^2}} \int_{-\infty}^{\infty} e^{j\psi} \, e^{-\psi^2/2\sigma_\psi^2} \, d\psi = e^{-\sigma_\psi^2/2} \tag{G.6}$$

and substitution of (G.5) gives

$$r(\tau) = \frac{A^2}{2} \exp(j2\pi f_0\tau) \exp\left(-\frac{(2\pi)^2 R_\mu |\tau|}{2}\right) \tag{G.7}$$

The Fourier transform of $r(\tau)$ is the spectral density

$$R(f) = \frac{A^2 R_\mu/2}{(\pi R_\mu)^2 + (f - f_0)^2} \tag{G.8}$$

which is the Lorentz spectrum. Its appearance is shown in Figure 4.15. The 3 dB bandwidth B_L of the spectrum is determined by the relation $R(f_0 \pm B_L/2) = R(f_0)/2$, which gives

$$B_L = 2\pi R_\mu \tag{G.9}$$

G.2 Filtered Phase Noise

G.2.1 *Approximate distribution*

Following Foschini and Vannucci (1988) we expand $\exp[j\theta(t)]$ in a Taylor series

$$e^{j\theta(t)} \approx 1 + j\theta(t) - \frac{1}{2}\theta^2(t) \tag{G.10}$$

For phase noise processed by an integrating filter

$$y = \frac{1}{T}\int_0^T e^{j\theta(t)} \, dt \approx 1 + j\zeta_1 - \zeta_2/2 \tag{G.11}$$

where

$$\zeta_1 = \frac{1}{T}\int_0^T \theta(t) \, dt \tag{G.12}$$

and

$$\zeta_2 = \frac{1}{T}\int_0^T \theta^2(t) \, dt \tag{G.13}$$

The square envelope of y is, neglecting higher order terms

$$|y|^2 \approx 1 - \zeta_2 + \zeta_1^2 \tag{G.14}$$

Expand $\theta(t)$ in the time interval $[0, T]$ into a Fourier series

$$\theta(t) = a_0 + \sum_{k=1}^{\infty} a_k \sqrt{2} \cos\left(\frac{k\pi t}{T}\right) \tag{G.15}$$

with random coefficients

$$a_0 = \frac{1}{T} \int_0^T \theta(t)\, dt = \zeta_1 \tag{G.16}$$

and

$$a_k = \frac{\sqrt{2}}{T} \int_0^T \theta(t) \cos\left(\frac{k\pi t}{T}\right) dt \tag{G.17}$$

Integration by parts yields

$$\frac{1}{T} \int_0^T \theta(t) \cos\left(\frac{k\pi t}{T}\right) dt = -2\pi \int_0^T \frac{\mu(t)}{k\pi} \sin\left(\frac{k\pi t}{T}\right) dt \tag{G.18}$$

where $\mu(t) = \theta'(t)/2\pi$ is frequency noise with spectral density $R_\mu = B_L/2\pi$. The quantities

$$X_k = 2\pi \int_0^T \mu(t) \sin\left(\frac{k\pi t}{T}\right) dt \tag{G.19}$$

are independent equally distributed zero mean Gaussian stochastic variables with variance

$$\sigma^2 = (2\pi)^2 \int_{-\infty}^{\infty} R_\mu \sin^2 \frac{k\pi t}{T}\, dt = (2\pi)^2 R_\mu \frac{T}{2} = \pi B_L T \tag{G.20}$$

From Parseval's relation follows that

$$\zeta_2 = \sum_{k=0}^{\infty} a_k^2 = a_0^2 + \sum_{k=1}^{\infty} \frac{2X_k^2}{(k\pi)^2} \tag{G.21}$$

The moment-generating function of $|\mathcal{Y}|^2$ is

$$\Psi_{\mathcal{Y}}(s) = E\{\exp[(1 - \zeta_2 + \zeta_1^2)s]\} = \exp(s) \prod_{k=1}^{\infty} E\left\{\exp\frac{-2X_k^2 s}{(k\pi)^2}\right\} \tag{G.22}$$

The averages in the right-hand side of (G.22) are of the form

$$E\{\exp(-ax^2)\} = \frac{1}{\sqrt{2\pi\sigma^2}} \int_{-\infty}^{\infty} \exp[-x^2/(2\sigma^2) - ax^2]\, dx = (1 + 2a\sigma^2)^{-1/2} \tag{G.23}$$

and

$$\Psi_{\mathcal{Y}}(s) = \exp(s) \prod_{k=1}^{\infty} \left(1 + \frac{4\sigma^2 s}{(k\pi)^2}\right)^{-1/2} \tag{G.24}$$

From Handbook of Mathematical Functions, Abramowitz and Stegun (1968), formula 4.5.68

$$\sinh z = z \prod_{k=1}^{\infty} \left(1 + \frac{z^2}{(k\pi)^2} \right)^{-1/2} \tag{G.25}$$

and

$$\Psi_{\mathcal{Y}}(s) = \exp(s) \left(\frac{\sqrt{2\beta s}}{\sinh \sqrt{2\beta s}} \right)^{1/2} \tag{G.26}$$

with $\beta = 2\pi B_L T$.

G.2.2   Moments

Recursive relations for the moments of filtered phase noise have been presented by Garrett et al. (1990), Dallal and Shamai (1994), Kaiser et al. (1993), and others. They can be used for numerical calculations of moments and to obtain analytic expressions for low order moments using a symbol manipulating computer program.

It is convenient to express the filtered phase noise in terms of the normalized phase noise process $\psi(s) = \theta(sT)$.

$$\mathcal{Y} = \frac{1}{T} \int_0^T e^{j\theta(t)} \, dt = \int_0^1 e^{j\psi(s)} \, ds \tag{G.27}$$

The stochastic process $\psi(s)$ has a zero mean Gaussian distribution with variance $\beta s = 2\pi B_L T s$. The mean of the square envelope of $\mathcal{Y}$ is

$$E\{|\mathcal{Y}|^2\} = \int_0^1 \int_0^1 E\left\{ e^{j[\psi(s) - \psi(t)]} \right\} ds \, dt = \int_0^1 \int_0^1 e^{\beta|s - t|/2} \, ds \, dt$$

$$= \frac{4}{\beta} \left(1 - \frac{1 - e^{-\beta/2}}{\beta/2} \right) \tag{G.28}$$

The mean square value of $|\mathcal{Y}|^2$ is

$$E\{|\mathcal{Y}|^4\} = \int_0^1 \int_0^1 \int_0^1 \int_0^1 E\left\{ e^{j[\psi(t) - \psi(s) + \psi(u) - \psi(v)]} \right\} ds \, dt \, du \, dv \tag{G.29}$$

The expectation is easy to carry out and the result is a combination of exponential functions. Integration over these gives

$$E\{|\mathcal{Y}|^4\} = \frac{8(783 - 270\beta + 36\beta^2 - 784e^{-\beta/2} - 120\beta e^{-\beta/2} + e^{-2\beta})}{9\beta^4} \tag{G.30}$$

The variance of $|\mathcal{Y}|^2$ is obtained by combining (G.30) and (G.28)

$$\text{Var}\{|\mathcal{Y}|^2\} = E\{|\mathcal{Y}|^4\} - (E\{|\mathcal{Y}|^2\})^2 \tag{G.31}$$

References

Abramowitz, M. and Stegun, I. A. (1968). *Handbook of Mathematical Functions*, National Bureau of Standards, Applied Mathematics Series, No 55.

Amano, K. and Iwamoto, Y. (1990). 'Optical Fiber Submarine Cable Systems,' *J. Lighwave Tech.*, vol. 8, pp. 595–609, April.

Andersson, T. (1990). 'Design of the Receiver in a Digital Fiberoptic Communication System,' *Internal Technical Report, Telecommunication Theory, Lund University, Sweden*, (in Swedish), September.

Azizoglu, M. and Humblet, P. A. (1995). Reference Transmission Schemes for Phase Noise Immunity,' *IEEE Trans. Comm.*, Vol. 43, pp. 1624–1635, February/March/April.

Bar-David, I. (1969). 'Communication under the Poisson Regime,' *IEEE Trans. Inform. Theory*, vol. IT-15, pp. 31–37, January.

Bell, D. A. (1960). *Electrical Noise, Fundamentals and Physical Mechanism*, Van Nostrand, London.

Bell Telephone Laboratories (1971). *Transmission Systems for Communications*, 4th ed., Bell Telephone Laboratories.

Bennett, W. R. (1960). *Electrical Noise*, McGraw-Hill, New York.

Bjarklev, A. (1993). *Optical Fiber Amplifiers: Design and System Applications*, Artec House, London.

Campbell, N. (1909). 'The study of discontinuous phenomena,' *Proc. Camb. Phil. Soc.*, vol. 15, pp. 117–136. February and 'Discontinuities in Light Emission,' *Proc. Camb. Phil. Soc.*, vol. 15, pp. 310–328, November.

Carratt, M., Reinaudo, C., Jocteur, R., and Trezeguet, J-P. (1987). 'Dispersion Shifted Fiber for Long Unrepeatered Submarine Systems,' *El. Comm.*, vol. 61, pp. 384–388.

Carrier, G. F., Krook, M., and Pearson, C. E. (1966). *Functions of a Complex Variable*, McGraw-Hill, New York.

Cheng, D. K. (1983). *Field and Wave Electromagnetics*, Addison-Wesley, Reading.

Cox, D. R. and Isham, V. (1980). *Point Processes*, Chapman and Hall, London.

Da Rocha, J. R. F. and O'Reilly, J. J. (1986). 'Linear Direct-Detection Fiber-Optic Receiver Optimization in the Presence of Intersymbol Interference,' *IEEE Trans. Comm.*, vol. COM-34, pp. 365–374, April.

Dallal, Y. E. and Shamai (Shitz), S. (1994). 'Power Moment Derivation for Noisy Phase Lightwave Systems,' *IEEE Trans. Inform. Theory*, vol. 40, pp. 2099–2103, November.

Davenport, W. B. and Root, W. L. (1958). *Random Signals and Noise*, McGraw-Hill, New York.

Davis, M. H. A. (1980). 'Capacity and Cutoff Rate for Poisson-Type Channels,' *IEEE Trans. Inform. Theory*, vol. IT-26, pp. 710–715, November.

Dugundji, J. (1958). 'Envelopes and Pre-Envelopes of Real Waveforms,' *IRE Trans. Inform. Theory*, vol. IT-4, pp. 53–57, March.

Edelcrantz, A. (1796). *Afhandling om Telegrapher*, (in Swedish), Johan Pehr Lindh (printer), Stockholm.

Einarsson, G. (1986). 'Pulse Broadening in Graded-Index Optical Fibers: Correction,' *Appl. Optics*, vol. 25, p. 1030, April.

Einarsson G. (1989). 'Detection of On-Off Modulated Optical Signals,' *Lecture Notes in Control and Information Sciences*, vol. 128 : Topics in Coding Theory, Springer-Verlag, Berlin.

Einarsson G. and Sundelin, M. (1995). 'Performance Analysis of Optical Receivers by Gaussian Approximation,' *J. Optical Comm.*, vol. 16, to appear.

Einarsson, G., Strandberg, J., and Tafur Monroy, I. (1995). 'Error Probability Evaluation of Optical Systems Disturbed by Phase Noise and Additive Noise,' *J. Lightwave Tech.*, vol. 13, pp. 1847–1852, September.

Feller, W. (1957). *An Introduction to Probability Theory and Its Applications*, vol. I, 2nd ed., Wiley, New York.

Fleming, J. W. (1976). 'Material and Mode Dispersion in $GeO_2 \cdot B_2O_3 \cdot SiO_2$ Glasses' *J. Amer. Ceramic Soc.*, vol. 59, p. 503–507, November - December.

Fleming, J. W. (1978). 'Material Dispersion in Lightguide Glasses' *Electronics Letters*, vol. 14, pp. 326–328, May.

Foschini, G. J., Greeenstein, L. J., and Vannucci, G. (1988a). 'Noncoherent Detection of Coherent Lightwave Signals Corrupted by Phase Noise' *IEEE Trans. Comm.*, vol. 36, pp. 306–314. March.

Foschini, G. J. and Vannucci, G. (1988b). 'Characterizing Filtered Light Waves Corrupted by Phase Noise,' *IEEE Trans. Inform. Theory*, vol. 34, pp. 1437–1448, November.

Foschini, G. J., Vannucci, G., and Greeenstein, L. J. (1989). 'Envelope Statistics for Filtered Optical Signals Corrupted by Phase Noise' *IEEE Trans. Comm.*, vol. 37, pp. 1293–1302, December.

Gagliardi, R. M. and S. Karp (1976). *Optical Communications*, Wiley, New York.

Gallager, R. G. (1968). *Information Theory and Reliable Communication*, Wiley, New York.

Garrett, I, Bond, D. J., Waite, J. B., Lettis, D. S. L., and Jacobsen, G. (1990). 'Impact of Phase Noise in Weakly Coherent Systems: A New and Accurate Approach,' *J. Lightwave Tech.*, vol. 8, pp. 329–337, March.

Gloge, D. (1971). 'Weakly Guiding Fibers,' *Appl. Optics*, vol. 10, pp. 2252–2258, October.

Gordon, J. P. and Mollenauer, L. F. (1991). 'Effects of Fiber Nonlinearities and Amplifier Spacing on Ultra-Long distance Transmission,' *J. Lightwave Tech.*, vol. 9, pp. 170–173, Febuary.

Gowar, J. (1984). *Optical Communication Systems*, Prentice-Hall, London.

Haus, H. A., Atkinson, W. R., Branch, G. M., Davenport, W. B., Fonger, W. H., Harris, W. A., Harrison, S. W., McLeod, W. W., Stodola, E. K., and Talpey, T. E. (1960). 'Representation of Noise in Linear Twoports' *Proc IRE*, vol. 48, pp. 69–74, January.

Hayaski, I., Panish, M. B., Foy, P. W., and Sumski, S. (1970). 'Junction Lasers which Operate Continuously at Room Temperature,' *Appl. Phys. Lett.*, vol. 17, pp. 109–111, August.

Hecht, J. (1985). 'Victorian experiments and optical communications,' *IEEE Spectrum*, vol. 22, pp. 69–73, February.

Heidemann, R., Wedding, B., and Veith, G. (1993). '10-GB/s Transmission and Beyond,' *Proc. IEEE*, vol. 81, pp. 1558–1567, November.

Helstrom, C. W. (1978). 'Approximate Evaluation of Detection Probabilities in Radar and Optical Communications,' *IEEE Trans. Aerospace Electr. Syst.*, vol. AES-14, pp. 630–640, July.

Helstrom, C. W. (1979). 'Performance Analysis of Optical Receivers by the Saddlepoint Approximation,' *IEEE Trans. Comm.*, vol. COM-27, pp. 186–191, January.

Helstrom, C. W. (1984). 'Computation of Output Electron Distributions in Avalanche Photodiodes,' *IEEE Trans. Electr. Devices*, vol. ED-31 pp. 955–958, July.

Helstrom, C. W. (1988). 'Computing the Performance of Optical Receivers with Avalanche Diode Detectors,' *IEEE Trans. Comm., vol. 36, pp. 61–66, January.*

Helstrom, C. W. (1992). 'Analysis of Avalanche Diode Receivers by Saddlepoint Integration,' *IEEE Trans. Comm.*, vol. 40, pp. 1327–1338, August.

Henry, C. H. (1986). 'Theory of Spontaneous Emission Noise in Open Resonators and its Application to Lasers and Optical Amplifiers,' *J. Lightwave Tech.*, vol. 4, pp. 288–297. March.

Henry, P. S. (1985). 'Lightwave Primer,' *IEEE J. Quantum Electronics*, vol. QE-21, pp. 1862–1879, December.

Henry, P. S. (1989). 'Error-rate performance of optical amplifiers,' *Optical Fiber Conference OFC'89, Houston, Texas*, Tech. Dig. Series, vol. 5, paper THK3.

Hodgkinson, T. G. (1987). 'Receiver Analysis for Synchronous Coherent Optical Fiber Transmission Systems,' *J. Lightwave Tech.*, vol. LT-5, pp. 573–585. April.

Holzmann, G. J., and Pehrson, B. (1994). 'The First Data Networks,' *Sc. American*, vol. 270, pp. 112–117, January.

Humblet, P. A., and M Azizoglu (1991). 'On the Bit Error Rate of Lightwave Systems with Optical Amplifiers,' *J. Lightwave Tech.*, vol. 9, pp. 1576–1582, November.

Jacobsen, G. (1994). *Noise in Digital Optical Transmission Systems*, Artech House, Boston.

Jones W. B.. (1988). *Optical Fiber Communication Systems.*, Holt, Rinehart and Winston, New York.

Kaiser, C. P., Shafi, M., and Smith, P. J. (1993). 'Analysis Methods for Optical Heterodyne DPSK Receivers Corrupted by Laser Phase Noise,' *J. Lightwave Tech.*, vol. 11, pp. 1820–1830, November.

Kanamori, H., Yokota, H., Tanaka, G., Watanabe, M., Ishiguro, Y., Yoshida, I., Kakii, T, Itoh, S., Asano, Y., and Tanaka S. (1986). 'Transmission Characteristics and Reliability of Pure-Silica-Core Single-Mode Fibers,' *J. Lightwave Techn.*, vol. LT-4, p. 1144–1149, August.

Kao, K.C. and Hockham, G. A. (1966). 'Dielectric-Fibre Surface Waveguides for Optical Frequencies,' *Proc. IEE*, vol. 113, pp. 1151–1158, July.

Kapron, F. P., Keck, D. B., and Maurer, R. D. (1970). 'Radiation Losses in Glass Optical Waveguides,' *Appl. Phys. Lett.*, vol. 17, pp. 423–425, November.

Kazovsky, L. G. (1985). 'Decision-Driven Phase-Locked Loop for Optical Homodyne Receivers: Performance Analysis and laser Linewidth Requirements' *J. Lightwave Tech.*, vol. LT-3, pp. 1238–1247, December.

Kimura, T. (1988). 'Factors Affecting Fiber-Optic Transmission Quality,' *J. Lightwave Tech.*, vol. 6, pp. 611–619, May.

Li, T. and Teich, M. C. (1991). 'Bit-Error Rate for a Lightwave Communication System Incorporating an Erbium-Doped Fibre Amplifier,' *Electronics Letters*, vol. 27, pp. 598–600, March.

Li, T. and Teich, M. C. (1992). 'Performance of a lightwave system incorporating a cascade of erbium-doped fiber amplifiers,' *Optics Communications*, vol. 91, pp. 41–45, July.

Li, T. and Teich, M. C. (1993). 'Photon Point Process for Traveling-Wave Laser Amplifiers,' *IEEE J. Quant. Electronics*, vol. 29, pp. 2568–2578, September.

Lichtman, E. (1993). 'Optimal amplifier spacing in ultralong lightwave systems,' *Electronics Letters*, vol. 29, pp. 2058–2060, November.

Mahlke, G., and Gössing, P. (1987). *Fiber Optic Cables*, Wiley, New York.

Malmgren, E. (1972). *Bilder ur svensk telehistoria, (in Swedish)*, Televerket, Farsta.

Mandel, L. (1967). 'Complex Representation of Optical Fields in Coherence Theory,' *J. Opt. Soc. Amer.*, vol. 57, pp. 613–617, May.

Marcuse, D. (1982). *Light Transmission Optics*, 2nd ed., Van Nostrand, New York.

Marcuse, D. (1990). 'Derivation of Analytic Expressions for the Bit-Error Probability in Lightwave Systems with Optical Amplifiers,' *J. Lightwave Tech.*, vol. 8, pp. 1816–1823, Dececember.

Marcuse, D. (1991a). 'Calculation of Bit-Error Probability for a Lightwave System with Optical Amplifiers and Post-Detection Gaussian Noise,' *J. Lightwave Tech.*, vol. 9, pp. 505–513, April.

Marcuse, D. (1991b). *Theory of Dielectric Optical Waveguides*, 2nd ed., Academic Press, New York.

Massey, J. L. (1981). 'Capacity, Cutoff Rate, and Channel Coding for a Direct-Detection Optical Channel,' *IEEE Trans. Comm.*, vol. COM-29, pp. 1615–1621, December.

Mazo, J. E. and Salz J. (1976). 'On Optical Data Communication via Direct Detection of Light Pulses,' *Bell Syst. Tech. J.*, vol. 55, pp. 347–369, March.

Messerschmitt, D. G. (1978). 'Minimum MSE Equalization of Digital Fiber Optic Systems,' *IEEE Trans. Comm.*, vol. COM-26, pp. 1110–1118, July.

Michaelis, A. R. (1965). *From Semaphore to Satellite*, International Telecommunication Union, Geneva.

Mollenauer, L. F., Mamyshev, P. V., and Neubelt, M. J. (1994). 'Measurement of timing jitter in filter-guided soliton transmission at 10 Gbits/s and achievement of 375 Gbits/s-Mm, error free, at 12.5 and 15 Gbits/s,' *Optics Lett.*, vol. 19, pp. 704–706, May.

Morse, P. M. and Feshbach, H. (1953). *Methods of Theoretical Physics*, McGraw-Hill, New York.

Murata, H. (1988). *Handbook of Optical Fibers and Cables*, Marcel Dekker, New York.

Nagel, S. R. (1989). 'Optical Fiber - The Expanding Medium,' *IEEE Circuits and Devices Magazine*, vol. 5, pp. 36–45, March.

Nakazawa, M., Suzuki, K., Yamada, E., Kubota, H., Kimura, Y., and Takaya, M. (1993). 'Experimental Demonstration of Soliton Data Transmission over Unlimited Distances with Soliton Control in Time and Frequency Domains,' *Electronics Letters*, vol. 29, pp. 729–730, April.

Neumann, E.-G. (1988). *Single-Mode Fibers*, Springer-Verlag, Berlin.

Nyquist, H. (1928). 'Certain Topics in Telegraph Transmission Theory,' *Trans. AIEE*, vol. 47, pp. 617–644, April.

Okoshi, T. (1982). *Optical Fibers*, Academic Press, Orlando.

Olshansky, R. and Keck, D. B. (1976). 'Pulse Broadening in Graded Index Optical Fibers,' *Appl. Optics*, vol. 15, pp. 483–491, February.

Olshansky, R., Lanzisera, V. A., and Hill, P. M. (1989). 'Subcarrier Multiplexed Lightwave Systems for Broad-Band Distribution,' *J. Lightwave Tech.*, vol. 7, pp. 1329–1342, September.

Olsson, N. A. (1990). 'Lightwave Systems With Optical Amplifiers,' *J. Lightwave Tech.*, vol. 7, pp. 1071–1082, July.

O'Mahony, M. J. (1988). 'Semiconductor Laser Optical Amplifiers for Use in Future Fiber Systems,' *J. Lightwave Tech.*, vol. 6, pp. 531–544, April.

Oyamada, K., and Okoshi, T. (1980). 'High-Accuracy Numerical Data on Propagation Characteristics of α-Power Graded-Core Fibers,' *IEEE Trans. Microwave Tech.*, vol. MTT-28, pp. 1113–1118, October.

Papoulis, A. (1962). *The Fourier Integral and its Applications*, McGraw-Hill, New York.

Papoulis, A. (1984). *Probability, Random Variables and Stochastic Processes*, 2nd ed., McGraw-Hill, New York.

Parzen, E. (1962). *Stochastic Processes*, Holden-Day, San Francisco.

Personick, S. D. (1971a). 'New Results on Avalanche Multiplication Statistics with Applications to Optical Detection,' *Bell Syst Tech J.*, vol. 50, pp. 167–189, January.

Personick, S. D. (1971b). 'Statistics of a General Class of Avalanche Detectors with Applications to Optical Communication,' *Bell Syst. Tech. J.*, vol. 50, pp. 3075–3095, December.

Personick, S.D. (1973). 'Receiver Design for Digital Fiber Optic Comunication Systems,' *Bell Syst. Tech. J.*, vol. 52, pp. 843–886, July-August.

Personick, S. D., Balaban, P., Bobsin, J. H., and Kumar, P. R. (1977). 'A detailed Comparison of Four Approaches to the Calculation of the Sensitivity of Optical Fiber System Receivers,' *IEEE Trans. Comm.*, vol. COM-25, pp. 541–548, May.

Proakis, J. G. (1989). *Digital Communications*, 2nd ed., McGraw-Hill, New York.

Reed, W. A., Cohen, L. G., and Shang, H-T. (1986). 'Tailoring Optical Characteristics of Dispersion-Shifted Lightguides for Applications Near 1.55 μm,' *AT&T Tech. J.*, vol. 65, pp. 105–122, September/October.

Ribeiro, R. F. S. and Ferreira da Rocha, J. R. (1994) 'Computationally Efficient Methods for Bit-Error Rate Evaluation in Weakly Coherent Optical Systems', *J. Lightwave Tech.*, vol 12, pp. 1842–1848, October.

Rice, S. O. (1944). 'Mathematical Analysis of Random Noise,' *Bell Syst. Tech. J.*, vol. 23, pp. 283–332, July and vol. 24 (1945). pp. 47–156, January.

Runge, P. K. (1992). 'Undersea Lightwave Systems,' *AT&T Tech. J.*, vol. 71, pp. 5–13, January/February.

Runge, P. K. and Trischitta, P. R. (1983). 'The SL Undersea Lightguide System,' *IEEE J. Selected Areas Comm.*, vol. SAC-1, pp. 459–466, April.

Runge, P. K. and Trischitta, P. R. (1984). 'The SL Undersea Lightwave System,' *IEEE J. Selected Areas Comm.*, vol. SAC-2, pp. 784–793, November.

Runge, P. K. and Trischitta, P. R. (eds) (1986). *Undersea Lightwave Communications*, IEEE Press, New York.

Saleh, B. (1978). *Photoelectron Statistics*, Springer-Verlag, Berlin.

Saleh, B. E. A., and M. C. Teich (1991). *Fundamentals of Photonics*, Wiley, New York.

Salz, J. (1973). 'Optimum Mean-Square Decision Feedback Equalization,' *Bell Syst Tech J.*, vol. 52, pp. 1341–1373, Oct.

Salz, J. (1985). 'Coherent Lightwave Communications' *AT&T Tech. J.*, vol. 64, pp. 2153–2209, December.

Schwartz, M., Bennett, W. R., and Stein, S. (1966). *Communication Systems and Techniques*, McGraw-Hill, New York.

Shimoda, K., Takahasi, H., and Townes, C. H. (1957). 'Fluctuations in Amplification of Quanta with Application to Maser Amplifiers,' *J. Phys. Soc. Japan*, vol. 12, pp. 686–700, June.

Shlomo, (Shitz), S. and Lapidoth, A. (1993). 'Bounds on the Capacity of a Spectrally Constrained Poisson Channel,' *IEEE Trans. Inform. Theory*, vol. 39, pp. 19–29, January.

Snyder, D. L. and M. I. Miller (1991). *Random Point Processes in Time and Space, Second edition*, Springer-Verlag, New York.

Suzuki, M., Edagawa, N., Taga, H., Tanaka, H., Yamamoto, S., and Akiba, S. (1994). 'Feasibility demonstration of 20 Gbit/s single channel soliton transmission over 11500 km using alternating-amplitude solitons,' *Electronics Lett.*, vol. 30, pp. 1083–1084, June.

Taga, H., Edagawa, N., Yoshida, Y., Yamamoto, S., and Wakabayashi, H. (1994). 'IM-DD Long-Distance Multichannel WDM Transmission Experiments Using Er-Doped Fiber Amplifiers,' *J. Lightwave Tech.*, vol. 12, pp. 1448–1454, August.

Thiennot, F. P., Pirio, F., and Thomine, J.-B. (1993). 'Optical Undersea Cable System

Trends,' *Proc. IEEE*, vol. 81, pp. 1610–1623, November.

Tikhonov, V. I. (1959). 'Noise Influence on Operation of Frequency Phase Adjustment Circuit,' *Automatika i Telemekh.*, vol. 20, pp. 1188–1196.

Tonguz, O. K. and R. E. Wagner (1991). 'Equivalence Between Preamplified Direct Detection and Heterodyne Receivers,' *IEEE Photonics Tech. Lett.*, vol. 3, pp. 835–837. September.

Tyndall, J. (1854). 'On some Phenomena connected with the Motion of Liquids,' *Proceedings of the Royal Institution of Great Britain*, vol. 1, pp. 446–448, May.

Van Trees, H. L. (1971). *Detection, Estimation, and Modulation Theory*, Wiley, New York.

Viterbi, A. J. (1966). *Principles of Coherent Communication*, McGraw-Hill, New York.

Viterbi, A. J. and Omura, J. K. (1979). *Principles of Digital Communication and Coding*, McGraw-Hill, New York.

Wagner, S. S. and Menendez, R. C. (1989). 'Evolutionary Architectures and Techniques for Video Distribution on Fiber,' *IEEE Comm. Mag.*, vol. 27, no. 12, pp. 17–25, December.

Way, W. I. (1989). 'Subcarrier Multiplexed Lightwave System Design Considerations For Subscriber Loop Applications,' *J. Lightwave Tech.*, vol. 7, pp. 1806–1818, November.

Werts, A. (1966). 'Propagation de la lumiére cohérente dans les fibres optiques,' *L'Onde Electrique*, vol. 46, pp. 967–980, September.

Wozencraft, J. M. and Jacobs, I. M. (1965). *Principles of Communication Engineering*, Wiley, New York.

Wyner, A. D. (1988). 'Capacity and Error Exponent for the Direct Detection Photon Channel - Part I and II,' *IEEE Trans. Inform. Theory*, vol. 34, pp. 1449–1471, November.

Yariv, A. (1985). *Optical Electronics, 3rd ed.*, Holt-Saunders, New York,

Index